Between Two Breaths

...the seasons of creation

Bob G. Shupe

THE BIRTH AND DEATH OF EVERYTHING
THROUGH THE EYES OF FAITH, SCIENCE, AND RELIGION

Self-published by Bob G. Shupe
P.O. Box 2344, Brentwood, Tennessee 37024 USA
bob@esptn.com
www.bobgshupe.com

ISBN: 979-8-9851675-0-4
Copyright: September 28, 2021, TXu 2-279-078
Cover and Illustration Design: Ghislain Viau, Creative Publishing Design
Content Editor: Jerry Herbert
Grammar Editor: Pamela Cangioli, Proofed to Perfection
Formatted: Damonza.com

OTHER BOOKS BY THIS AUTHOR

The Bitter Pill, where did my benefits go?

First book on Healthcare Issues written and copyright 2006, four years prior to the passage of the Affordable Care Act. Many of Mr. Shupe's concerns would become reality.

Published by Wheatmark, 610 East Delano Street, Suite 104, Tucson, Arizona 85705, *www.wheatmark.com* .

ISBN-10: 1-58736-684-3

ISBN-13: 978-1-58736-684-0

LCCN: 2006931431

The Little Brown Church in the Vale, when did the lights go out?

Written and copyright 2008. A look at the dying 21st-century denominational movement and maybe it's only last hope.

Published by Wheatmark, 610 East Delano Street, Suite 104, Tucson, Arizona 85705, *www.wheatmark.com*

ISBN: 978-1-60494-137-1

LCCN: 2008929188

Ultimate Point of Vulnerability

Mr. Shupe's second book on healthcare is in the form of a "faction" novel written and copyrighted in 2009. A cure for cancer is discovered in the 1960s, only to be covered up by a secret syndicate, The Foundation, for fear the cure will wreak havoc on the U.S. economy.

Decades later, The Foundation, which has now gained power over nearly every aspect of American life, is trying to manage the economic crisis brought on by terrorist attacks and the same national health care system they had put in motion years before. Now, just as economic chaos looms, it is learned that China has the long-withheld cure and intends to release it, which could further devastate the U.S. economy and its sovereignty.

Will the U.S. be able to recover? Or has The Foundation taken Americans to the brink of no return? Is this America's Ultimate Point of Vulnerability, a "perfect storm" of divergent elements—crashing together—overwhelming enough to destroy the most powerful country in the world?

Published by Wheatmark, 610 East Delano Street, Suite 104, Tucson, Arizona 85705, *www.wheatmark.com*

ISBN: 978-1-60494-308-5

LCCN: 2009929259

Also available on Kindle

Behind the Healthcare Cost Curtain... there is an answer!

Think for a moment about where you get your information regarding the healthcare cost crisis. Now ask yourself what motivation you think each of those sources has to give you information. Going deeper, what motivation did the sources which your sources cited have for providing that information? If you do a little research, you will find most everyone—including the federal government—is asking for ways to fix this broken system. They are asking those providing healthcare services, procedures, and products, and those selling insurance and related services. Who are they *not* asking? You… and me?

It's important to understand who sells healthcare as a business, for profit and not for profit (non-profit). Even more critical is, who has the greatest influence over this process: those who are directly paying for healthcare, those who sell it, or those who need it?

Most U.S. consumers, when hearing the word *healthcare* freely used by politicians, the media, and the health care industry, have no idea they are hearing a word not yet fully accepted within literary circles or most sources we use to validate our language. As you read this book, it is critical you keep this concept in the foreground. As a simple example, when you hear about the "increased cost of insurance," is the source referring to the cost of healthcare or the cost of health care? This distinction is critical. Changing or creating words with which we communicate a problem or a solution to an uninformed public can be deceiving. I encourage you to stop accepting problems and start demanding solutions using terms you understand.

Self-published by Bob G. Shupe

P.O. Box 2344, Brentwood, TN 37211 USA

bob@esptn.com

ISBN: 978-0-578-50232-8

Copyright: TXu 2-132-553

Also available on Kindle

ABOUT THE AUTHOR

Mr. Shupe is very involved in his church and is a past Moderator of the Nashville Presbytery of the Cumberland Presbyterian Church. He is an active elder at his church and past Minister of Music, serving in that position for forty-eight years. He attended Free Will Baptist Bible College and then attended and graduated from Belmont University in 1973. He is the father of two, grandfather of two, and the husband of Valerie Shupe for the past fifty-two years.

Mr. Shupe, as a teenager, felt called to the Christian mission field which led him to the higher education he pursued. After serving for a year and a half at the First Baptist Church in Portland, Tennessee as Minister of Music and Youth, he felt a different call. After a two-and-a-half-year search, he discovered the insurance industry. This industry allowed him to provide for his family and fund and provide time for several mission projects throughout the remainder of his forty-three-year career. Some of those were working for eight years on a Sudanese Church project for refugees in the U.S.; music programs in Colombia, South America; and raising funds for various other mission projects while directing music for many years in a Cumberland Presbyterian Church.

Other music projects included co-writing a historic musical about the birth of the Cumberland Presbyterian Church in 1810 during the Second Great Awakening in the Cumberland Valley. The research for this Great Awakening project and his tenure teaching a Bible Discussion Group for several years gave birth to the idea for this, his fifth book entitled, *Between Two Breaths: The Seasons of Creation.* As the book developed it became apparent that his original idea had vastly expanded, giving rise to a second subtitle: *The Birth and Death of Everything Through the Eyes of Faith, Science, and Religion.*

In addition to serving as past moderator of the Nashville Presbytery of the Cumberland Presbyterian Church, he currently serves on the Presbyterial Board of Missions working with several churches on mission outreach. Mr. Shupe's love for history came out of research into the formation of several, now-large denominations that were birthed during the Second Great Awakening in the late 1700s and early 1800s. His love and appreciation for science was a result of his family moving to Florida in 1959 to a little town about twenty miles south of Cape Canaveral (for a while Cape Kennedy). He spent his young life from the third-grade through high school graduation in the middle of one of the greatest scientific advancements in history. Many of his friends' parents worked in the space program or for companies who supplied goods and services to the program. Mr. Shupe remembers standing on the steps of his elementary school watching the contrail of Alan Shepherd's space capsule soar overhead, making him the first man in the U.S. to reach space. A few years later he was sitting in his bedroom watching TV when news broke about three astronauts who were burned alive while trapped in their capsule during a training exercise. That happened less than twenty miles from where he was sitting.

All of this background was the impetus for his creation of this book.

CONTENTS

FALL | 411

Breathe on me, Breath of God,
fill me with life anew,
that I may love what thou dost love,
and do what thou wouldst do.

Edwin Hatch
1835-1889

Educated at Pembroke College, Oxford, Edward Hatch ministered in an Anglican parish in the slums of east London before accepting a position teaching Classics at Trinity College in Quebec. He was recognized as an authority on the early church due to his widely acknowledged Bampton Lectures, "On the Organization of Early Christian Churches."

In 1876, he penned the hymn "Breathe on Me, Breath of God." Although simple, it reflects Hatch's deep knowledge of Scripture. Its inspiration is John 20:21-22: "*Jesus said to them again, 'Peace be with you. As the Father has sent me, even, so send I you.' And when he had said this, he breathed on them, and said to them, 'Receive the Holy Spirit.'*" (RSV)

For those who wish to appreciate the intricacies of text and tune relationships, TRENTHAM is the only Short Meter tune (6.6.8.6) in *The United Methodist Hymnal* that works well with this text. Most Short Meter tunes are set with iambic texts (beginning on a weak beat). This hymn, however, requires a tune that begins on a strong beat as each stanza starts with an imperative verb invoking God to "Breathe on me."[1]

DEDICATION

Compared to acquaintances I have made over seventy-two years, I have known Tyler Spradling a short time—a little over twenty years. This book and its overriding message are dedicated to this servant-leader. This dedication in no way indicates Tyler may be in total agreement with my writing; rather, this book is dedicated to his ministry with youth and the open discussions he has had with those who may not always agree with him—which is the reason for this book.

As you will discover in this book, Earth time is relative, not defining. Compared to my total life, I have learned much from observing this man's life and example. Like many, Tyler started his career with a college education that on the surface seems to have had nothing to do with his current vocation. That is true of many college graduates, including me. Although, at the base of every college degree, one learns the importance of self-reliance and organization, I am not sure it helped Tyler in those areas. I believe he had those worked out before he attended his first post-secondary education class.

What have I learned from Tyler Spradling? Selflessness. Other than God and his family, nothing seems to take priority over serving those who are truly in need—of anything. Frugality. For years I wasn't sure he owned any clothes bought first-hand, especially pants. Those were purchased at a thrift store and often met an expected fate. They were cut off at the knee – no hem. Servant-leader. Tyler's God-led vocation, for several years, has been in youth ministry. While he holds a youth gathering each week on Wednesday evening, attending will only give you the tip of what happens between him and those he serves-leads.

A few examples of Tyler's approach to ministry I have observed include:

- More often than not, when a parent attending an activity or the youth participant looks into the crowd, they will find Tyler and, if appropriate, wearing cut-off pants. He considers part of his ministry being directly involved in the whole life of the student, not just on Wednesday evenings.
- For as long as I have known him, Tyler and his youth have organized a thirty-hour famine raising money to feed hungry children around the world. Those efforts, by a small group of teenagers and their leader, have raised well over $100,000. The group also goes without food for thirty hours during the event.
- Tyler finds kids in the community or they hear their friends talking about their experience and find the group. Each year the church holds a graduation dedication for those graduating at various levels of their education. The church is always amazed at who comes forward because they may have never seen the person on Sunday. The reason? These kids want to be the church, not just sit in it.
- Each year the youth give almost a week of their time to participate in a mission project helping people who are recovering from disaster in and out of state. Tyler offers shares of stock in the youth at varying levels to raise funds for the trip. When someone purchases stock (they sell out every year) that stockholder gets a daily report from that youth about what they did and experienced. One day of that mission week is set aside for the participants to have fun. At the end of the mission trip the youth prepare and serve the stockholders a meal and talk about their experiences.

While many pastors and adult leaders could, but do not, learn from these and other examples, there are other intangibles I am aware of in this man's life. For instance, I know that Tyler, out of respect for his future bride's parents, went to their place of business and asked the parents for permission to marry their daughter. What does this have to do with the subject matter of this book? Everything. Many in the faith community have yet to figure out what their faith is really about. Most unbelievers in the science community and other unbelievers have been unable to connect with someone of faith who is willing to listen to them without judgment. Maybe both communities should consider wearing used, cut-off pants.

Oh, did I mention, Tyler is my son-in-law?

Thank you, Tyler.

The Father-in-law Author

FOREWORD

Between Two Breaths… the Seasons of Creation is an expansive book on theology and science covering Genesis to Revelation. Bob asks, "Who is right, faith or science?" He notes that Christians feel they are in a position where they have to reject science to accept the Bible. His book takes four major themes—Theology/God, Christology/Christ, Ecclesiology/The Church, and Eschatology/The coming of Christ—which he presents as four seasons, and then examines each area in the light of faith and science. His goal is to create a platform for dialogue and discussion about what the Bible says and what science believes. I believe this book provides a way for Christians to think about their Christian faith in the context of a science-based world. Bob gives his readers ways to think about their faith and not reject commonly held secular views about our world and universe. *Between Two Breaths* gives its readers new ways to think about the stories of the Bible, presenting new understanding about Christian faith and the challenges of science.

Rev. Dr. Lynndon L. Thomas,

Director of Global Missions

Cumberland Presbyterian Church

ACKNOWLEDGEMENTS

Above all else on this created Earth, I am thankful for my wife of fifty-two years. In addition to being my bride she has been my business partner, my sister-in-Christ, encourager, and inspiration. Thank you, sweetheart, for allowing me to leave pieces of my brain all over the country in the form of books. I love you.

Writers employ a myriad of different techniques when compiling thoughts and material. I have always applied what could be considered the "fire-hose" method. Once I begin to formulate ideas, I simply start writing them down. Often, those ideas and thoughts come in large packages. If I don't write them down, I will certainly forget them or, at a minimum, lose the flow of thought. To capture my thoughts, I just keep writing. A sixty-word sentence, with multiple commas and other incorrect punctuation, is not out of the question. I realized early as an author with input from professional editors, this method had two serious flaws: content continuity and grammar suffer.

In the past I have relied on professional editors to rearrange my thoughts and correct my grammar. I did use a professional grammar editor on this work; however, I was blessed to have a close friend who just happens to be, in my opinion, a genius when constructing paragraphs, content, and thought.

My friend, Jerry Herbert, agreed late in this process to join me in molding this book into the project I envisioned. Instead of trying to discuss content by email with an out-of-state editor, Jerry and I alternated inviting each other to our homes to spend two to three hours, sometimes weekly, to iron out content. Thanks to Jerry, I am confident this writing can be used by unbeliever and believer alike to add substance to their discussions. Thank you, Jerry, for assisting in this monumental project!

Jerry and his lovely wife, Lee, spent several successful years in the printing business as small business owners prior to retiring. He and his wife are quite active in a local community church where Jerry teaches a Bible discussion group.

I also want to thank past and present members of my Sunday morning Bible Discussion Group, *The Truth Seekers Class*, for being an important part of my week. They have allowed me to explore with them thoughts and Scriptures often forgotten or never fully developed in discussions or sermons. Only this group of people would allow an eight-week study to last for a year and a half… and they continue to show up! The secret to growth and continuity was our effort to have members leave the class and instead become teachers of other classes. These folks and friends like Jerry and Lee were my teachers. If you want to learn something… teach it.

INTRODUCTION

One breath… each of us drew our first one at birth, and we will someday draw our last. Ironically, the eventful lives lived between the date we are born and the date we die will be represented on our grave markers by a simple dash.

The average person at rest takes about sixteen breaths per minute, 960 breaths per hour, 23,040 breaths per day, and 8,409,600 per year. If they live to be eighty, they will, in the course of their lifetime, have taken somewhere around 673 million of them. Those of us who live in a country like the United States—which guarantees "Life, Liberty, and the Pursuit of Happiness"—consider ourselves blessed. Yet none of these rights and privileges can sustain our life on this Earth if we are not able to draw that next breath.

Allow me to overstate this premise since it is so key to this entire writing. A human can live for years, sometimes a lifetime, without liberty or freedom. Countless wars have been fought around the world to liberate such groups and nations. Many died of oppression and brutality, but many lived, despite the odds. A person can live for more than three weeks without food. That same average person can last only three to four days without water.[2] Consider that, after five to ten minutes without breathing air, your brain is likely to sustain irreversible damage and even if you are revived, you will likely live in a vegetative state the remainder of your life, possibly sustained only by a feeding tube and artificial breathing machine.[3]

God created Adam and Eve to enjoy forever a world that contained the Tree of Life. But by one fateful choice, they assured the day that a last breath would be exhaled by the final human being ever to live on this Earth.

My purpose in writing this is to provide a Christian audience with "food for thought" regarding difficult questions that can arise when witnessing to a non-believer, specifically as it relates to Christian beliefs and the scientific community. While there can be vast differences of opinion separating us, an open mind leads to open dialogue.

For organizational purposes, I have separated the continuing course of creation into seasons:

- **Winter** – God before and during the Genesis creation, and all time after the Creation including the 400 years between the Old and New Testaments and up to the birth of Christ
- **Spring** – The birth of Christ through His ascension
- **Summer** – The development of the New Testament Church through Christ's third coming (yes, there will be three)
- **Fall** – Christ's coming back at the end of the original Creation and beyond for eternity… or infinity

From the Christian perspective, the long-awaited coming of Jesus Christ into the world is the beginning of everything. However, there were many events leading up to His coming, and there have been many since. As the preponderance of questions posed by non-believers focus on the time Jesus spent with us here on Earth, and the time immediately following, this work has much to say on that relatively brief period.

It is my prayer that some will find this writing to be useful as a preparation for discussion opportunities that may arise. My intent is not to provide my version of sound doctrine or scriptural interpretation. We need to know that it is okay to ask God questions and to use all of God's image breathed into our oldest ancestor and passed down to all of us.

This book begins with a hymn by Edwin Hatch, "Breath on me, Breath of God." Inspired by John 20:21-22, Jesus breathed on those gathered when He spoke to them. This writing will take you on a journey of timeless interruptions in God's creation story. You see, it is not about me, you, or the world. Rather, it is about His plan for eternity. It is majestic. It is beyond comprehension. It is larger than life itself, and it all began, as it will end, with… one breath.

WINTER

God before and during the Genesis creation, and all time after the creation including the 400 years between the Old and New Testaments and up to the birth of Christ

CHAPTER 1
PRE-CREATION

MANY IN THE faith community begin their story with Genesis 1:1: "In the beginning..." Many in the science community cite a single moment when the first indication of life occurred. But what about the instant before either of these indications? Based on the positions of both science and faith, each proposes either an infinite or eternal past prior to the birth of life as we know it today. Perhaps we could get these two communities to agree there is commonality in their insistence on starting in the middle of a very long story. To the faith community, the journey we are about to traverse in this book is simple. To the science community, it is complicated because the route keeps changing.

Throughout this book you are going to be asked to consider things that do not fit the traditional faith explanations or the scientific understanding you may have been taught previously. Perhaps you just stumbled upon this book and are searching. Regardless of past experience, I hope as you read this book you will deeply consider the truths that lie in your understanding of the things around you.

The first thing I am going to do is ask you to accept briefly that, before all the formless void spoken of in Scripture—or the scientific explanation of life's origins—there *was* something there. The great I Am, the Trinity... and there was evil. Where all of these came from, where they live or exist, what they look like, what work they perform, or how they came into being can only be speculated upon. All I am asking you to consider is that they existed before scientific or faith-based origins. The simple act of accepting this can change your perspective in regard to Scripture and science. It can also drive you insane if you dwell on it.

There are some things that are almost impossible to understand without a scientific explanation, such as the air under a multi-ton airplane wing at thirty-five thousand feet, or how the human eye works. Then there are some things which *are* impossible to understand, which Science is not equipped to explain.

If you are willing to accept this notion of a Pre-Creation reality, let me stir your imagination with some questions you absolutely cannot answer.

- Where did God come from?
- What was God's physical perspective during creation?
- What does the Trinity look like?
- If we call God the Father, is there a Mother?
- Is Christ God's *only* Son?
- Accepting that God said during Creation, "Let *us* make man in *our* image," did whoever "Us" referred to include a female?

Why are these, and other questions, important? According to a Jewish online publication, 10 percent of Protestants, 21 percent of Roman Catholics, and 52 percent of Jews do not believe in God. Within these groups, while 79 percent of Americans believe there is a God, only 66 percent are absolutely certain. Eighty-four percent of women believe in God, compared with 73 percent of men. Ninety-one percent of African-Americans believe in God, compared with 81 percent of Hispanics and 78 percent of whites. Eighty-seven percent of Republicans believe in God, compared with 78 percent of Democrats and 75 percent of Independents. And finally, 82 percent of those with no college education believe in God, compared with 73 percent of those who went to college.[4]

Chances are, if you are a professing Christian, sooner or later you are going to encounter someone who doesn't believe there is a God. This person may quite possibly bring up one or more of the six questions posed above. If you have thoughtfully considered these questions, you will be much better prepared to engage in a meaningful discussion with them. Let's take a look at this list.

Where did God come from?

In order to speculate on where God came from, we have to go back to the beginning of time, correct? But what if the concept of time has nothing to do with it? In an article titled, "Where did God come from?" written by, Joshua M. Spaulding—in the online publication Eternalanswers.org—the author states, "It is a fact that something cannot come from nothing. Newton's Third Law, '*For every action, there is an equal and opposite reaction*' is an observable fact. If something could come from nothing, then we should be able to produce a rose with no ingredients at all. This is obviously not possible. We need ingredients in order to make something—anything."

He goes on to state this question comes from a false understanding of time, and as a result, the question itself is illogical. "Asking where God came from is illogical because it presumes that God was created. God does not exist within time and He is not a creature. This is not just faith, this is fact. We know this because we know that from nothing, nothing comes."[5] To his point, in Genesis 1:1, "In the beginning… ," God did not start with nothing—He started with something that had no form and was void of anything, in particular, Life. Something similar occurs when we, who are void of anything spiritually, are made into new creatures: "Therefore, if anyone is in Christ, the new creation has come: The old has gone, the new is here!"[6]

There may have been some form of time where God existed before He created the heavens and the Earth, but since time, by its nature, is measured according to its surroundings, God's time, or whatever it may be called, would have no relevance to our existence. In the beginning, God didn't call it *time*. He called it *day and night*. Our time was established when God set everything related to us, and His creation, in motion.

The Hebrew word for day is *yom* in the Book of Genesis. This word appears in Scripture more than 1,400 times. Editors of Abarim

Publications, founders of the Center for Rational Theology located in Belgrade, Serbia, assert the Hebrew use of this word in Genesis was to denote the action occurring in a day, not the length of day. Meaning, God determined part of the day would be light and the other part darkness. Left unsaid is that 21st-century minds define the word as having twenty-four hours, because a day in 2021 contains twenty-four hours. To define a day having twenty-four hours in the first chapter of Genesis assumes the length of a day has always been the same length. [7] We will pick this discussion back up later in a more relevant discussion.

As stated previously, Science cannot explain everything. The concept of a creator is dismissed because there is no *cause.* So, the scientific community theorizes that life is a result of happenstance, an accident, a "Big Bang" (which by scientific standards was neither big nor a bang)[8], or distorted explanations of the true science of evolution.

So, the question—Where did God come from?—is a logical issue only if approached from a human point of view; that's how we figure stuff out. Actually, if we could determine where God came from, there would be no need to worship Him. He would just be another creation, created somewhere else by something else that was either without cause or was created by something else. The question can only be answered by someone who accepts that God is the Alpha and Omega[9]. In this case, the answer is stated in Scripture: "I Am."[10]

What was God's physical perspective during Creation?

Most discussions about creation pit faith and science against each other as two opposing forces: the unstoppable force hitting the immovable object. Those listening to these debates are led to believe that there is no option but to choose one or the other. However, I believe there is a third choice that both Scripture and Science support. Supporting this third choice is the physical evidence that is literally all around us, our universe. We cannot dismiss the fact that the universe has been around for a very long time—roughly 13.8 billion years, give or take a day or two.[11] This is in direct contrast to some in the faith community who hold that, according to Scripture, the Earth was created between six to ten thousand years ago. But what if both are correct? I can only move the discussion forward with questions and observations, not dogma that cannot be proven. The faith-based explanation has its roots in Scripture, while the scientific explanation will always follow the scientific method. We can examine these two viewpoints more closely by addressing the question posed in this section.

A scientific evaluation of anything would begin with the following: identify the problem or question, gather relevant data, form a hypothesis based on the data, test the hypothesis, observe the results, draw a conclusion, then try to replicate the results. Let's see what that might look like.

The question or problem would be that no place or matter existed on Earth for God or *a god* to be present on, or in, when He spoke the heavens and the Earth into existence. From this, it could be concluded that God must have been somewhere else when He created the heavens and the Earth. Where was that place? The next step would be to gather data or relevant information to form a hypothesis. All the scientific community's method has offered is their understanding of where molecules and elements originated that are contained in what they can touch or see. There is no other place or matter, unless it is billions of light-years outside the observable solar system.

If it isn't part of our Earth, our time wouldn't affect it and it could not be measured. Coming to a conclusion would be difficult. The biggest blow to their process is that the beginning of creation could not be replicated. In fact, it never has been. The only conclusion left inside the scientific process would have to be that God could not have created the heavens and the Earth while present, because there was no cause, and therefore... there is no god. Their conclusion is that we exist on this Earth today through a naturally occurring process—sky, water, and land somehow have always been right where they are today, at least for 13.8 billion Earth years.

When the faith community comes to this question, they have an answer for creation. Their answer, based on faith, says there was a God who brought this all into existence from nothing, or at least from a very dark, formless ball of something. However, they too have no answer concerning the question as to the location of the God who, in the beginning, created everything from nothing.

Those who follow science and faith have a common process enemy known as "the moment of singularity." Simply put for all to understand, this term means "a beginning." Science has it harder than the faith community, since in recent time science has stated that a singularity is impossible, as it would have to suspend many of the time-tested rules of the universe for a singularity to have been present at some point in the distant past.

Science has developed the Big Bang theory—some event that caused the formation of the very first molecule, cell, bacteria, etc. However, it must still answer where those ingredients, and that event, came from to create the cause. It seems that every direction science has taken keeps interfering with other laws that it long ago accepted as fact.

Back to the third choice. Scripture says God created the heavens and the Earth from a formless, lifeless heaven and Earth. It does not say a totally empty space. We will later see that even empty space is made of… matter. Scripture says he placed two great lights in the heavens to rule the day and the night. Could it be that they were already there? Nothing in the Genesis creation account supports creation or changes to the rest of the universe. The universe could have already been in place for the previous 13.8-billion-years; calculated to fit within our understanding of time measured in… Earth years.

Who is right, faith or science? Based on physical evidence, I think both can be right—but with lingering, unsolvable equations. Science is still struggling with explaining where life began in a 13.8-billion-year-old universe. Science has also discovered little to support how this all came into being billions of years ago without changing its findings based on the next discovery or hypothesis. If at some point science does come up with a solid, replicable answer, then they can start working on what was in place prior to the 13.8-billion-year period they just figured out.

The faith community isn't struggling so much with the larger explanation as they are with more literal interpretations of when creation took place. Using Scripture, some believe creation took place 10,000 earth years ago. There are several different schools of thought within the faith-based community, including Muslim beliefs. The major difference between faith and science is that faith has nothing to prove; you must just believe and accept. That is why they are considered faith-based. Still, there are issues they don't talk about very much, namely 2,000-Earth-year-old archeological finds, including very old dinosaur bones and temple building civilizations dating back to 9,000 BC. The most recent find is a species named *Nyasasaurus parringtoni*, roughly 243-million-years-old. The find was announced by University of Washington paleontologist Sterling Nesbitt and colleagues in *Biology Letters*.[12]

So, where was God and the rest of the Trinity standing, sitting, or lying when He (They) created the heavens and the Earth? I think He (They) were wherever they are today, except that Christ later sends the Holy Spirit to Earth to be our comforter. I think both faith and science have a humanity lifetime left to answer several other intriguing but, as far as our present and future is concerned, irrelevant issues. The more they both discover, the more they will realize just how little they know… and how close they have been all along.

What does the Trinity look like?

To answer this question, we must first dig deeper. Is the question asking for a physical description? Is the question asking for the communal relationship between its three parts? Is the question asking how three beings, defined as one, process individual results? Or does the question spring from a deep mistrust of Christianity in general? This basic concept is difficult to get one's head around. Since the term is so critical to the understanding of the faith community, it should be clarified that, although the concept of the Trinity is clearly outlined and described in the Bible, the actual word, *Trinity*, appears nowhere in Scripture. [13]

Since many people are visual, let's start there. In an online publication, *EveryStudent.com, a Safe Place to Explore Questions About Life and God*, it is suggested that the Trinity is first viewed by humans as we see all else, in a three-dimensional world. Somehow, the mention of a trinity conjures up a picture of three entangled people, hopelessly inseparable, only able to accomplish a task with their mind or by awkwardly maneuvering about to arrive at a different location. Let's back up to the part about three-dimensional: *EveryStudent.com* gives the following illustration. The Trinity is not 1+1+1=3, but 1X1X1=1.[14]

Others have tried to explain the Trinity as being similar to ice, water, and steam. They are the same, but appear very differently depending on temperature. The same idea works with an egg: eggshell, egg white, egg yolk—individual elements, yet collectively an egg. These explanations are seemingly feeble attempts to define our Creator. However, they are examples from our natural world, which is the only viewpoint we possess. Point in fact, when humans venture into outer space they must alter the environment to be like Earth or they die. At 28,000 to 30,000 feet up, the rest of the universe turns on us. From there on out, as far as we can go, everything is designed to kill us. At 50,000 feet, even with an artificial breathing apparatus, we cannot survive unless we are wearing a pressurized suit to keep the atmosphere from destroying us, from inside out.

With this limited understanding it is difficult, if not impossible, to describe the concept of three-in-one. So, should we just give up? No, but we need to have an explanation prepared for an encounter with someone who is searching.

What about the relationship between the three Spirits? [15] It has been pointed out that it seems supremely egotistical for a God to create an entire human species just so He can receive their praise and glory. Although an expected view of the created, it could have merit if it were true. That is why it is important to understand the relational value of the Trinity. Instead of ego, let's look at a couple of other, more relevant terms directly relating to our faith: holiness and righteousness.

If the Trinity didn't exist, and there was only a God, how could it be possible for that singular God to be holy, or righteous? Holy and righteous to whom—himself? No, that is a concept even mere mortals can understand. If you are the only person in the world, can you be the richest person in the world, can you be the most famous, the most talented, the smartest person in the world? You may argue that you would be, since you are the only person breathing; but what would it matter? It wouldn't. Now let's bring in the concept of the Trinity.

The Trinity, the Three-in-One Godhead, the I Am, creates the need, not to be holy, but to be accountable to each other so that they deserve to be called holy and righteous, to have others to love, and cry over, to be responsible for, to make plans, to create the Good News, to share whatever is created and to work together and sacrifice to save it. This may be just a glimpse at the overpowering promise of the Trinity and more important, why the Trinity is… the Trinity.

What does the Trinity look like? Read the creation story and in particular about man's first breath. Read about Christ's last breath on the cross and what happened around Him. Read about how the Holy Spirit fell upon those at the resurrection and Christ's promise to send a comforter while He would be away for a while. What does the Trinity look like? Don't blink! You will miss it…

If we call God the Father, is there a Mother?

"No. Could have been. Why do you ask? Makes sense. Depends on who you ask. We were created in the image of the Trinity which includes a father with an only child. Hey, Jesus didn't have an earthly father!" Many times, these and other responses are given to this question and similar derivatives. It is an inquiry based on a logical assumption. It is particularly relevant in the 21st century, as women finally begin to approach equal status with men.

Scripture places a great deal of emphasis on Christ being the only Heavenly Son of God. Though he has an earthly mother, there is no mention whatsoever of how He was conceived before the creation event, or even if He was conceived, before the world began.

The following opening paragraph appeared in an NBCNEWS.com article: "God had a wife, Asherah, whom the Book of Kings suggests was worshipped alongside Yahweh in his temple in Israel, according to an Oxford scholar."[16] The article later states the conclusions reached are based on 1 Kings 15 and several partial inscriptions on discovered ancient pottery fragments. The authors assert that it could have been based on wide-spread polytheism prior to 586 B.C. when an elite community within Judea was exiled to Babylon and the Temple in Jerusalem was destroyed. This event was followed by a move to monotheism: one God.

If there was a special woman in God's life, she may have died, if that is even possible. The Mormon Church's first prophet, Joseph Smith, introduced the concept of a heavenly mother. Their belief is based on a few different ideas, but the main one is simply logical: God is the father, Christ is His only Son. It therefore follows that there must have been a heavenly mother.[17]

Is it possible, assuming there was a heavenly mother, that she was sent to Earth and became Mary, the earthly mother of Jesus? This is probably more difficult to believe from a logical standpoint than the concept of the Immaculate Conception (covered later in detail). If this is the subject of discussion with a non-believer, perhaps the best response is, "I don't think this can be answered right now, but I'll ask God when I see Him."

Some points can be made, however, since they are scriptural. Jesus was especially drawn to the behavior of children. He knew that the most important aspect of a child's growth centered around a father… and a mother. This comes straight from the Fifth Commandment of his Father. Jesus was also very respectful of women, who in that age were treated as inferior by men. He was especially attentive to his earthly mother, Mary.

Two other scriptural facts stand out when considering this question: Instead of marrying, Christ chose the church to be His bride.[18] He also was only one of three humans in history without a biological father. As with many things God does, these do not make sense to us. And that is the whole point. We cannot think like God. We cannot compare anything we do to His actions or decisions. You see, there have been aliens on the Earth and in the heavens. There will be again. If you believe in God, you believe in beings from somewhere else besides the Earth. Christ is one of these. He wasn't born as we are on Earth. So, maybe He wasn't born in Heaven according to Earth's procreation process?

One last twist: Who on Earth didn't have an earthly mother? Adam and Eve. We know that to form a human, DNA is needed to instruct their creation from conception. We also know men determine the sex of a child when conceived and developed in the womb. Apparently, Adam was not determined by another man's X or Y chromosome. Obviously, neither was

Eve. God made those decisions. More completely stated, these two individuals didn't have an earthly father *or* a mother. Unless we ask in Heaven, we may never know the DNA origin of Christ. The beauty of relationships, whether the Trinity, close friends, or a husband and wife holding their newborn baby, is not about their DNA. The beauty is simply the love between those individuals. Not accepting the loving embrace of faith in the Trinity is the greatest hope lost by humanity.

Is Christ God's only Son?

"*For God so loved the world that He gave His only begotten Son, that whoever believes in Him should not perish but have everlasting life.*"[19] The definition of begotten: "A somewhat old-fashioned adjective, *begotten* is the past participle of the verb *beget*, which means to father or produce as offspring."[20] It is more than interesting that the use of this word, while not inferring a woman was not involved in Christ's birth, certainly doesn't emphasize Christ was born of a man and a woman.

"The Greek translation of "only begotten" is *monogenes*, a word used nine times in the New Testament that can mean "one-of-a-kind or unique." (As interpreted by compellingtruth.org). [21]

Another widely used translation, the New International Version (NIV), doesn't seem to accept this definition: "*For God so loved the world that he gave his one and only Son, that whoever believes in him shall not perish but have eternal life.*"[22] This seems to simply state that Christ was the *only* Son God ever had. It does not, however, imply that the begetting process itself was unique or one-of-a-kind. Rather, one more easily can infer from it that Christ was the result of procreation, or a physical relationship between a man and a woman.

While Jesus' earthly family would have been considered blended, there isn't a word or title given to a family where there is no mother; at least, not yet. I suppose you could consider God and Christ part of a Trinity family. I suggest to you the following Scripture: "*How will this be," Mary asked the angel, "since I am a virgin?" The angel answered, "The Holy Spirit will come on you, and the power of the Most High will overshadow you. So the holy one to be born will be called the Son of God.*"[23] Although not as skilled as today's physician, this Scripture was written by a 1st-century physician, Luke. I wonder how Luke processed this when he first heard this information? What would your Ob/Gyn say if you told them you were pregnant and no man was involved in any way? Counseling, maybe?

All translations aside, if the Trinity exists as spoken of in the Bible, Christ was God's only Son. If you get someone to accept the reality of the Trinity, this question should be quite easy to explain.

Does the phrase "our image" in the Creation story include a female creator?

This question is created from earlier Scripture:

"Then God said, '*Let us make mankind in our image, in our likeness, so that they may rule over the fish in the sea and the birds in the sky, over the livestock and all the wild animals, and over all the creatures that move along the ground.' So God created mankind in his own image, in the image of God he created them; male and female he created them.*[24]"

These two Scriptures are a bit of a challenge to follow. Verse 26 says, "in our image" and in Verse 27, "in His own image." So, which one is it?

Both verses refer to the creation of mankind. Verse 27 expands on the definition of mankind by stating, "male and female He created them." Verse 27 also gives us a clue as to what God actually looks like and what His image represents. Verse 26 suggests male and female will be made in the collective image of the Trinity, those who were likely present on the sixth day of creation.

There are several points to be made regarding who is being referred to in verse 26. Here are a few possibilities:

- Looking at the Trinity, one might see a changing image reflecting not just man and woman but all races of people.
- One of the Trinity is a female, perhaps the Holy Spirit, since the 1st-century Christians saw Christ as a man and He is called the Son and God. God is always referred to as the Father, definitely a masculine reference.

- There were more beings present at creation than the Trinity, perhaps angels. The only issue with this is that almost always when angels are mentioned in the Bible, they are referred to in the masculine form. The term in both Greek and Hebrew is the masculine form of the word. One exception is Zechariah 5:5-11.[25]
- And then, perhaps Christ's Mother, God's wife, was present?

Back to square one. In Genesis 2, the activities of the sixth day of creation are expanded upon, but very little is added regarding the nature or purpose of the man and the woman in the creation story. So much more could have been discussed—humankind's ability to reproduce, think, see, hear, remember, and expand its shell to accommodate the growth of organs, and all other human emotions and traits, including faith, trust, hope, love, and charity. But someone else was present at creation.

In Ezekiel 28, the first ten verses seem to speak of a human ruler. However, beginning in verse 11 to verse 19, it is implied that his ruler is Satan, or a human king possessed by Satan.[26] A similar pattern is unveiled in Isaiah 14:12-17. I will discuss these later in the book.

Revelations 12:7-12 recounts a story about a war in Heaven between Satan and His forces against God's Kingdom. The passage refers to Satan's desire to supplant God and his attempted ascent to the very throne of God. He and His forces were cast down to the Earth.[27] We are not told exactly when this occurred but it must have been during creation or prior to it since the Serpent, the Devil, the Evil One shows up in the garden as Adam and Eve are getting settled in.

An interesting comment is made at the end of verse 12: "... For the devil has come down to you, having great wrath, because he knows that he has a short time."[28] We know the Earth was formed at least 6,000, 10,000, 204,000, or even 4.5 billion years ago, according to the Pope (in a huffpost.com article in 2017). [29] Any of these dates would make God's time in Revelation 12:12b seem much different than the six-day creation event. That verse says that Satan's time on the Earth will be a short time. In addition to knowing Satan was on the Earth at least near the creation story, he will also be on the Earth when it ends at Christ's second coming (or third—see later explanation) as stated in Matthew 25:41 (NIV): "*Then he will say to those on his left, 'Depart from me, you who are cursed, into the eternal fire prepared for the devil and his angels.'"* [30]

You might wonder what all of this discussion about Satan has to do with a female being present during the creation of mankind. If the war in Heaven happened after the creation of mankind, is it possible Satan was present? It is just a thought, but what better way for Satan to try to get back at God for rejecting Him than to corrupt God's humankind creation? Present or not at creation, Satan had a motive.

Many Other Questions...

There are more questions about the existence and mind of God than we could ever include here. Some of those might be:

- Does God breathe the same kind of air as we do?
- Where does God live—in the Heaven where Christ went to prepare a place for us, or some other place?
- Why did God decide to create something new and different?
- If God has foreknowledge, why would He create a humanity he knew was going to fail?

Three of the many things God created and breathed into us included a sense of curiosity, wonder, and experimentation. If we are curious about how something is made, then why would we not also be curious about spiritual things? Just as important, why would some also doubt what Scripture says, even to the point of completely rejecting it? At the end of the day... however long that really is... there are some things we will likely never know. I have heard some faithful Christians say we will find out when we get to Heaven. And then again, like leaving your favorite watch behind on Earth when you leave for Heaven, all of this probably won't matter much anyway.

CHAPTER 2

CREATION AS WE KNOW IT

Then the Lord God formed a man from the dust of the ground and breathed into his nostrils the breath of life, and the man became a living being.

GENESIS 2:7 (NIV)

ONE OF THE major differences between science and religion is time. Science is always confined by time—billions and billions of years. However, time has no relationship or relevance between God and Man. Most Christians, when thinking about the beginning of time, go directly to the Creation Story in Genesis. They go specifically to Genesis 2:7 regarding the beginning of mankind. However, in a discussion concerning creation, we need to go back further, much further—in fact, so far back that time becomes irrelevant.

To take this journey, you may need to release any long-held notions regarding History and Scripture. In fact, it might push what you accept as the limits of your belief. To begin, we will see how far back creation goes. Later, we will also examine how far into the future it may go.

Begin by taking off your watch and closing your calendar. These are worthless tools on this journey. Once you leave Earth, neither of these is relevant. As an example, science has discovered that those who leave the Earth and enter outer space visiting the International Space Station are younger when they return to Earth than if they had never left. Although they're only thirteen milliseconds younger after a year in space, it is still possible to alter time as it relates to our earthly human bodies.[31]

Here is a simple, literal recap of the creation of the heavens and the Earth according to Genesis, chapters 1 and 2:

Genesis 1: 1-31

Verse 1—It begins with, "*In the beginning... ,*" but does not expound on this. We must assume it means the beginning of, "*the heavens and the earth.*" Do "*the heavens*" include the vast universe discoverable by telescope? Was that also created at this Genesis 1:1 moment—or was it, like the voided, formless Earth, already in existence? Scripture says it was without form and void, but it does not say that it wasn't there.

Verse 2—The earth was a dark, water world without form and was void of anything living or dead. We discover here that God was not alone. His Holy Spirit was there hovering just above the dark water.

Verse 3—This is the first recorded instance where God speaks. It isn't clear as to whom He is speaking; perhaps to Himself or to a substance called light, or to someone called "light." Whomever or whatever, it was obeyed and it appeared. There is more clarity added by reading John 1:1.

Verse 4—God was pleased with the light that He added to His creation. He then caused the Earth to spin, thus creating day and night. This was the beginning of the future need for watches and calendars. It is not clear why the sun (probably created in verse 1) was not already shining on the earth. It could have been blocked by another planet or by dark clouds in the atmosphere. Although it is allegorical, John 1:4-5 fills in missing information about this situation.

Verse 5—God gave the light and darkness names: day and night. He declared that each day would have a morning and evening. This seems to suggest that a twenty-four-hour day is immediately established, but we can only assume this to be true. If it is true, it would have no bearing on the length or span anywhere else than within the heavens and the Earth. We can only speculate if there is day and night in a place like Heaven, how long it might be, or if there is even a need for varying lengths.

Verse 6—This verse, taken by itself, could suggest God caused dry land to form, which by separating the water provided a stage for creations to come. Skipping to verse 8 helps clarify that there was apparently a need to pull some of the water up into the heavens, likely forming clouds. Dry land does not appear until verse 9. There are several ways this could have occurred when He spoke.

Verse 7—Just confirms verse 6 was completed.

Verse 8—God names the firmament heaven and declares an end to the second day. Note that God does not label this action as "*good.*"

Verse 9—It appears God places all the remaining water on the Earth into one place. This suggests it was initially located in several different places. This action caused the dry land to appear. There is a bit of a physics challenge here. Simply pushing all the water into one place would not cause the land to rise out of the water. Water always seeks its lowest level. However, moving that much water weight could have triggered massive earthquakes and volcanoes, which subsequently pushed land upward from the mantle. Science tells us that there are tectonic plates floating on the Earth's inner mantle. It has been postulated that these plates have moved great distances since their beginning. Perhaps, in the time described in this verse before the plates separated, all the land was grouped in one place and the ocean in another. This could have occurred during an earlier time.

Verse 10—It is here that God calls the waters "seas" and the dry land "earth." Unlike verse 8, God calls this part of creation, "*good.*"

We should note that nothing is yet breathing in the heaven, or on the Earth, or in the sea.

Verse 11—God speaks into existence grass, herbs, plants, and fruit trees. He did not plant seeds but created them already fully grown. They would accomplish two things: it would produce seed to multiply and convert carbon monoxide into breathable oxygen.

Verse 12—This verse validates verse 11. God acknowledges this creation to be "*good.*"

Verse 13—God proclaims the third day to have an evening and a morning, a beginning and an end. He may have had to mention this again if the Trinity had to tweak the Earth's rotation from verse 4.

Verse 14—Interestingly, God again says there is to be light in the firmament of the heavens to divide night and day. Those lights could only be the sun during the day and the moon, and possibly stars, at night. This would suggest that our solar system was also fully formed at this time. Science suggests that our moon was formed before this through a collision with another planet. The creation of the seasons at this point provided humans with what they would need later to develop the concept of time.

Verse 15—Validates that the light in verse 14 is to give light on the Earth. This light is further defined in John 1.

Verse 16—After speaking these earlier lights into existence, God now identifies the sun and the moon, calling them "*two great lights.*" He seems to now clarify that the heavens mentioned earlier included stars. Could the earlier creation of these two great lights possibly have been fine tuning in response to the result of the rotation of the earth and its effect on gravity?

Verse 17—Validates that placing light on the Earth is extremely important.

Verse 18—Validates the purpose for the light, and that this part of creation is good.

Verse 19—God proclaims the evening and morning of the fourth day.

Verse 20—God speaks fish and birds into the creation, which begin to use the oxygen in the air and water previously created in order to sustain these new life forms. There is no mention of His breathing life into these creatures.

Verse 21—God creates even larger fish and birds. He sees that what He has done is good.

Verse 22—This is the first time God blesses His creation, telling them to multiply and fill the Earth.

Verse 23—God proclaims the evening and morning of the fifth day.

Verse 24—God creates living creatures that roam the Earth, each according to its kind. This term, "*according to its kind,*" is used extensively throughout this first chapter of Genesis. This may suggest that "*its kind*" may have existed somewhere else at some point. Or... perhaps God literally designed what they looked like based solely on His creative imagination.

Verse 25—Validates verse 24. God saw it to be another good creation.

Verse 26—This verse is the first place where the use of the plural is obvious with the terms "*Let Us...* ," "*in Our...* ,"and "*Our likeness...* "He certainly was not alone during this effort.

Verse 27—This verse, however, is singular. "*God created man in His own image; in the image of God He created him...* " Then, almost as an afterthought, "*male and female He created them.*" This human creation will be fleshed out later in the second chapter of Genesis. This verse raises more questions than it answers.

Did God, in the beginning, look like humans of today? Did He possess arms, legs, feet, hands, body, head, hair, beard? Does His body function like ours? Must He eat? Digest? Sleep? Speak? Breathe air like ours? There must have been a space of time between the creations that occur in verses 24 and 25 and the creation of man and woman. Later we find out that, after He created man, the man complained of loneliness, and so God created woman. He created her by using one of Adam's ribs, suggesting cloning. At a minimum, a DNA code was applied.

Verse 28—God again blesses the man and woman and gives them dominion over His entire creation. As with the beasts of the field, He instructs them to be fruitful and multiply. Often it is noted that the woman is given the pains of childbirth only after she is escorted out of the garden. If this is true, then it follows that childbirth in the garden would have been painless. Had the fall not occurred, there would have been childbirth, since God instructed Adam and Eve to be fruitful and multiply.

Verse 29—God validates His gift of every living thing to the man and woman, but apparently as vegetarians.

Verse 30—All the living creatures were also to eat plants and fruit, not each other or humans.

Verse 31—Then God saw everything He had made, even the things He previously did not acknowledge as positive, and said it was all indeed very good. He then declared the end of the sixth evening and day.

Genesis 2:1-25

Verse 1—Proclaims all of creation is complete… finished.

Verse 2—God was tired from all the work He alone had completed, so He rested on the seventh day, creating the Sabbath. So, as perfect and mighty as God is, He can be overworked.

Verse 3—God blessed the seventh day and sanctified it as a day of rest. This concept may be difficult to fathom. Was this a physical or mental tiredness? We may never know.

Verse 4—The author proclaims the previous account to be the history of creation and then goes back to define certain sections. It should be noted that, considering when this account was written, its descriptions are strikingly in agreement with modern-day science, except for conflicting calendars.

Verse 5—Returns to the day before chapter 1, verse 11. It is emphasized that God, at this time, had not caused it to rain from the firmament He separated and placed above, nor was there any man to till the ground. Apparently, no need for rain.

Verse 6—For some reason it is mentioned that, instead of rain, the whole Earth was watered by a mist that went up, not down. It sounds similar to modern in-ground irrigation systems. This fact is later referred to regarding the unprecedented rain detailed in the flood that eventually fell on Noah.

Verse 7—This verse seems to just appear without reason, except that it sets up God's desire to plant a Victory Garden in East Eden in the following verse, where He places the man, not the woman. It is here that we initially are told of the first breath of humanity. It was this breath alone that caused man to be a living being—which hadn't happened before, when he was formed from the Earth. Speaking of being formed, did God just bend down and spit into the dust of the Earth to make that living human body? Or was the body fully formed, just lying there, waiting on life? This is all we are told about how man became alive. What would Adam's reaction have been when God's breath filled his lungs? Was he instantly aware? Was God's breath enough to activate every muscle motor, brain memory, eye focus, and stereo hearing? Did he have to be trained to care for the Earth, the animals, and the Garden? Do you think Adam thought, at this time, about the last breath he would take?

Verse 8—God creates a Garden and places the man inside.

Verse 9—God creates everything in the Garden, including the Tree of Life and the Tree of the Knowledge of Good and Evil.

Verses 10 through 14—Directions to the Garden and minerals that are contained in four rivers nearby. Sounds like an Indiana Jones treasure map, doesn't it?

Verse 15—Here, it is mentioned once again that God placed Adam in the Garden to tend and keep it.

Verses 16 and 17—God forbids Adam from eating the fruit of the Tree of Knowledge of Good and Evil. If Adam does, he will surely die. This may have been the first time Adam thought about the last breath he might take. Even today, the distance between these two breaths can be short, and stop suddenly.

Verse 18—It is at this point, after telling Adam the one thing he could not do, that God looked on Adam and realized His living creature was lonely. God decided to make Adam a comparable helper, meaning a person with a few minor variations.

Verse 19—Adam was given the honor of naming every animal and bird God created, as he thought they should be called.

Verse 20—God had not yet created the woman. It was during this naming process that Adam discovers there is no one who looks and acts like him. (Refer to verse 18 above)

Verse 21—God invents anesthesia and puts Adam to sleep. God operates, removes a rib, and sews him up.

Verse 22—God fashions the rib into a woman—no explanation—and brings her to Adam, who is recovering from surgery.

Verse 23—Apparently God tells Adam how He created Eve because Adam proclaims her as bone of his bone and flesh of his flesh—in other words, a perfect DNA match. God also allowed Adam to give Eve her name, calling her "woman" because she was taken out of man. These verses leave out the sophisticated details of the most complicated thing yet to be created or invented on this planet: a human being. Science addresses this by speculating on billions and billions of years of evolution and primordial soup. There have even been suggestions that bacteria flew in on comets.

Verse 24—Here we find the oldest reference to the marriage of a man and a woman. That they were created from the same body leads to the suggestion that once married, they should again become one flesh.

Verse 25—Neither had any need for clothes. Yet.

John 1:1-50

Verse 1—Here we find added clarification about the plural used in Genesis 1. Not only was the Holy Spirit present, so was Christ the Son of God (the Word). If this is properly understood, it is Christ's first appearance on the Earth, not at His birth. Earlier, I mention Christ's third appearance when He returns at the end of earthly time. This reference gives that impression.

Verse 2—Validation of verse 1.

Verse 3—It appears that, although Genesis proclaims creation was spoken into being by God, the reference in John 1 is to Christ—that everything was made through Christ, and none of this would have been possible if not for Christ. He was instrumental in designing and creating the very world and living beings that would one day kill him.

Verse 4—It seems all of the life created by God's spoken word resided in Christ, including the light.

Verse 5—There is a vague reference here to evil, or Satan, as darkness. The light, Christ, was shone into the darkness, but it was not comprehended by the darkness. Meaning possibly that Satan did not understand what was happening, nor why. It also suggests, by reference, that Satan was also present on the formless void called Earth before it ever took on its Trinity-inspired shape. So how far back does Satan's story go? How did he get to this location, and why? Later, my friend. Be patient.

Verses 6 through 8—Introduction of Jesus' cousin, John the Baptist, and the cousin's prophesied calling to proclaim the coming of a Messiah.

Verse 9—John the Baptist's acknowledgement that Christ was the true Light given to everyone entering this world. In other words, when someone is first born, and takes their first breath, they are made aware of who Christ is. This knowledge is stored in them to digest as they grow, eventually urging them to accept this light as true and receive saving grace for their soul before they take their last breath.

Verse 10—Acknowledgement that Christ was in the world when it was created and validation that the world was made through Him. Interestingly, the world at the time did not know Christ, or in Satan's case did not recognize Him. This raises a question: When God was walking in the Garden, did He walk alone or was He accompanied by Christ and the Holy Spirit? This also might give us insight as to what the Trinity looked like in person—perhaps one being with three different vestiges? Or... maybe not.

Verses 11 through 13—These verses transition to the purpose of Christ's second appearance on Earth. Interestingly, Scripture refers to Christ's next return as the Second Coming. It will actually be His third coming—creation, born as a man, and the rapture.

Verse 14—It is here that Christ mirrors Adam in taking His first breath when born of Mary as a human. Christ's last breath on the cross, after He sighs the words in Luke 23:46 NKJV—"*...Father, into thy hands I commend my spirit...*"—will be a return to the power and might of creation itself. The anguish of the Trinity when Christ died on the cross likely caused the earthquake and storm mentioned in the retelling of those events. It seemed to be as close as the original creation ever came to breaking into pieces and returning to a place without form and void.

Verses 15 through 50—These verses are the introduction to Christ's ministry over the next three-year period, including the establishment of this ministry as well as an invitation for certain people to follow Him.

Anytime someone reads Scripture, they should understand their responsibility to the original manuscript. What does that mean? As most folks know, there are numerous versions of the Christian Bible. BibleGateWay.com is a popular online site containing about fifty-nine English versions, and globally the full Bible has been translated into more than 680 different languages. In addition, the New Testament has been translated into 1,534 languages and other parts of the Bible into another 1,133.[32] Some pastors, who have studied Kione Greek and/or Hebrew, will translate sections on their own. The versions used by these translating pastors may be closer to the true meaning of a verse than many of the other, more popular versions.

Looking back to ancient times and moving forward to the present, members of the Jewish faith lay certain other texts beside the Hebrew Bible, or the Old Testament, to validate and better define its meaning. Among these are the Mishnah, written in 22 AD, explaining Jewish law code that up until that time was largely oral tradition. Also used is the Talmud, a collection of teachings and commentary on Jewish law. The Talmud's first edition was finalized around the 3rd century, followed by a second edition near the 5th century. Another accompanying book was the Midrash, composed between 400 and 1200 AD. [33]

All ancient languages assign very narrow meanings to some words, while other words may have a variety of meanings. Only by comparing the use of a word in multiple Scriptures under different and similar circumstances can the reader or listener best determine the message to be gained. This understanding applies nowhere better than in the passages just listed in Genesis and John.

The creation story is first told to us in Genesis 1 and 2, which are among the oldest recorded passages in Scripture. It is almost a certainty that whoever wrote these verses was not present when the events unfolded. For that reason, these passages were written using the words and experiences of the person writing the words. They were not a firsthand account. This does not weaken the assertion that all Scripture is inspired by God and is, in fact, the Word of God. Regardless of the circumstances, other than the Ten Commandments—which were written personally by God, in stone—a human was the author. Employing a written language that was early in its development, the author had to record what he heard from differing versions of oral sources that might have been passed down for several thousand years.

Let me give you my uninspired version of the Scripture verses listed previously.

The Trinity either came into being at some point and from somewhere, or perhaps they just simply always were. At some point, the Trinity decided to create a very orderly, complex world populated with all manner of creatures and humans. There was abundant vegetation that existed on dry land surrounded by bodies of water. The creatures and humans were able to move about and rest according to available light. The time this took is listed as six days with a seventh day of rest.

After the creation was finished, the Trinity established a garden for the humans to tend and to provide them with food. In that garden, the Trinity grew a tree of knowledge. Eating from this tree would give the humans the ability to determine right from wrong. They were told not to eat the fruit of that tree. They did, with some coaxing from Satan, the Serpent. This led to the humans being expelled from the garden into the rest of the Earth, which was not nearly as hospitable. They would now have to experience pain and death. It is unclear as to whether the Trinity initially intended for humankind to take in that first breath provided by God, never to breathe their last.

By combining the first two books of Genesis with the first book of John, we can see that the Trinity—God, Christ, and the Holy Spirit—were all three present and active during this event. Before they took action, nothing was breathing yet on this formless, void body floating in space.

As scattered as this beginning may have been, it was our beginning, and according to other Scripture, it will be where civilization eventually draws its last breath.

The Winter of Creation is far from over.

CHAPTER 3

THE FRAGILE CREATION OR BILLIONS OF COINCIDENTAL EVENTS HAPPENING AT RANDOM WITH CAUSE

In the beginning *God created the heavens and the earth.*[34]

In that case, it (Book of Thomas) would be the only first-century Gnostic Christian text—virtually all other Christian texts whose goal is gnosis are from the second or third centuries. Many second-century Christian texts had gnosis as their goal.[35]

Jesus said, "If they say to you, 'Where have you come from?' say to them, 'We have come from the light, from the place where the light came into being by itself, established [itself], and appeared in their image.' If they say to you, 'Is it you?' say, 'We are its children, and we are the chosen of the living Father.' If they ask you, 'What is the evidence of your Father in you?' say to them, 'It is motion and rest.'"[36]

Instead of developing from complex molecules such as RNA, life might have begun with smaller molecules interacting with each other in cycles of reactions. These might have been contained in simple capsules akin to cell membranes, and over time more complex molecules that performed these reactions better than the smaller ones could have evolved, scenarios dubbed "metabolism-first" models, as opposed to the "gene-first" model of the "RNA world" hypothesis. Oh, and if you thought all that was mysterious, consider this: Scientists admit they don't even have a good definition of life![37]

Did people and dinosaurs live at the same time? No! After the dinosaurs died out, nearly 65 million years passed before people appeared on Earth. However, small mammals (including shrew-sized primates) were alive at the time of the dinosaurs. Many scientists who study dinosaurs (vertebrate paleontologists) now think that birds are direct descendants of one line of carnivorous dinosaurs, and some consider that they in fact represent modern living dinosaurs. This theory remains under discussion and shows that there is still much we don't know about dinosaurs.

Where did dinosaurs live? Paleontologists now have evidence that dinosaurs lived on all of the continents. At the beginning of the age of dinosaurs (during the Triassic Period, about 230 million years ago) the continents we now know were arranged together as a single supercontinent called Pangea. During the 165 million years of dinosaur existence this supercontinent slowly broke apart. Its pieces then spread across the globe into a nearly modern arrangement by a process called plate tectonics. Volcanoes, earthquakes, mountain building, and seafloor spreading are all part of plate tectonics, and this process is still changing our modern Earth.[38]

The first is a personal stance: "I don't know if God exists or not." This is often called weak agnosticism. The second is a universal claim: "We cannot know if God exists or not." This is often called strong agnosticism. It is actually a claim to knowledge. It makes the claim that there isn't enough evidence for anyone to know whether God exists or not.[39]

Atheism has its own creation myth, writes Spencer, and it goes like this: non-belief is the love child of reason and science—a human advance arising from scientific and philosophical progress in Europe, particularly movements like the Copernican Revolution in the 16th century, the scientific revolution in the 17th and the Darwinian in the 19th.

> *Spencer describes this myth: "Gradually, wonderfully, the human race matured, with every confident scientific step forward pushing our infantile, crumbling ideas of the divine closer to oblivion," he writes. This myth is "true enough to be believable, (but) it is not true enough to be true." Why does any of this matter? Because these days the foundation myth of atheism is widely accepted in the West and has implications. Atheism, so the story goes, is rational, while religion is embarrassingly irrational; atheism represents clear thinking human progress, while religion is regressive, primitive superstition.*[40]

TODAY, IN OUR 21st century, and throughout history, two things have gnawed at the core of understanding. These two opposing views, like good and evil, are seemingly monolithic and can never share the same space. What are they? Science and faith… and then there is religion… later.

The opening passages of this chapter are a small collection of views from people of faith: Gnostics, science, agnostics, and atheists. Between the cracks of these classifications live many more views, some well thought out and others simply extreme in both directions. All parties attempt to defend their positions with varying degrees of success based on several criteria. The reason these groups still exist and are increasing is because of two basic principles. The first is the failure of each to offer undeniable evidence. This failure only increases the number and desire of those who find the topic fascinating. The second is a basic distrust by both groups of anything that does not support their position.

Let's take a look at some of these issues.

Helpful Definitions

As we get deeper into this discussion about just how large this creation is, there are some terms that should be defined as you make this journey. These terms are often thrown out in reports, articles, and discussions. Sometimes they are used incorrectly or in a confusing way. Perhaps this list will be helpful to you, as it was to me in doing this research. You may wish to earmark these pages for future reference. [41]

Asteroid—Asteroids are chunks of rock or metal that used to be parts of other small planets that have broken off due to a collision in space and now orbit the sun. The name asteroid means "star-like" referring to their emission of light. Asteroids can be lumped into orbital groupings called families and form into an asteroid belt.

Big Bang—The Big Bang Theory is a model describing the origin of the universe. It explains how the universe expanded from a high-density, high-temperature state into the cosmos we see today. Although there are alternative theories, this is the most widely recognized theory of how the universe began.

Black Hole—Made famous by Einstein's theory of general relativity, a black hole is a small, dense core remnant of a dead star. Since the density of this core is more than about three times more than the density of the sun, the strong force of gravity produces a black hole.

Comet—Besides being a beautiful phenomenon that's rarely visible from Earth, comets are icy, frozen balls of gas, rock, and dust. Comets gain their trademark bright glow by emitting gas as their orbit passes close to the sun, sometimes even resulting in a glowing tail.

Constellation—One of the first things we learn about space is its many constellations. From Orion and his belt to the mythical Seven Sisters and the big and little dipper, constellations map out a group of stars that form shapes in the sky. Constellations were first used by explorers to map direction as well as a tool for storytelling.

Dark Energy—Like the cosmic mystery its name evokes, dark energy is hard to define. The easiest way to think of it is a property of space that accounts for an expanding universe. The universe's "empty space" may still have its own, dark energy. However, scientists are still trying to pin down the properties of this dark energy that makes up most of the universe.

Dark Matter—Somewhat similar to dark energy, dark matter also covers a large chunk of the universe and remains quite mysterious to scientists. The main difference is that dark matter explains more of how groups of objects in the universe function together. Though scientists have a greater understanding of what dark matter isn't than what it is, a leading theory posits dark matter is made from exotic particles like axions or Weakly Interacting Massive Particles (WIMPS).

Escape Velocity—Escape velocity is the speed that an object needs in order to escape a planet or moon's gravitational pull. For example, the escape velocity for a spaceship to leave the surface of Earth is about 25,000 miles per hour.

Galaxy—We all know that we live in the Milky Way galaxy, but what exactly is a galaxy? A galaxy is a wide-ranging mix of dust, gas, stars, and various solar systems brought together by a singular gravitational pull.

Gravity—Based on the Latin word for "weight," gravity is a natural phenomenon where things with mass or energy are drawn toward its center. The force of gravity is what keeps us planted on Earth's surface and also keeps the Earth orbiting the sun.

Light-Year—A light-year is as simple a concept as its name suggests. It is an astronomical distance that measures the distance that light travels in a year. One light-year roughly equates to six trillion miles.

Meteor—A meteor is a rocky object or asteroid that vaporizes when entering the Earth's atmosphere. They're often called "shooting stars" and provide a beautiful glowing tail as they enter the atmosphere. Next time you see one, be sure to make a wish upon a meteor!

Milky Way—The Earth's galactic home, the Milky Way, is a spiral galaxy that is constantly rotating. The Milky Way contains a high amount of dust and gas wrapped around the central galactic bulge. The galactic bulge contains a dense amount of dust, gas, and stars. And within that bulge exists a large black hole. Our solar system lies about 30,000 light-years from the galaxy's center, or 30,000 times 6,000,000,000,000 miles!

Oort Cloud—Named after astronomer Jan Oort whose theory about a cloud comprising a shell of icy objects existing beyond the Kuiper Belt, or the region that exists in our solar system beyond Neptune, could be responsible for the long-term comets visible from Earth (like Halley's). The cloud is theorized to comprise the remains of disc materials of planets and the sun.

Orbit—In short, an orbit is the gravitationally curved trajectory of an object in space. In other words, an orbit is the circular motion in which objects in space travel due to gravity's pull. For example, the moon orbits around the Earth and the Earth orbits around the sun in a circular motion. An object that orbits another is called its satellite.

Solar System—A solar system is a gravitationally bound system of a sun and the objects that orbit it. That includes the bodies that orbit the objects orbiting the sun. For example, our solar system contains the sun, the planets that orbit it, as well as moons and other objects that orbit them.

Star—Other than serving as beautiful guiding lights in our sky, stars are astronomical objects that are composed of gas that pulls itself together with its own gravitational pull. Nuclear fusion at the star's core creates the light that stars emit.

The Theory of Relativity—The theory of relativity was developed by Albert Einstein in 1905 and includes both theories of special relativity and general relativity. The theory determines that the laws of physics are the same for all non-accelerating observers and was monumental in introducing a framework for space and time.

Universe—And lastly, we have landed at the universe. This loaded definition has flummoxed scientists for generations. A universe contains billions of galaxies and even more stars along with their planets, moons, asteroids, comets, dust, and gas, all of which exists swirling around in the massive thing we call space.

Our Home Is Fragile

The *"In the beginning"* believed and taught by people of faith often leaves gigantic holes in their timeline. This position often appears to be intentional, as seemingly undeniable evidence is simply ignored. As a person of faith, I prefer to address these issues head-on. First, let's look at what was central to the Biblical Creation account: the heavens and the Earth.

The Genesis Earth in the faith example is easy to identify. It is a formless mass, void of anything other than… a formless mass with certain deep places which were dark. The undefined object is the *"heaven."* Which heaven, what heaven, that just touching the Earth, extending how far, in which direction? Since the account mentions the creation of presumably the sun and the moon to initiate day and night, we could assume it includes our solar system containing only the other eight or nine planets—depending on whom you ask—orbiting around the main light source. If this is true, then the Creation would also include several layers of space as you move away from the Earth. It isn't accurate to say "as you move up from the Earth," because once you leave the Earth, up, down, sideways, east, west, north, south, like time, are irrelevant.

Early observers had only their eyes to search the skies. They surely assumed everything they could see was all the same big creation, including objects beyond our solar system. They had no reason, or tools, to think otherwise. Today, however, 21st -century man knows that our solar system is but an infinitesimal grain in the—invent a number, one doesn't exist—seeable universe. Question: Did God create everything we can now see through our technology, as well as endless more which we cannot?

Let's come back to Earth…

As large as our planet is, compared to our relative understanding, it is one of the most fragile elements in the knowable universe. Let's list a few.[42]

- Our solar system could not exist without everything around it.
- Simply, the tropospheric ozone is an atmospheric layer adjacent to the Earth. If its amount would decrease even slightly, then most animals would begin to experience respiratory failure.

- The Milky Way is unique, because this galaxy is the only one (currently known) in which there is life.
- If the Earth's continents were concentrated to a greater extent in the southern hemisphere, then sea salt in the air would not be enough to stabilize the surface temperature of our planet.
- The atmosphere of the Earth has a thickness of about 298 miles, but its bulk is within 9.23 miles, or between 48,000 to 50,000 feet of the surface. For comparison, commercial airlines routinely cruise between 30,000 and 40,000 feet. If atmospheric pressure were less, then liquid water would evaporate too easily and condense too rarely to sustain life. Without an artificial source of oxygen and a pressurized suit, a human would die quickly between 48,000 and 50,000 feet above the surface of the Earth. Think of a location roughly nine miles from where you are sitting. That distance turned vertical is all that separates all of humanity from extinction. Now consider that the circumference of the Earth is 24,900 miles. Looking at an object that large makes a nine-mile-high atmosphere seem very thin and easily manipulated.
- According to scientists, the Milky Way is one of the most massive and brightest galaxies in the universe and is the ideal size to sustain life. It is not too large, since the gravitational effect of gas and stars would disrupt the orbit of the sun, and not too small, in which interstellar gas would not be enough to maintain the formation of stars for a time sufficient for life to arise.
- The triatomic hydrogen cation, also known as protonated molecular hydrogen or H_3 +, is one of the most common ions in the universe. If this ion were less than now, the simple molecules necessary for the formation of planets and the chemistry of life would never have formed.
- If the angle of Earth's inclination was more, or less, as on most other planets of the solar system, the temperature differences on the surface would be too large to support a variety of life forms, including human life.

- If, for some reason, the amount of iron in the ocean and soil rises (compared to the current level), then iron poisoning will destroy highly developed life forms. If the iron content decreases, many of the nutrients needed to support life will disappear. If there were less sulfur, then the formation of a solid inner core would begin too early, which would lead to its increase too quickly, also violating the magnetic field.
- The solar system in which we live is very far from the center of the galaxy where there is a large black hole. Closer in, many stars have faded and there are also many supernovae that kill life. In addition, the solar system is located between the spiral arms of the galaxy, away from supernovae. This location also provides an excellent overview for exploring and observing our galaxy and the universe.
- Even if we take gravity for granted, the truth is that without gravity, human life would not exist. The amount of gravity on the Earth's surface is perfect for human life as compared to other planets within our solar system. Earth's distance from other planets creates our gravity and would be significantly altered with small adjustments in distance. It is generally accepted that gravity is a major factor in biological changes, such as the evolution of species.
- All life forms on Earth are carbon (since all life forms consist of water and breathe oxygen in one form or another). Scientists believe that these elements appeared much later in the stars. The early stars were very massive, but this did not last long. If these elements were not created, then life would not have arisen.
- Thanks to the rotation of the Earth, the sun rises from the horizon every morning, and sets in the evening. If it were not for rotation, it would be unbearably hot on one side of the world, and everything would freeze on the other. Life would be impossible anywhere on the planet. Rotation also contributes to our level of gravity.

- The moon has a critical, positive effect on the Earth. Scientists now know that the Earth originally had a rotation period of eight hours. Such a fast rotation period led to the fact that the wind speed near the surface was over 497 mph. The gravitational pull of the moon over the past four billion years has increased the Earth's rotation period to roughly twenty-four hours.
- If the Earth did not have a strong and relatively stable magnetic field, then all life would be literally fried by cosmic rays and solar storms. The bad news is that every 300,000 years, the magnetic poles change places. This reversal process takes thousands of years.
- The Earth's moon is four hundred times smaller than the Sun, while it is also four hundred times closer to the Earth than the Sun. As a result, during a lunar eclipse, the moon perfectly closes off the light from the Sun. This makes the Earth the only place in the solar system from which to observe this solar eclipse. Solar eclipses have helped science test some of the laws of physics, including Einstein's theory of relativity.
- The huge size of Jupiter makes it a wonderful natural shield for the Earth from many asteroids and comets—it attracts this space debris to itself. In addition, planets like Jupiter are extremely rare in the universe.
- Earth does not exist in a vacuum. The Cosmos in the solar system is full of asteroids, comets, dust, and traces of gas. Even now, small cosmic stones fall to Earth daily. Large space objects crash into the planet often enough that scientists must keep a constant eye on them. In the very first years of our planet's formation, collisions with comets and asteroids brought water and other important chemicals to the Earth, which subsequently made life possible.
- The Earth revolves around the Sun at the "right distance." This is the so-called "habitable zone" or "Goldilocks' Zone," where water can exist in liquid form, which is the basic requirement for life.

- The Moon and Sun are where they are for a reason. If either of them was slightly closer, or further away, the Earth would suffer catastrophic circumstances. Both the Sun and the Moon are responsible for the Earth's ocean tides. The Moon more so than the Sun. The results of the Sun moving closer or further away would have more to do with temperature. A closer Moon would cause gigantic tides, some so large they would bury New York City and other coastline cities. The Sun moving closer or further away would either bake the Earth or freeze the Earth. Either way, life would cease to exist.

Is it possible that all of this and more just happened? If so, for what cause? If so, at what odds? If so, how long did it take? Understanding the size of everything combined in our existence, isn't it likely this accident would have happened many more times? If so, where are they? You are right, they may be out there beyond our sight. This may also just be a way of understanding eternity...

The Vastness of Space

No one knows if the universe is infinitely large, or even if ours is the only universe that exists. And other parts of the universe, very far away, might be quite different from the universe closer to home. Simply put, some of what we see today, or at night, by the naked eye or through sophisticated satellites or telescopes, has taken the light of what we observe nearly 14 billion years to reach us. This light is the youngest because it is a snapshot of our newborn universe, long before the first stars and galaxies formed. Closer to home, the Milky Way is a huge city of stars, so big that even at the speed of light, it would take 100,000 light-years to travel across it.[43] While defining some of this terminology, remember 100,000 light-years is 100,000 times 6,000,000,000,000 miles.

Today, science's best estimate is that the Andromeda galaxy is 2.54 million light-years away. Because distances within galaxies are measured using the light within them, it is currently impossible to measure the space between galaxies because there is no light, just dark matter.

The most red-shifted light we can detect in the observable Universe suggests that light has reached us from galaxies that are 13.8 billion years old. Try not to confuse how old something is with the light-years it would take to get there. Also, the amount of time it takes for the light of an object to reach us is not the age of the object. We don't know how long the object was there before it emitted light, nor do we know how long it will be in existence, because none of us will likely live long enough to see the light disappear.[44]

So, what is a light-year? A light-year is a measure of distance, not time. It is the total distance that a beam of light, moving in a straight line, travels in one year. To obtain an idea of the size of a light-year, take the circumference of the Earth (24,900 miles), lay it out in a straight line, multiply the length of the line by 7.5 (the corresponding distance is one light-second), then place 31.6 million similar lines end to end. The resulting distance is almost 6 trillion (6,000,000,000,000) miles! Now multiply that by 2.54 million and that will give you a rough idea of how far it is to the Andromeda Galaxy. Not your average weekend trip.[45]

These numbers, as unimaginable as they are, should give an idea of the vastness of space. It is just too much for a human to comprehend. And somewhere in this vast, unmeasurable infinity lies a grain of sand on the beach of the universe: the Earth, surrounded by its heaven. Then visualize yourself standing somewhere on that planet, surrounded by everything else. As a point of reference, it takes the International Space Station (ISS), flying at an altitude of 220 miles above the Earth's surface at a speed of 17,227 miles per hour, ninety-two minutes to circle back to the place where you are standing.[46]

When you are sitting perfectly still, contemplating the meaning of life, the Earth is spinning around on its axis at a speed of 1,038 mph at the equator in a twenty-four-hour period. Just 80 degrees north or south of that point, the speed slows to 180 mph.

The Earth, as a planet, is circling around the sun at a speed of 66,660 mph to complete one orbit in the 365.2425 days that make up our Earth year.[47]

Now ask yourself: "Am I an accident of nature?" This exercise could lead you to depression and anxiety over a human's uselessness, when compared to the enormity of the universe. That would be where faith and science tear apart.

Not that you don't have enough to consider, but did God create both the solar system in which we live and everything beyond it? If not, he probably knew who did. If it wasn't Him, there must be more than one god.

And Then There Were Dinosaurs

For much of my life I have, as a Christian, struggled with this question: "What about the dinosaur?" I was raised to believe that since the Bible said six days, everything was created in six days… and God rested on the seventh. Therefore, you rest on Saturday or Sunday, depending on which faith you profess.

Squaring this belief with the scientific evidence of dinosaurs created tension in my life every time it came up in a secular setting. Over time, I came to deal with it by accepting the explanation that a day in the Old Testament period was of a different length than it is today. I would later learn that this is the case, but in the wrong direction. (Again, that science thing!). Science postulates that a day at the beginning of existence was shorter, not longer. This theory suggests that days were only between five to seven hours, due to a collision with another large planet. The result is believed to have formed our moon. The gravitational effect of the moon—and also the Sun, to some degree—caused the Earth to slow its rotation to where it is today.[48]

Well, that didn't help solve the Christian timeline of 10,000 years versus the evidence of hundreds of thousands of years since dinosaurs roamed the Earth. I genuinely had a problem for which there seemed to be no answer. So how did I handle it, as a mature adult? I ignored it.

In a well-written essay by Richard F. Carlson, retired research professor of physics at the University of Redlands and a PhD. in physics and an MA in theology and biblical studies from Fuller, we find his assessment of this ageless tug between faith and science:

> *We read in the Bible about creation, and all of us are aware of at least some aspects of that same creation as understood by natural science.*

> *The accounts of creation in these sources do not use the same words and do not have the same overall areas of concern, but we can expect that at least the accounts should not conflict. The book of nature described by science should not contradict the book of nature as read from the Bible. But for the past century and a half, too much conflict has existed.*[49]

His phrase "*...at least the accounts should not conflict*" screamed at me from the text. Yes, that is what I have reasoned for decades. Growing up as a Christian, I am confident I wasn't the only one troubled by the inability to settle this conflict.

My early choices for a profession while growing up were not only fireman, James Bond, and major league baseball player, but also archaeologist and paleontologist. However, I ruled out dating rocks and bones when I discovered just how old the scientific disciplines claimed some of them were. As a Christian, I couldn't do that... not with a straight face. Chief among the dating thing was determining the age of dinosaur bones. Someone needed to fix whatever dating device they were using!

As the maturity of my reasoning caught up with my growing body, I began to see God's handiwork in a new perspective. I always thought that the conflict between science and faith was the result of bad science. It had not occurred to me that God never intended for me to understand everything. Imagine that! It isn't human nature to accept that there are things you will never know, much less understand. I began to wonder if perhaps God intended for this to remain a mystery, and subsequently a continuing conflict between science and faith. Perhaps it was His will for both communities to be ever searching and ever revealing new things to us. One of these things, I have discovered, is that there is very little factual conflict between the two. The conflicts lie in that both rely on a great deal of speculation when arriving at conclusions. A good example is the faith-based and scientific explanations as to how life began—neither can be proven nor disproven.

As you may have guessed, I watch a great many programs on the Science Channel. As I listen to the narrator and watch an artist's rendering of what is going on light-years away, I am always fascinated by remarks that could just have easily ended in, "...and that's how God created the universe" instead of, "...based on what we know, we believe this is how it happened... for now, until we discover something else that changes everything." To give written proof to any naysayers that I have lost my faith, let me go ahead and say this: Some Christian-based presentations which conclude with, "and this is the way it is because God created the heavens and the Earth" could just as easily end with, "...based on what we know, we believe this is how this happened... for now, until we discover something else that changes everything."

Which brings us back to dinosaurs. It quite clearly turns out I was wrong, and the devices used to date rocks and bones are accurate—depending on whom you ask. What do you think the following terms have to do with dating the age of the universe, the Earth, people, and dinosaurs?

> *Fossils, radiometric dating, isotopes, carbon, uranium, atoms, neutrons, protons, nucleus, parent, daughter, half-life, mass-spectrometer, carbon 14-dating, uranium-235, uranium-238, potassium-40, igneous rocks, sedimentary rocks, volcanic ash, igneous brackets, zircon crystals, amino acid, magnetic fields, lasers...* [50]

These terms, and others, are at the base of all disputes between faith and science. When you are staring at, or even better, holding a bone in your hands, it is very hard for anyone to deny the bone exists. The question is, "...how old is it?" While faith has reason to doubt some forms of dating, it also has a bit of a struggle proving that the Earth is 10,000 years old, beyond what is written in the Bible.

Some Christians argue dinosaurs could not have existed because God would not create such creatures capable of wreaking havoc upon human civilization. Enormous amounts of archeological evidence counter this argument, but we should also note the Bible mentions God created several other ferocious animals.[51]

Jonah 1:17 describes a great fish swallowing a man.

Are dinosaurs mentioned in the Bible?

Job 40:15-24 describes a behemoth creature with a tail like a cedar and an herbivore diet.

In Job 41, a leviathan is recorded with armor-plated skin that no hook can pierce. Some label it as a crocodile. Psalm 104:26 indicates it lived in the water. Isaiah 27:1 indicates it is a twisting serpent, while Job 41:18-19 indicates the creature literally breathed fire. It is important to note that from a scientific perspective, the first dinosaur bones were not discovered until 1819—not exactly an ancient event.

For instance, using this same scientific method, ten archeological finds in 2018 alone bolstered evidence for events and people mentioned in Scripture. [52]

The Science of It All

The scientific community has deduced that dinosaurs first appeared on the Earth 240 to 247 million years ago, and for about 175 million years inhabited much of the Earth's surface. An extinction event, the source of which is still widely debated, occurred around 65 million years ago, wiping out all but the avian dinosaurs. Many scientific communities believe birds are a surviving form of dinosaur.

Paleontologist Richard Owen coined the term *dinosaur* in 1842 when he examined the bones of three very large beasts. The word dinosaur is derived from the Greek *deinos*, meaning "terrible" or "fearfully great," and *sauros*, meaning "lizard" or "reptile."[53]

During the time of the dinosaurs, all of the land on the Earth was connected in one large area science calls Pangaea. The land mass was surrounded by one very large ocean. This description fits well with the Biblical description of the Earth prior to God's Spirit moving on the surface of the Earth during Creation. The only problem is that this took place about 190 million years before the Biblical account, according to some Biblical scholars. [54]

So, if dinosaurs lived some 240 million years ago and lasted for about 175 million years, where does science place humans in the timeline? Although living things began to show up 250 million years ago, the Earth is said to be 4.5 billion years old. Notice the "b" in the word following 4.5. That is an important distinction related to this paragraph. Apparently, science will only give a name to a period so far back. There comes a point when periods just don't do justice to how far back history goes. I guess you could call it reverse eternity? The title for this 250 million period is called the Prehistoric Period; the word is self-defining.

This period is followed by the Stone Age, the Bronze Age, and the Iron Age. All four of these periods cover a span from roughly 250 million years ago through about 1200 BC. That means these four periods ended about 3,200 years ago and about 1,200 years before the birth of Christ.[55] Or, Before the Common Era (BCE), as some would argue. This stone age era is marked by the use of tools by our early human ancestors (who evolved around 300,000 BC.) and the eventual transformation from a culture of hunting and gathering to farming and food production.

In the Paleolithic period (roughly 2.5 million years ago to 10,000 BC), early humans lived in caves or simple huts or tepees and were hunters and gatherers. The end of this period marked the end of the last Ice Age, which resulted in the extinction of many large mammals. The subsequent rising sea levels and climate change eventually caused man to migrate. This Stone Age era ran up to about 3000 BC. Then humans discovered bronze and figured out how to make things out of it that improved and protected their lives—the Bronze Age. To give some reference, Egyptian pyramids and hieroglyphics were developed during this era. This period lasted until about 1300 BC when yet another discovery was made: iron. Although the Iron Age was the shortest of these four periods, advances in many civilized ideas seem to escalate.[56]

The longer mankind continues to have an appetite for discovery, and technology continues to produce more sophisticated instruments to feed those appetites, new discoveries are going to continue to bend our understanding of really ancient history and the humans who lived and thrived during these periods. I will cite two discoveries, both of which are changing many things civilization has held as unchangeable for decades.

Amanda Borschel-Dan, writes on *The Times of Israel* website on July 16, 2019:

> *An unprecedentedly vast Neolithic settlement—the largest ever discovered in Israel and the Levant, say archaeologists—is currently being excavated ahead of highway construction five kilometers from Jerusalem, it was announced on Tuesday.*

> *The 9,000-year-old site, located near the town of Motza, is the "Big Bang" for prehistory settlement research due to its size and the preservation of its material culture, said Jacob Vardi, co-director of the excavations at Motza on behalf of the Antiquities Authority. "It's a game changer, a site that will drastically shift what we know about the Neolithic era," said Vardi. Already some international scholars are beginning to realize the existence of the site may necessitate revisions to their work, he said.*

This article continues to document a city of some 3,000 people who were quite advanced in building, husbandry, agriculture, and forging advanced tools. Also found were public halls and places for worship containing statues and other artifacts. All of which was very unexpected for a 9,000-year-old civilization.[57]

Sandra Scham, writing for *Archaeology*, a publication of the Archaeological Institute of America in their November/December 2008 edition, reported on another amazing, history-changing discovery in Turkey. The site is described as:

> *The press here is fond of calling the site "the Turkish Stonehenge," but the comparison hardly does justice to this 25-acre arrangement of at least seven stone circles. The first structures at Göbekli Tepe were built as early as 10,000 BC, predating their famous British counterpart by about 7,000 years.*
>
> *Schmidt and his colleagues estimate that at least 500 people were required to hew the 10-to-50-ton stone pillars from local quarries, move them from as far as a quarter-mile away, and erect them. How did Stone Age people achieve the level of organization necessary to do this? Hauptmann speculates that an elite class of religious leaders supervised the work and later controlled the rituals that took place at the site. If so, this would be the oldest known evidence for a priestly caste—much earlier than when social distinctions became evident at other Near Eastern sites.*
>
> *Archaeologist Klaus Schmidt downplays extravagant spiritual interpretations of* Göbekli Tepe, *such as the idea, made popular in the*

press, that the site is the inspiration for the Biblical Garden of Eden. But he does agree that it was a sanctuary of profound significance in the Neolithic world. He sees it as a key site in understanding the transition from hunting and gathering to agriculture, and from tribal to regional religion. [58]

So, did man and dinosaurs exist at the same time? According to the United States Geological Survey (USGS) under the Department of the Interior, *"No! After the dinosaurs died out, nearly 65 million years passed before people appeared on Earth."* The closing sentence from their website states, *"This theory remains under discussion and shows that there is still much we don't know about dinosaurs."*[59] As you can see, science doesn't have a wrap on this movie just yet. I would postulate that neither does the faith community.

Summing Up the Beginning

At the end of the discussion, does it really matter if the dinosaurs and humans lived during the same time? When attempting to bring science and faith to a point of agreement, which involves ironing out conflicts in both group's assumptions, we really just need to figure out the time discrepancy. Give and take on this issue would assist both groups in settling most of their other differences. It might be more practical for a scientist to accept a Creator if things happened over a few million years instead of a twenty-four-hour day. Then there is the small challenge getting some scientists to believe there is a God.

These thoughts and my statement at the end of the previous subheading may be considered heresy to some in the faith community. The scientific community may suggest I am putting words in its mouth. To me, this approach does not diminish God's power or divinity; instead, it enhances God's ability to create unthinkable, unimaginable, fantastical realities. Or, you can just create your own word, because I can't think of any human word, in any language or dialect, that properly describes such divine abilities.

As much as our carnal beings have tried to figure out—through scientific process and endless searching through scientific experimentation—where we came from and when it happened, we only open another door that leads to yet another never-ending search. An example of this is the story of the creation of man from the faith point of view, but also from what we know about a human being.

The Scripture simply says in Genesis 2:7, *"Then the Lord God formed a man from the dust of the ground…"* Let's think about that for a moment. The next thing we are told is, *"…and breathed into his nostrils the breath of life, and the man became a living being."* The only two elements mentioned are *dust of the ground* and presumably air when the word *breath* is used. Two elements. Both were previously created. However, one came from outside God's being and the other came from inside: His breath. The water [60] and oxygen contained in that life-giving breath from the Creator, combined with the dust of the Earth, and the human race came into existence. Just like that.

We can marvel endlessly over the mind-boggling size of our known universe, and how everything seems to work like gears and motors in a very complicated grandfather clock. But to create a human that has yet to exist from dust and a single breath of air, that would be the money shot, as they say in my hobby of photography.

Think about Adam's body for a moment before that first breath. There he was, lifeless. He was perfect, the very image of those who were gathered for the event. But what was contained inside that lifeless body? Not to pretend I know all the names of everything inside my body, I will keep this simple. Adam's lifeless body contained a brain with all of the billions of connections that would allow him to experience emotions, figure computer-rattling equations, create art-music-literature-languages and countless other things. It contained ears that turn vibrations into intelligible sounds, recorded and remembered by the brain. Two precious eyes dominate the face with the ability to receive light waves in various forms to create 5G (latest version as of this writing) stereo images recorded, recognized, and stored by the brain. A mouth that is the beginning of the digestive system and sustainer of life which can also form sounds to communicate-sing-laugh-kiss-drink-eat-and speak words worked out by the brain. In a woman's breasts and body, milk to feed newborn children and reproductive organs.

Hands with opposing thumbs. Feet that allow the body to stand, walk, run, balance, and swim. Arms and legs connecting hands and feet connected to the brain for coordination. Internal organs to pump blood, fight infection, filter and direct the digestive system, muscles-tendons-bones-ligaments to give the body strength and shape, skin to cover it all that was resilient-waterproof-strong-durable-ever-replacing, and a source of cooling and heating. DNA, RNA, cells, microscopic bacteria, and millions of other things I am not smart enough to know.

Last but not least, inside this lifeless body are the lungs that process oxygen and distribute it to the entire body, without which, in as little as two minutes, he dies. It seems that this lifeless, impossibly complex body was waiting on one breath of air to become a living human being. At that moment all of the previously mentioned parts of this marvelously created body sprang to life and continued to function… without any more help from their creators.

… and the faith community and science want to argue over how long it took for this to happen. Frankly, what does it matter? Does crowding this into a twenty-four-hour day make it any more miraculous? Does stretching the process out over a few million years make it more believable? Truth is, to me, science has not, nor will it ever, create life from nothing. You can tell they have given up on that approach when they start figuring out how a living cell arrived on Earth from somewhere out there. Even if they do, who created those cells?

The faith community may want to lighten up on the idea of forcing a human-conceived timeline onto God's creation story. Am I a heretic because I am rooting for the scientific community to prove beyond a doubt how life originated? I want to put all of this aside and recognize God for who He is and what He has done and is doing, despite humankind's best efforts to mess it up. My aim is not to diminish *what* God is, but to recognize *who* He is.

So... Let's Talk About God

Earlier we spent a good deal of time talking about the Trinity—the Godhead, three-in-one. It is hard enough describing the Trinity, much less defining it. Instead, let's spend some time talking about God and several things we have been taught about Him. As with much of this study, you will just have to let some things go... if you truly believe we were made in His image.

Let's start with this one: Was creation God's full-time job? Did He have other things to do? Before creation, what did God do in Heaven? What was a day like in the life of God? Some folks imagine God sitting on a throne with all of the believers standing and sitting around Him, singing, praising, and having a great time. Well, is that it? So, God just sits there and we all look at Him for eternity? If not, what else does God do, what else is He in charge of that takes up a good portion of His 100,000-million-hour long day? Oops, no watches in Heaven. I guess the twenty-four-hour thing is over, or at a minimum, the eight-hour work day.

So, what about before creation, what else was God in charge of wherever He resided? These are great questions, but can they be answered? I think so.

From the website Jesus Created Ministries you will find ten things they believed existed or were planned by God before the Creation event took place:[61] Here are a few.

- **Before Creation, the Godhead existed as the Father, Son, and Holy Spirit.**
    - Psalm 90:2 "*Before the mountains were brought forth, or ever You had formed the earth and the world, even from everlasting to everlasting, You are God.*"
    - John 1:1-2 "*In the beginning was the Word, and the Word was with God, and the Word was God. 2He was in the beginning with God. 3All things were made through Him, and without Him nothing was made that was made.*"
    - Hebrews 9:14 "*How much more shall the blood of Christ, who through the eternal Spirit offered Himself without spot to God.*"

- **Before Creation, God chose the saints for salvation.**
  - Ephesians 1:4-5 *"Just as He chose us in Him before the foundation of the world, that we should be holy and without blame before Him in love, 5 having predestined us to adoption as sons by Jesus Christ to Himself, according to the good pleasure of His will."*
  - 2 Timothy 1:9 *"Who has saved us and called us with a holy calling, not according to our works, but according to His own purpose and grace which was given to us in Christ Jesus before time began."*

I should also mention that this group believes with certainty that the Earth was created between 5,000 and 10,000 years ago. Some of their doctrine seems to be in defense of predestination. If that is true, one of the significant things which occurred prior to Creation was compiling a list of those who, through the foreknowledge of God, were selected for salvation and entry into Heaven; a tougher position to defend than many scientific speculations.

Turning to a more personal theme, conjecturing as to who else might have been with God before Creation, Dr. Michael S. Heiser, PhD. in Hebrew Bible and Semitic Languages at the University of Wisconsin-Madison, answers in his provocative and enlightening book, *The Unseen Realm: Recovering the Supernatural Worldview of the Bible.*

> *Children often ask, 'What was there before God made the world?' The answer most adults would give is that God was there. That's true, but incomplete. God had company, and I'm not talking about the other members of the Trinity.*

Dr. Heiser points to Job 38:4-7:

"Where were you at my laying the foundation of the earth?
Tell me, if you possess understanding.
Who determined its measurement? Yes, you do know.
Or who stretched the measuring line upon it?

On what were its bases sunk?
Or who laid its cornerstone,
when the morning stars were singing together
and all the sons of God shouted for joy?"

The Dr. rightly asks, "Who were these sons of God"? He believes they were not humans, who had not yet been created. You might think they were angels, but he doesn't believe that is accurate. Based on his knowledge of ancient Hebrew belief, he mentions that there is, as in earthly Hebrew family structure, hierarchies. He suggests that this is also true where God came from; in his understanding, Heaven. There were angels and then there were archangels. While angels were often messengers, archangels serve God with more important tasks. All of these angels, regardless of name, are non-human and created by God. Dr. Keiser suggests that God had, and therefore has, a *Divine Family.* Since the term, *begotten*, means to be "caused by and born of a man," these other beings could be half-brothers to Christ, like adopted children.[62]

It seems the more we read, the larger the crowd at Creation. It makes sense by our human standards. Why would someone go to all this work and then allow only a few people to watch? Creation would have been a spectacular event, worthy of an immense audience. So, who else might have been, maybe not in the crowd, but created before the Genesis Creation?

Biblical evidence exists in several places denoting not only who Satan is but where He came from and where He was sent. Isaiah 14:12 gives us a clue: "*How you are fallen from heaven, O Lucifer, son of the morning*!" Luke 10:18, "[Jesus] said to them, '*I saw Satan fall like lightning from heaven*'" Ezekiel 28:14, *"You were on the holy mountain of God."* The rest of this passage, Ezekiel 28:12-15, describes Satan to His fall. Revelation 12:7-9 is a short look at a very large event in Heaven before the Genesis Creation account. We find that a "War" broke out in Heaven after the Devil, Satan, Lucifer decided to ascend up to the throne of God, to be like God. It didn't go well for Satan or the angels He convinced to follow Him. The Archangel Michael came to the defense of God, eventually driving Satan and all His followers out of Heaven and casting them down to the Earth.

Satan's arrival on the Earth would logically have occurred after it was created, since it was called "the Earth" and not some other name for a dark formless void. However, why would God intentionally send Satan to His newly created, perfect world? Oh, to tempt His newly created couple in the Garden. With this information, we are left to wonder if the war happened before Creation of after. It also makes sense that Satan would have been sent to a cold, dark, formless void as punishment, only later to be used by God to test the will of His newly created humans. Thus, Satan would not only have been present for Creation, he would have been right in the middle of it.

Whatever timeframe surrounds human existence, it cannot compete with an infinity of events prior to human's creation, nor the other side of infinity when Christ returns for the third time. What about after creation - did God's new, full-time job become overseeing His creation and fixing what breaks? Wait... are you saying God may have made a mistake? Yes, more than one. On purpose? Maybe.

Other Aspects of the Man and Woman Creation

Let's start at the beginning or our story, we humans. We know of at least one issue prior to Creation involving the creation and appointment of a certain archangel in Heaven that didn't work out so well. Isn't it amazing how, even when God seems to make the wrong call, it turns out perfect? Just call it a God thing. Remember, our purpose is to better communicate with those who reject faith, not so much those who choose natural answers.

Let's take a look at what happened right out of the gate in Eden. Perfect atmosphere, perfect clothes—didn't need any. Perfect plot of land, perfect crops. Beautiful scenery. No pain, disease, or hard labor. No decisions to make... wait a minute, that's the problem, the glitch. Decisions. Why would God allow this perfect couple to make decisions, especially between right and wrong? But there was no right and wrong. There was just a tree of knowledge of good and evil... and the perfect couple was told not to eat its fruit. There was no decision, just a command—not a suggestion. Well, maybe more of a parent's unsolicited chatter about what will happen if you touch the hot stove... there will be consequences.

Do you see where this is going? It's not a reality TV program, but it does have a very ugly ending. God made a decision, at some point before this garden scene, to create an archangel. A most perfect, beautiful, intelligent archangel. That decision didn't go well. It seems free will has always been a part of everyone's DNA. After this perfect archangel went off track, God had him driven down to the Earth, where he shows up in a perfect garden occupied by a perfect couple. This outcast angel doesn't force a decision on the perfect woman, he just creates doubt. It is interesting that a big deal is made out of casting the perfect couple out of the perfect garden, but it seems the once-perfect archangel gets off with a slap on the wrist… at least for then.

Back to the good world… outside the perfect garden. Remember the Creation story, where God would create something and then proclaim it "good." I wonder if He said that after careful consideration of what result He saw after the Creation. I also wonder if, after creating a perfect, most beautiful, intelligent archangel, he paused and said, "…and it was good"? Although he said on several occasions during Creation what He created was good, He didn't say it was perfect.

After being expelled from the garden, Adam and Eve had children, beginning with two boys. After they grew up, each chose an occupation. One became upset that his occupation didn't appear as worthy as that of his brother's, so he took a "good" rock and crushed the head of his seemingly imperfect brother. We, of course, rightly blame this on Satan. However, Satan didn't willingly descend down to the Earth from Heaven, He was driven there. The murdering brother, born from perfectly created parents, made in the image of God, used one of the emotions he was born with and killed another human being.

I told you previously I would be called a heretic, right? Actually, I am just anticipating what a non-believer might bring up in a conversation with you. Hopefully, you will find some help in these pages when that time does come.

What the Atheist, Agnostic, or Naturalist Would Like to Know but, Is Too Wise to Ask

What follows is a summary of a few questions I have heard, read, dreamed, or thought of while thinking about my faith. I believe it is okay to ask God any question. Remember, curiosity was breathed into humans along with all of the many other traits of the Trinity during Creation.

- Wasn't Creation a full-time job; didn't God have Heaven to run?
- God defined normality by Creation.[63]
- How long was man created before taking on life?
- Does God breathe oxygen as we do, or did He just adapt to the air He created?
- Will we breathe the same type air in Heaven as on Earth?
- What was in the first DNA code in Adam and Eve?
- Did God breathe life into animals, CO_2 into plants? There is no mention of how they became alive.
- Does God breathe? He must, if he breathed life into mankind.
- Is the air breathed by God the Holy Spirit?
- Is there air in Heaven?
- What was Adam's first breath like?
- Without water, life cannot exist. Where did it originally come from?
- Why is water so plentiful throughout our known universe but, life is so scarce?
- Does science assume water was just here?
- Are there other creations?
- Who else may have been present at Creation?
- What else was created prior to the heavens and the Earth?

This chapter's purpose was to introduce you to a multitude of views, opinions, theories, and skepticisms held by many humans. In making these introductions, I have attempted—at a minimum—to either make you angry or place enough doubt in your mind to cause you to reflect inwardly. Whatever your resulting determinations, please remember that when communicating with someone who doesn't believe as you do, the last thing they want to hear is an intractable refusal to dialogue. An open discussion with plenty of give and take leads to mutual respect. Respect leads to trust. Trust leads to openness. If you are willing to have a conversation on the logical and material benefits of science without being defensive, your unbelieving friend may be as willing to consider that which you have to share regarding the spiritual benefits you have received in your walk as a believer.

Not to beat this to death, but this is exactly what must happen between the science and faith communities. To accomplish this, it may require all of us to spend more time with those who do not agree with our viewpoint.

CHAPTER 4
THE MISSING DEAD SEA SCROLLS

Genealogies Versus the Uncaused Cause

January 24, 1936

Dear Phyllis,

I will attempt to reply to your question as simply as I can. Here is my answer:

Scientists believe that every occurrence, including the affairs of human beings, is due to the laws of nature. Therefore, a scientist cannot be inclined to believe that the course of events can be influenced by prayer, that is, by a supernaturally manifested wish.

However, we must concede that our actual knowledge of these forces is imperfect, so that in the end the belief in the existence of a final, ultimate spirit rests on a kind of faith. Such belief remains widespread even with the current achievements in science.

But also, everyone who is seriously involved in the pursuit of science becomes convinced that some spirit is manifest in the laws of the universe, one that is vastly superior to that of man. In this way the pursuit of science leads to a religious feeling of a special sort, which is surely quite different from the religiosity of someone more naive.

With cordial greetings,

your A. Einstein [64]

"...science without religion is lame, religion without science is blind."

Einstein – from his Essay, "Science and Religion" published 1954 [65]

In the same way that the scientific community continually strives for that elusive, perfect, factual truth, there is much the faith community does not know about the writings of the Old and New Testaments. Within the Old Testament alone it is estimated by Biblical scholars that some 2,000 years are missing between Chapters 2 and 11 of the Book of Genesis. Though not to be discussed here, there are also about eighteen years missing in the New Testament regarding the childhood and young adulthood of Christ. Those in the faith community ponder the following questions: Are these periods missing, were they never written, or was it God's divine inspiration of Scripture that we should never know?

Something we do know is that there were multiple books written during or after those contained in the Bible we recognize today. Whole texts and tens of thousands of fragments have been discovered. Some of these discoveries match what has already been incorporated into the Bible, while others are slightly different. There are also complete writings that do not make an appearance at all in our traditional Bible. Some of these have raised questions regarding the contents of accepted Scripture.

The most controversial of the additional writings were discovered in late 1946 or early 1947 by Bedouin teenagers who were tending their goats and sheep near the ancient settlement of Qumran, located on the northwest shore of the Dead Sea in what is now known as the West Bank. One of the young boys tossed a rock into a cave and heard something shatter. He and his companions later entered the cave and found a collection of large clay jars, seven of which contained leather and papyrus scrolls. Thus began a worldwide pursuit of intense study related to the best-selling book in history.

The cache, purchased later by an antiquities dealer, ultimately ended up in the hands of various scholars who estimated that the texts were upwards of 2,000 years old. Later, Bedouin treasure hunters and archaeologists unearthed tens of thousands of additional scroll fragments from ten nearby caves. Together, they make up between 800 to 900 manuscripts. [66]

For the faith community, the discovery of these new scrolls and fragments was the equivalent of a major scientific breakthrough that puts into question everything that had been previously believed on a particular subject. While these new writings didn't necessarily call into question the existence of God, some did suggest that long-held belief in the divinity of Christ was overstated. Tests and identification of these findings continue today and will likely not be complete for many years to come.

And what about the missing records in the book of Genesis? In a *Biblical Chronology and Dating of the Early Bible* by Curt Sewell, we are told:

"There are three ancient text versions of the Old Testament—the Septuagint, the Masoretic, and the Samaritan Pentateuch. Although scholars say that all agree on the important doctrines, there are noticeable differences between them. Of particular interest to us now are the different numbers in the genealogies that are given in the 5th and 11th chapters of Genesis, which record the number of years from the Creation until the birth of Abraham." [67]

It is important to note that Mr. Sewell has some harsh things to say about archeological techniques in dating rocks, bones, and other historical artifacts. He is quite adamant that the history of the Earth cannot extend back further than about 5,000 to 10,000 years. He cannot accept, based on his and other chronological research, that the age of the earth is 4.5 billion years old. He suggests dinosaurs were somehow part of God's creation.

Not to be overlooked in his article is the quote above about three separate texts used to translate the Old Testament. He also admits there were no surviving writings of eye witnesses to the first eleven chapters of Genesis, nor could there have been if writing had not yet been invented. He also mentions discrepancies in genealogy between Genesis 5 and 11. While all of the "begets" in the first few books of the Old Testament are very boring to read, they are nonetheless extremely important.

When it comes to clearly spelling out the age of the Earth, all sides involved—evolutionists, creationists, liberals, and conservatives—claim to have the most credible measuring stick. This is counter-productive. For some reason, all sides feel it is important to date the Earth in a way that validates how they view history. But unlike the winner of a war where the winner gets to rewrite the history of the events, this conflict will never have a winner. Why? Because one side is trying to validate a theory that keeps changing, while the other side stands on faith alone—a non-starter when it comes to theory.

The unspoken reason that the debate regarding the age of the Earth can never be resolved is due to the wide chasm between the stated dates of origin. Resolving a deferential between 10,000 years and 4,500,000,000 years would require a miracle, which the science community does not accept as reality. As I am writing a book on this subject, permit me to take a stab at a settlement.

I realize that as soon as some read this, they are going to insist that this is not a new thought. I agree, it isn't. In fact, I will give you a reference. It is called the *Gap Theory.* [68]

This reference is from an online site called GotQuestions.org. In the About Us section of the site, we find the following statement:

> *GotQuestions.org is a ministry of dedicated and trained servants who have a desire to assist others in their understanding of God, Scripture, salvation, and other spiritual topics. We are Christian, Protestant, evangelical, theologically conservative, and non-denominational. We view ourselves as a para-church ministry, coming alongside the church to help people find answers to their spiritually related questions.*

In the last paragraph of this article, the writer calls the Gap Theory a "strained reconciliation" on the part of some to reconcile "old earth, evolutionary theories with the book of Genesis." As you know by now, I don't have a problem asking questions of my readers, or God. Jesus made it clear that children were important to Him and, I assume, equally what they thought. I am comfortable in bringing a childlike innocence to my understanding of the difference between Genesis 1:1-2 and 1:3-5. My understanding of this matter is not to prove anything… just to understand.

First, my non-theologian, non-scientist observation is that verses 1 and 2 of Genesis chapter 1, is a description of the Trinity's need to create something from nothing. Here is the NKJV version of verses 1 through 5:

Verse 1, "In the beginning God created the heavens and the earth."

Verse 2, "The earth was without form, and void; and darkness was on the face of the deep. And the Spirit of God was hovering over the face of the waters."

Verse 3, "Then God said, 'Let there be light'; and there was light."

Verse 4, "And God saw the light, that it was good; and God divided the light from the darkness."

Verse 5, "God called the light Day, and the darkness He called Night. So, the evening and the morning were the first day."

Verse 1 contains a very important word. After it is used, it can no longer be associated with anything else that follows. That word is "beginning." In Hebrew, the opening phrase, "In the beginning," is based on *bereshit*. The word can mean: in the beginning, at the start, or at the head of. In Jewish tradition, the word can refer to either the first weekly Torah portion in the annual Jewish cycle of Torah reading or to the first book of the Torah itself called, *Sefer Bereshit*. Jewish tradition also suggests that this opening statement is not of a first-person perspective of some man's understanding of God, but from an omniscient third-person perspective, a voice that reveals the Glorious Power that created the entire cosmos by means of His Word. [69]

When you "begin" a project of any kind, you may have gathered what was needed to complete the project, made extensive plans for how the project would proceed, gathered all the personnel required as well as thousands of other details. However, all of this preparation is not the actual *beginning* of the physical work.

As an example, construction of a house does not actually begin until that first scoop of dirt is turned to lay the foundation. However, a lot of preparation and planning has already taken place, including the ordering of materials, to get to that point. The same was true with Creation—except those materials were created out of nothing.

From several other Scriptures we understand that the Trinity lives in Heaven, the place where Christ ascended to, "…to go to prepare a place for us." That place to which Christ ascended was where there previously had been a war, after which Satan was thrown down to the Earth—this prior event must also be figured into this *beginning* equation. It seems that the Trinity came into our universe, likely from Heaven, after creating the Earth. Nothing says this happened immediately, however. This would have been the one and only "*In the beginning...*" Notice that heavens are plural and Earth is singular. There is nothing further said about the condition of the earthly heavens at this time, but it was worth Scripture mentioning that the Earth was a mess and that there was water, over which there was some reason for God's Spirit to be hovering.

Beginning in verse 3 we have creative action. Unlike the heavens and the Earth, which were created from nothing in verses 1 and 2, God for the first time begins in verse 3 changing what He created in verse 3. It is possible that the creation of the heavens and the Earth in verses 1 and 2 were not necessarily part of Day 1. The intent of verses 3 through 5 are very specific and clear. All five of these verses are in the proper order, but don't seem to share time. A lot could have happened between verse 2 and 3, maybe even billions of years. Through verse 2, all God has created is an imperfect world. Notice He *didn't* say right after creating a universe, our planet, and the heavens above—no small feat—that it was "good." That declaration does not appear until verse 5.

Here's the deal. To me, it really doesn't matter if Creation was 10,000 years ago or 250 million years ago. It was before I was born, and before accurate, long-term storage of written records existed. Science isn't concerned with genealogy recorded in a book of faith, but the faith community does have to admit that very large animals with bones appearing to be significantly older than other animal bones, and bones of people who don't look a lot like 21st -century bodies, have been around since before any recorded history.

Scripture doesn't specifically tell us what the time-line should look like. Why? It is a book of faith. If God had placed a timeline in His inspired Bible, etched with the same finger He used to write the Ten Commandments for Moses on stone tablets, would it have affected our path to salvation? How? It would have given us something else to worship. Instead, He gave us science to work with the faith community to better understand His creation and how everything works together for mankind's benefit.

The faith community cedes nothing to the scientific community by simply admitting that there is evidence of things that existed on this Earth which do not have to fit their interpretation of genealogies contained in Scripture. It doesn't create any errors in the Bible. Rather than insisting on the Creation story timeline, Bible scholars may be better served by insisting that science, in the absence of accepting Biblical awareness, show evidence that all of this "something" we call the universe simply came from… something. By doing so, the science community would only be agreeing with the faith community, beginning with Genesis 1:3. That would leave the science community with a clear assignment to prove where their *beginning stuff* first came from. So far, their efforts have centered on two possibilities yet to produce replicable results: singularity and the Big Bang theory.

Such a collaboration would require each to make a concession. The faith community, while not conceding the twenty-four-hour, six-day Creation Scripture, would accept that the billions of years of Earth history postulated by science could have occurred between Genesis 1:2 and Genesis 1:3.

The hardest part would be for the science community to accept that there is a God.

CHAPTER 5
THE STORY OF A PERFECT GOD

... and a few of His mistakes

Therefore you shall be perfect,

just as your Father in heaven is perfect.

MATTHEW 5:48 (NKJV)

The universe is perfect —

eerily, uncannily perfect—

AS A SETTING FOR CREATING LIFE.

LIVESCIENCE.COM, BY RAFI LETZTER, FEBRUARY 12, 2018

IF WE ARE going to discuss perfection, we should first get a handle on the root word "perfect." Or, since the above reference came from the New Testament book of Matthew, the meaning of the Kione Greek word used in this passage, *teleioi*. Scholars agree that the use of this word by Matthew as a nominative plural masculine adjective at the end of both phrases doesn't mean what most think. This Greek word means complete or completeness—*not* "without imperfection." [70] Restated, "therefore you shall be complete, just as your Father in Heaven is complete."

Does this make God any less of a creator? No. It just means that what God attempts, He completes.

His command to us in Matthew means that we should also be complete in what we do, including our relationship with him. Much as Scripture commands that we be either hot or cold and not lukewarm, lest He spit us out of His mouth, God desires that we be complete. [71] Since man cannot be perfect, this interpretation of perfect from Greek identifies something a human can achieve. Since there can be no deviation in something considered perfect, what about science?

It took more than 350 years for the Catholic Church to *apologize* (in 1992!) for imprisoning the great heliocentrist astronomer Galileo Galilei and forcing him to recant his description of the solar system. Maimonides, the medieval Spanish-Egyptian Jewish philosopher, took that geocentrism to heart, arguing that even the ancient Hebrew Bible described a world where everything revolved around our planet. This position was defended as recently as 1975 by Rabbi Menachem Mendel Schneerson, the Lubavitcher Rebbe, employing Albert Einstein's theory of relativity.

But scientific research has demonstrated that we are able to exist on Earth solely because of the precise makeup and workmanship of our universe, and this to an extent that the ancient philosophers couldn't have imagined:

> *The universe is perfect—eerily, uncannily perfect—as a setting for creating life. All sorts of physical constants—the speed of light, the charge of an electron, the ratios of the four fundamental forces—seem fine-tuned to create a universe where life as we know it could emerge.*

As an illustration of just one example of this, writer Anil Ananthaswamy explained for PBS:

> *[The neutron] is 1.00137841870 times heavier than the proton [a bare hydrogen nucleus], which is what allows it [a neutron] to decay into a proton, electron, and neutrino—a process that determined the relative abundances of hydrogen and helium after the Big Bang and gave us a universe dominated by hydrogen. If the neutron-to-proton mass ratio were even slightly different, we would be living in a*

very different universe: one, perhaps, with far too much helium, in which stars would have burned out too quickly for life to evolve, or one in which protons decayed into neutrons rather than the other way around, leaving the universe without atoms. So, in fact, we wouldn't be living here at all—we wouldn't exist.

That is, even as tiny a number as the mass of a neutron—the subatomic particle inside all atomic nuclei except that of hydrogen—is perfectly calibrated to allow worlds like Earth to emerge and survive over long spans. This, the thinking goes, is evidence that our universe exists only because there are thinking beings here to observe it. [72]

Further complicating a pure scientific approach are recent issues raised by, of all groups, the scientific community. The follow quote was posted on now.northropgrumman.com by Rick Robinson on November 29, 2020: "The two greatest achievements of modern physics—quantum mechanics and Einstein's theory of general relativity—share one serious problem: They don't agree. As a result, physicists find themselves in the awkward position of seeking to explain the universe using two theories that are not quite able to explain or account for each other." The article goes on to elaborate:

Most of the time, these very different approaches to nature do not get in each other's way. Einstein's relativity theory, in spite of its popular association with nuclear energy, is mainly used to explain how the universe works at large scale: the physics of baseballs, planets, galaxies. Quantum mechanics, in contrast, deals with the very small scale: atomic nuclei and smaller, down to the most fundamental building blocks of matter, quarks.

When the mathematical technique called renormalization is used to demonstrate how this unity is applied to gravity, it doesn't work. In the quantum mechanics framework, gravity should be associated with theoretical particles called gravitons. But gravitons, by reshaping space, render renormalization calculations endlessly complicated, says Space.com.

According to physicist Netta Engelhardt, quoted at Space.com, introducing gravitons into standard renormalization calculations "effectively means that you need an infinite number of experiments to determine anything. That's not a realistic theory." [73]

More than when and how creation came about, maybe explaining what happened before that should be what faith and science openly discuss?

Science as an Abstract

One of my favorite cable channels is the Science Channel. One of the most interesting programs to me involves the examination of the Earth's surface through the cameras of orbiting satellites. The satellite will find something unusual on the surface of the Earth, then provide a scientific description of what the image might represent and how it was formed. Frequently, a scientist describing the image will comment about its shape. Without fail, if the lines of the object are straight or at odd, symmetrical angles, the scientist will remark that it must be "man-made." His reasoning is that nothing in nature forms a perfectly straight line. Let's linger here a bit.

If it is true that no straight lines are created by nature, it nevertheless seems to do a great job when it comes to circles and spheres. Most of the things that the Science Channel scientists were observing were stationary, such as land masses, forests, rocks, and water. Things in motion, on the other hand, spin into a neat sphere. Why is that? The pull of gravity. Drop an object into the water and it will create a perfect circle as the force of the falling object is absorbed and spreads out across the stricken water surface—every action will have an equal reaction. My point? This thing we call life was not an accident. To believe that life was created by inanimate nature seems counter to science.

While little of organic matter forms a straight line, all of it is unimaginably complex whether great or really, really small. The majority of life forces are perfect—they have to be perfect. Yes, there are anomalies, times when life is born with a glitch. The results of these are defects or altered consciousness. Often, these glitches are carried into new generations and occasionally show up generations after previous genetic misfires. This points to perfect beginnings for these life forms, not evolving, overcoming disastrous beginnings. Back to straight lines.

Although a well-kept secret, some know the answer to life, the universe, and everything in it. It's 1/137, the universal constant. But according to an article written by Michael Brooks in *NewScientist*, October 6, 2018: "Controversial hints suggest this number might not be the universal constant we had assumed, instead varying subtly over time and space." [74] The writer goes on to say:

> *Change this number by a smidgen, and you change the universe. Increase it too much, and protons repel each other so strongly that small atomic nuclei can't hold together. Go a bit further and nuclear fusion factories within stars grind to a halt and can no longer produce carbon, the element on which life is based. Make alpha much smaller, and molecular bonds fall apart at lower temperatures, altering many processes essential to life.*

Okay, I am serious this time; back to straight lines. Earlier, we were comparing the size of stuff, say the Earth compared to Jupiter, or you compare the smallest microparticle embedded in your brain to a watermelon. To get to the basics, let's go from two extremes we factually know about… well, almost fact.

The smallest living organism known to man—according to the Guinness folks, who keep up with these kinds of things—is a *Nanoarchaeum Equitans*. This minuscule microbe, discovered in 2002, was in a hydrothermal vent on the sea floor off the coast of Iceland. Its cells are only 400 nanometers across. In addition, its genome is only 490,885 nucleotide bases long, which makes it the smallest non-viral genome ever sequenced.[75]

By the way, 400 nanometers is one billionth of a meter times 400. For perspective, a human hair is 75,000 nanometers in diameter. A human blood cell is 6,000 to 8,000 nanometers. The Ebola Virus is about 1,500 nanometers long by 50 nanometers wide. [76]

Let's compare this *Nanoarchaeum Equitans* to the largest known (well, theorized) collection of stuff almost at the edge of our ability to see, using electronic telescopes stationed in outer space. To measure this very large thing we will have to use time, not a yardstick. It isn't one thing, but a collection of things. It is just on the other side of the edge of our universe. Yes, apparently our universe has an edge... created by nature. Yes, humans used to think that it was possible to get in a ship and sail off the edge of the Earth.

Science says 13.8 billion years ago, the universe as we know it began with a "Big Bang." Over a very long time, space itself expanded from the explosion. Matter was manipulated by gravitational attraction, resulting in what we look at today as we peer into the heavens on a clear, cold night. As we have not yet developed the ability to see outside our universe, stars and all other objects giving off light fade to black. But, like a camera aimed at a dark spot in the sky with its shutter left open, cradled in a device to compensate for the rotation of the earth, there are uncountable other bodies of light beyond our ability to observe with the naked eye; a total, of some 46 billion light-years away.

Forty-six billion light-years. Just how far is that? Well, here are some facts: Because of the vastness of space, science uses the distance from the Earth to the sun as a unit of measure. That distance is right at 93 million miles. This is considered one *astronomical unit*, or AU. Perhaps you may wonder why light-years, a better-known unit of measure, are not used. Here's why: Try to turn off the light at your bedroom door and then get into bed before it turns dark. That's how fast the speed of light is. The definition of one light-year is the distance that light travels in a vacuum in one Earth year. One light-year equals 6 trillion miles, or 62,239 AU! Our solar system is just simply *too small* a place to be measured in light-years. (It's all relative!).

Our ability to see farther into space—to answer some of the darkest unknowns—was made possible to some degree when the US launched the Hubble Space telescope into orbit. After a frightening glitch in the satellite was repaired with a spacewalk, things, wonderful things, came into view. Theories were proven and fantastic discoveries made. The newest instrument designed to peer even deeper into the unknown is the Giant Magellan Telescope with its seven round mirrors arranged like a honeycomb measuring eighty feet in diameter. With its advanced computer hardware and software, the new telescope, sitting on Earth at the Campanas Observatory in Chile, allows humans to see ten times further into space than the Hubble Telescope sitting in outer space, and with significantly sharper images. [77]

Recently, with the development of the ability to observe objects in space emitting red ultraviolet radio waves, science has accepted the number of objects is innumerable. [78] [79] For perspective, if that is possible when speaking about these kinds of distances and volumes, if a galaxy's light takes 13 billion light-years to first reach our eyes, science estimates that object is about one-billion years old. [80] The soon-to-be-launched James Webb telescope will look between 13 and 14 billion years into the past. [81]

Armed with this knowledge, let's embark on a little journey. If we could propel straight up from the Earth indefinitely, away from the sun, this is what we would find: The first roughly nine miles of the journey would consist of moving through the layer of atmosphere that allows humans to continue to live and occupy the Earth without artificial assistance. Having left the atmosphere, a quick 238,900 miles later, we are passing by the moon. In 29 AU (2.8 billion miles), we reach Neptune's orbit. By now, we have been on this journey for many, many human lifetimes, but light that left the sun a little over four hours ago in light-years just caught up with us! Outside Neptune's orbit is the Kuiper Belt, a virtually empty ring that has icy bodies, most of them smaller than Pluto, making slow orbits around the sun.

Beyond the fringes of the Kuiper Belt is the Oort Cloud. Unlike the orbits of the planets and the Kuiper Belt, which are relatively flat like a disk, the Oort Cloud is a giant, spherical shell surrounding the sun, planets, and Kuiper Belt Objects. It's like a big bubble with thick walls around our solar system, made of icy pieces of space debris the size of mountains and sometimes larger. This is where some comets come from. This Oort Cloud is massive, with the inner edge thought to be between 2,000 to 5,000 AU from the Sun, and the outer edge between 10,000 to 100,000 AU from the Sun. If the 100,000 AU estimate is correct, it would take about 1.5 light-years to get through it. As we leave the Oort cloud, it will be almost three more light-years before we reach the nearest star system,

Alpha Centauri. We must now travel another 25,000 light-years to reach the outer edges of our Milky Way Galaxy. And after this? Well, it has been estimated that there are 120 billion more galaxies in our observable universe. [82 83 84]

So, we reach the end of our journey, having passed trillions of objects. However, there is more out there, and it is called *dark matter*. Many may think what surrounds us on Earth and in space is just empty, transparent air. Nope, it is also dark matter. This is the most mysterious, non-interacting substance in the Universe. Its gravitational effects are necessary to explain the rotation of galaxies, the motions of clusters, and the largest scale-structure in the entire Universe. [85]

The description of this "journey" was necessary to get you to consider earlier questions about science and faith. Take a little time, mark your place and close the book, and think about the relationship between the size of a *Nanoarchaeum Equitans* at 400 nanometers and the inconceivable vast outer edge of the Oort Cloud. And… beyond that. Once you have considered this, ask yourself the question: "Was this all by chance, and did it happen in six days?"

In an opinion piece written on April 27, 2014, Amir Aczel wrote in *Time Magazine* the following closing comment to his article, "Why Science Does Not Disprove God:"

> *Science and religion are two sides of the same deep human impulse to understand the world, to know our place in it, and to marvel at the wonder of life and the infinite cosmos we are surrounded by. Let's keep them that way, and not let one attempt to usurp the role of the other.* [86]

Now consider again the previously described distance from the Earth to the outer edge of the Oort Cloud. Somewhere in all that vastness stands a body, a human… you. Feeling a little insignificant? It's okay.

CHAPTER 6
POST CREATION

... an immeasurable length of time

Now Cain talked with Abel his brother; and it came to pass, when they were in the field, that Cain rose against Abel his brother and killed him.

GENESIS 4:8

EARLIER, WE DISCUSSED the story of Creation in quite a bit of detail, and how the first breath of man came about. Another first-and last-breath narrative quickly follows: the story of Cain and Abel. After Creation, Abel was the second human to be born of woman and the first human on Earth to die, as recorded in the Bible. His death came at the hands of the first human on Earth to be born naturally, Abel's brother, Cain.

Abel's first breath, unlike that of his father, Adam, would have been an involuntary gasp for air after exiting the womb. We must remember, both Cain and Abel were born in an imperfect world, not the Garden. Cain and Abel's first breaths were likely no different than those of a 21st-century baby, except the 21st-century child's breath typically would be drawn in the sterile environment of a hospital. Abel's first breath was probably a little easier for the parents to handle than Cain's. Remember, this had never happened before Cain's birth. New parents in the 21st century think they are overwhelmed with the thought of bringing a human life into the world for the first time. What about a set of new parents delivering a newborn themselves with no one to assist, or even an instruction manual? This was the ultimate natural childbirth! While not recorded, I imagine Adam's first reaction was, "Now what do I do?

Because they were literally the first children, Cain and Abel would have pretty much been on their own figuring out the whole childhood thing. According to Scripture, each child discovered what they liked and did best, and adopted it as their trade. There was probably a lot of trial and error.

The fourth chapter of Genesis is a bit confusing and seems to cover an enormous amount of time. According to scriptural records, either time was measured differently in this chapter or some of these folks just lived a really long time. There are some age-old questions raised within the verses of chapter 4, not the least of which is: Where did Cain's wife come from, and how did Cain build a city if there was no one to live in it? Again… the time thing. These pose no dilemma for science, as there were hundreds of thousands of years for all this to happen, right? On the other hand, the faith community should be prepared with answers to these questions. If asked, what would be your response?

Let's return to Abel as he lay dying from the hand of Cain. Were Abel's last breaths gasps, or did he die instantly, with no pain? Was he staring up at his brother wondering why Cain would do such a thing? Had they witnessed the death of an animal or perhaps even a person before? What we can be certain of is that both of them would have been acutely aware that this was wrong. They had probably heard stories told by their parents while sitting under the stars at night about walking in the Garden in the cool night air and enjoying a perfect world. All of that must have seemed so strange to them, having never been a part of that experience or being allowed back into that place protected by God's guardians, the cherubim.

We can only imagine that Abel's last breath would have contained the moist scent of soil freshly turned by his brother, Cain. At some point, Abel exhaled one last time. Where were their parents during all of this? Were they even aware that such a thing could happen? We know that they are still alive, because later it is recorded that they have another son, Seth… yet another first breath.

A Very Long Mid-Creation Winter

Scene 1

According to Scripture, somewhere around 3000 BC, some 1,000 years after Cain, Lamech was born. He was the son of Methuselah, who lived 969 years. Lamech was also the father of Noah, who himself lived to be 950. After Noah was 500 years old, he became the father of Shem, Ham, and Japheth. Their story begins in the sixth chapter of Genesis, and by all accounts, things had been going downhill ever since that first tragic murder long ago in Cain's field.

Things were so bad that God decided to put an end to His creation. God was angry. Obviously, He had passed this trait on to Adam when He created man and woman in His own image. It is difficult to find in these chapters the love for humankind expressed much later in the timeline by His Son, Jesus Christ. But in Genesis 6:8, we find that Noah found Grace in the eyes of the Lord. Seemingly by default, Noah's family found that grace as well.

From the time of the Creation story until Noah, many first breaths—and subsequently last breaths—had occurred on the Earth, both human and animal. Now a mighty, angry God was about to end them all. Even the heroic, renowned men of old could not prevent the destruction that was about to be unleashed from above and below the Earth. All that drew breath were about to draw their last. For humans, this form of death would be one that is universally dreaded: drowning. The animals would be spared the anticipated horror of this, but would die, nonetheless. In the end, there would be an acceptance by humans of the inevitable and the release of one's last breath, held as long as possible. Recorded in verse 22 of chapter 7 of Genesis are these chilling, stinging words: *"All in whose nostrils was the breath of the spirit of life, all that was on dry land, died."*

Floating among the bloated, dead bodies of humans and animals was a boat, built by a man in excess of 500 years of age, with his wife, his three sons, and their wives and future brides. It was to be God's first reboot, a punctuation mark on the continued failing of His first attempt at creating a world that would praise and honor Him. I wonder what the conversations must have been between God, His Son, and the Holy Spirit after the rains stopped. In my mind, I can't help but think that it could not have gone well. It seems the compromise was a rainbow.

Scene 2

Through Noah's sons, the Earth eventually repopulated. Some 900 years later, around 2100 BC, Abram came onto the scene. Later called Abraham, the son of Terah, he moved to Haran in the land of Canaan.

Chapter 12 of Genesis now begins a more relatable timeline continuing on to the birth of Jesus. Lifespans from this chapter forward also moderate to modern-day life expectancies. It should be noted, however, that folks were still having babies at age ninety and 100—that is rather uncommon in our day. There is also no mention as to why earlier humans were living hundreds of years and now struggling to live past 100.

The faith community argues that God chose to have people live longer and that conditions were different on earth in the early stages, thus allowing the longer lifespans. However, as sin progressed, it weakened our ability to sustain life. Science argues that the length of life is guided by our genes, and that no evidence exists for a human to be born with a gene allowing him to live for almost a thousand years. Science, however, has discovered that certain alterations to the genes of fruit flies and some worms can increase their normal life by about 35 percent. These genes are aptly called… Methuselah genes. What else? [87]

Abraham may not have had his genes changed, but he did have his name changed by God. Names are changed throughout the Bible—this was a familiar trait of Jesus. He called James and John the "sons of thunder" and Simon Peter was named "the rock." Don't overlook the fact that Christ's name was also changed to Jesus! These name changes had significance, unlike today. This was not the same as calling someone Bubba instead of by their real name. Let's take a quick look at why the change in Abraham's name was significant.

The name Abram in Hebrew, using English letters, is spelled, Avram. Abraham, using English letters, is spelled Avraham. In Hebrew, this new name results from adding the Hebrew letters, in English, "heh." In Hebrew, these three letters represent "breath of God."

God proclaims in Genesis 17:5, *"No longer shall your name be called Abram, but your name shall be Abraham; for I have made you a father of many nations."* This name change signified to Abraham that magnificent events were about to unfold. Just as Adam was breathed into life by the breath of God, a future people would come into existence through the lineage of Abraham (meaning "breath of God") and Sarah. From these people would come Jesus, the Christ.

It should be mentioned that another story is told between Noah and Abraham—the account of the Tower of Babel. People thrived and spread throughout the Earth after Noah. One group settled in an area that would become known as Babel. The first part of chapter 11 of Genesis tells their story, which begins with the construction of a massive tower.

These people apparently had masonry skills, and set about using them to construct a tower. This was not just any tower, but a tower that made a statement! In verse 4 of chapter 11 of Genesis, they say, *"...let us make a name for ourselves, lest we be scattered abroad over the face of the whole earth."* Apparently, God felt the reason they were able to accomplish such an impressive feat was because they all spoke the same language. Following in verse 6, God says to His heavenly host, *"Indeed the people are one and they all have one language, and this is what they begin to do; now nothing that they propose to do will be withheld from them."* In verse 7, He decides, *"Come, let Us go down and there confuse their language, that they may not understand one another's speech."*

This first language was likely created by God, along with breathing life into Adam. Scripture speaks of discussions between Adam, Eve, and Satan. Also, Adam was charged with naming all the creatures God created. Unless this was done silently, there must have been some form of speech. This theory is labeled *Adamic*. Theorists speculate that the Adamic language would have been corrupted after the fall. It is thought that the confounding of languages at the Tower of Babel was a simple transliteration of the original language by its creator. Surely the One Who created it could simply change the language of those involved in this project reaching to the Heavens. If this is correct, all future languages could have evolved from variations in the original Adamic vocabulary. [88]

Although this tower incident precedes Abraham by at least a few generations, the separation must not have been long, because several of the areas mentioned in the Babel story match those in the Abraham account.

Before leaving Abraham, a few observations:

Following the flood, Noah's clan developed into many groups of peoples, one of which again drew God's concern and intervention in the telling of the story of the tower of Babel. That is probably the reason He took such a personal role in the grooming of Abraham by changing his name, testing him with the Isaac sacrifice, and calling Abraham something that is said of no one else in Scripture: *a friend of God.*

The other thing is the statement God made after He and His friends came down to see the Tower of Babel: "*...now nothing that they propose to do will be withheld from them.*" The New International version (NIV) says, "*...then nothing they plan to do will be impossible for them.*"

Consider these excerpts:

> *Unprecedented, deep understanding of the evolution of life will become possible as well. Complete knowledge of the history of life is unachievable because we will never have access to extinct evolutionary intermediates, but reconstruction by comparison of extant genomes will yield detailed information on these long-gone life forms. This might seem a somewhat abstract goal, but "Nothing in biology makes sense except in the light of evolution." Such understanding will revolutionize biological science.*
>
> *Up to now, all true achievements of genome analysis, be it inference of key evolutionary events or discovery of new enzymes for genome editing, have been driven by a combination of automatic computation and human expertise. This approach is being challenged and will become completely impractical in the new era of total sequencing. Because of the sheer number of sequenced genomes, all analysis will have to be fully automated.*
>
> *If this sounds like a call for replacement of human expertise with artificial intelligence, it is. To master the giga-databases of the future, new generations of AI are essential, as is hardware such as quantum computers.*
>
> *Sequencing the entire range of Earth's biodiversity is not a pipe dream*

> *anymore. In fact, it might become tangible reality within the lifetime of the current generation of scientists. However, to turn this wealth of data into valuable information that can help society, as opposed to useless noise, tremendous—and not entirely foreseeable—advances in computer science and technology are essential. Such are the paradoxes of the information age.* [89]

If God and His friends had an issue with the tower of Babel, do you think He just may be somewhat interested in the development of artificial intelligence (AI)?

Scene 3

This scene moves us closer to the coming event that will change the measurement of linear time forever. We move the BC clock backward to about 2066 BC, the time of Isaac—the son of Abram. We should also pause and consider the son of Sarah's maidservant, Hagar. This lad's name was Ishmael.

Prior to Isaac's unexpected birth, Abram's inheritance was to go to Eliezer of Damascus. Because she had been unable to provide him with an heir, Sarah urged Abram to sleep with her maidservant. God was disappointed in Abram for agreeing to this. Down the road, it created a lot of stress in Abram's family, as Sarah regretted her decision. Also, God was now faced with the prospect of dealing with Ishmael, whom He did not intend to be Abram's heir.

Abram was eighty-six when Ishmael was born. When Abram turned ninety-nine, Ishmael was thirteen and becoming a young man. It was then that God would provide a way for both of them to fulfill God's purpose in their lives and the lives of the thousands they would touch. It involved a couple of name changes. God informed Abram that another son would be conceived and born of his wife, Sarai. Before the lecture, God informed Abram his name would be changed to Abraham, *the breath of God*, and Sarai would forever forward be called Sarah, *a mother of nations*. There was one other small detail, a yet unknown preventive surgery called, circumcision. This surgery was to be performed on all newborn males on the eighth day of life. Furthermore, He mandated that all living males were to immediately undergo this procedure, regardless of age. Ouch!

It is believed that this command, related to a covenant established by God between Himself, Abraham, and all of Abraham's *"descendants after him throughout their generations"* Exodus 17:10, was the first time in history that the circumcision rite was ever performed.

Scientists have discovered that on the eighth day after birth, the blood-clotting elements Prothrombin and Vitamin K are at their highest levels. This means that on this day, the bleeding stops faster than on any other day of life. It is also less painful for a newborn. [90] This was not the case for the rest! However, Scripture does not record any complications, except perhaps for some screaming.

In addition to the benefits of clotting on the eighth day, science states that circumcision prevents urinary tract infections and cancer in later years. [91] Although today this may not be significant, it certainly was over 2,000 years ago, when hygiene was not a priority. This surgery is one of the undeniable commonalities between science and faith. The difference is that science cannot claim it was discovered by a scientist and passed through the medical maze to become a "best practice," FDA-approved procedure.

As will be mentioned in later chapters, this circumcision covenant will create division and hatred when the gospel is preached to the Gentiles after Christ's resurrection. The Apostle Paul particularly will suffer persecution in his attempt to change the 2,000-year-old tradition. This issue alone will almost tear apart the New Testament churches in Jerusalem and Antioch.

Covered earlier, Abraham's generations were to number more than the stars above his head. Although his lineage would achieve that goal promised by God, that number of lives after Christ's resurrection opened up His promise to include everyone—not just Jewish believers—who were born on Earth; past, present, and future. With the Gospel opened up to everyone, God's earlier promise will take on an even greater significance!

In Genesis 35:29, Isaac, the heir of Abraham, draws his last breath at the age of 180. He accomplished much in his lifetime, including the birth of two sons, Jacob and Esau. As is often the case in the Old Testament, here again we find two brothers who do not get along. Jacob steals Esau's birthright and flees to avoid his brother's wrath, although Esau ultimately forgives him. Perhaps this was because Jacob's Divine assignment was to move to Egypt in order to begin the establishment of what would become known as the twelve tribes of Israel, each tribe named for one of his twelve sons. It is during this time that we encounter yet another name change: Jacob's name is changed to Israel.

The name change was the result of an injury during a wrestling match between Jacob and an angel that he would identify as the face of God. Despite the great pain inflicted upon him, Jacob refuses to give up. The angel at some point touches Jacob's hip, dislocating the socket and almost crippling him. Though limping, this does not deter Jacob. The angel tells him that because of his character, he will no longer be called Jacob, but Israel, "*because you have struggled with God and with men, and have prevailed.*" [92]

Talk about a dysfunctional family! For details, read Genesis chapter 49. Jacob's step-uncle Ismael's gift was to establish twelve kings which would become great leaders, but they were not the intended path from Abraham to Christ. The move to Egypt would set the stage for a historical series of events a few generations later. Jacob draws his last breath after pulling his feet into his bed in Genesis 49, verse 33. He had requested to be carried from Egypt back to the Land of Canaan, to be buried in a cave purchased by Abraham. It was the burial place for Abraham, Sarah, Isaac, Isaac's wife Rebekah, and Jacob's first wife, Leah.

Dying in Egypt, Jacob was embalmed in the usual Egyptian custom, a process that took forty days. Then Jacob's sons, along with a large contingent of Egyptian people, soldiers, and dignitaries made the journey back to Canaan, where Jacob was buried in the family cave along with Abraham and other close relatives. Afterwards, Joseph and the entire traveling troop returned to Egypt. However, this friendly alliance between the Jews and Egypt would last only until a new pharaoh is installed. Joseph dies at 110 years old, is embalmed and buried in Egypt. [93]

In the thousands of years since Creation, man's interactions with God have driven one point home about their creator: God is a circuitous planner. The straightest line between two ancestral points is not a straight line. Unlike human creation… isn't it? In about 1898 to 1806 BC, Jacob's son

Joseph would be a transition in Jewish history followed by a complete change in attitude of the Egyptians toward his race. God was guiding His chosen people toward a manger in Bethlehem, 1,600 years down the road. A blessed event of "Biblical proportion" would take place. Here, an exhausted mother, a stressed-out father, a few animals, maybe a shepherd or two, and a beautiful, literally perfect newborn baby was just on the other side of breathing His first breath of air. Air He helped create a very, very long time ago.

What follows now is a story about another ark, but one without rain. This small ark will protect a young baby boy from the same fate Jesus will escape at His birth: the annihilation of all male babies, but this time... not just the first-born.

Scene 4

It is about 1525 BC. Relations between the Egyptians and the Jewish Nation have been peaceful, accepting, and cooperative thanks to Joseph's stature with the Egyptian people. But Joseph passed away, times changed, and a new guy took over.

The new guy was the incoming King of Egypt, also known as Pharaoh. The opening chapter of Exodus wastes no time setting the stage for this scene in the unforgiving sands and heat of Egypt. It begins by telling us that eleven sons of Jacob came to Egypt with their father. His twelfth son, Joseph, was already there. Again, from these sons will come the twelve tribes of Israel. In addition to the twelve brothers, there surely were sisters, but tradition did not allow women to be important. Apparently, it was also not part of Gods plan. There is no reason I can find other than that this was the tradition of the time. Truth is, America didn't begin to seriously work on the oppression of women until the mid-twentieth century. Other countries have yet to figure it out.

Scripture tells us right up front that Jacob, all his children, and that generation had died. The new Pharaoh, in verse 8 of Exodus 1, is introduced: *"Now there arose a new king over Egypt, who did not know Joseph."* So, Joseph, what have you done for me lately? The new king made an early observation of a well-known fact, that there were more Jews in Egypt than

Egyptians! His concern? This vast number of people could turn on them and take over. His next statement, like many new guys, shows his inexperience and displays his need to control. In verse 10 of Exodus 1, he states, *"Come, let us deal wisely with them, lest they multiply, and it happen, in the event of war, that they also join our enemies and fight against us, and so go up out of the land."*

Let's take one of our millions of lifetime breaths here, as the new Pharaoh should have done, and look at the big picture. Although these folks were in truth slaves or, maybe better stated, the object of indentured servitude, they were content. Their masters provided food and shelter, a somewhat normal family life, and it was peaceful, although difficult for every Jew. The king had obviously never been in charge of getting work done. If he had, he would have realized to cut back on the number of workers was not going to turn out well, or for that matter, increase production. His plan? Work them so hard that they would be too tired to procreate. This obviously wasn't going to work, given human nature.

Instead of less births, there were more! Next, he insisted the maidservants who delivered the babies put an end to all male children who were born to Jewish women. The orders were not followed. When the maidservants were asked what happened, they reported that Jewish women were so energetic they had their children quickly before the maidservants could arrive. His response was to tell them to just throw the newborn male babies in the river. Time for the second ark story.

Before moving into one of the most important series of events during this Winter of Creation, I want to pause and remember the unknown number of innocent babies who were allowed only a few short breaths of air between their first and last as water filled their lungs and they found themselves in the loving arms of Jesus. I imagine it must have been emotionally exhausting to Christ knowing these babies would meet the same fate as others when He would be born to create an opportunity for Him to enter the world He helped create.

These precious babies often only get a mention when these verses are used in a sermon or Bible study. Innocent babies whose lives are taken at any time in history deserve to have their own short stories recognized and honored as individuals, not just part of a story. Amen.

Scene 5

Imagine… after getting up every morning, eating breakfast with your family, going to work, doing an honest day's work that is appreciated by your employer, then returning home, playing with your children, eating dinner together and thanking God for His blessings, then retiring for a good night's sleep with no concerns for danger and then everything suddenly begins to change… everything. Change started slowly, but very decidedly downward. It is first realized at work. Your employer is demanding increased hours and production with no increase in reward. You are required to arrive earlier and stay later. Those who are injured on the job are ignored or forced back to work. It affects your family life and everything about your daily living.

Rumors begin to fly, and then are validated. It seems that the only survivors of new births are female. The male babies not only die, but their parents never get to see them. They are disposed of somehow with no final hug or kiss on the forehead, no chance to see what they look like or to name them. You are deprived of the opportunity to follow your centuries-old burial traditions. And then… it gets worse!

This occurs by order of the new guy, the king, the pharaoh… all out of fear that this tribe of people are becoming too large to control. Repression turns to oppression; indentured servitude turns to pure slavery. But the Jewish nation continues to grow, babies continue to be born, male and female. Why? Compassion. It is difficult for us to process how that same breath of God that was breathed into his creation in the Garden of Eden could produce such opposing human characteristics.

Chapter 2 of Exodus starts out with a simple sentence that is much loved by those who trace genealogies: *"And a man of the house of Levi went and took as his wife a daughter of Levi."* It does not immediately give you their names. At this point, that isn't as important as their family lineage. They both are descendants from Levi, one of the twelve tribes of Israel (Jacob). The reason this is important is because of who the baby is, and more importantly, what he is about to become. Jewish history and God's plan for eternity is riding on what is about to happen.

The father of this baby was one of those men going to work every day in the earlier example. His name was Amram. He was also the father of two other individuals key to God's plan, Miriam and Aaron, the sister and brother of another child to be born of Amram and his wife, Jochebed. Levi would have been the great-grandfather of these three children.

Amram had been pushed to life's edge at work, and Jochebed suffered the stress of holding their family together. Nonetheless, there must have been times of intimacy, a time to escape the drudgery of daily life and simply enjoy companionship. They must have known this might lead to a pregnancy and ultimately to a very tough decision in the event that the child was a male. After about nine months, their fears were realized. That fear, and her love for this male child, pushed Jochebed to hide the infant for three months. This was enough time to nurse the child to the point where he was thriving.

It became difficult to continue hiding the child, causing Jochebed to build a very small ark made of bulrushes covered with asphalt and pitch. She placed the ark in the reeds by the Nile River bank near the bathing place of the Pharaoh's daughter. The fearful mother then charged her daughter, Miriam, to watch over the child from a distance. As hoped, Pharaoh's daughter, named Bithia, did indeed find the child. When Miriam saw her compassion as she wept over the abandoned baby boy, she approached her and asked if, considering the child was Hebrew, she should go and find a Hebrew woman to care for him. Bithia agreed, and asked the young Hebrew girl to find someone. Miriam brought back her mother, Jochebed, and introduced her to Bithia.

Amram and Jochebed suffered tremendously through this ordeal, but perhaps the greatest hardship to overcome was having their baby named by someone else. The baby was to be called Moses, which meant *"Because I drew him out of the water."* Exodus 2:10 Although Moses would be known by many other names during his lifetime, it is interesting that one of the most important human links in the lineage leading to the birth of Christ is not known through history by his Jewish name.

Moses' other names were: Jekuthiel (by his mother), Heber (by his father), Jered (by Miriam), Avi Zanoah (by Aaron), Avi Gedor (by Kohath), Avi Soco (by his wet-nurse), Shemaiah ben Nethanel (by people of Israel.) [94]

Moses is raised by an Egyptian stepmother, with distant influence from his real mother. He is given the education of a prince. As an adult, he is charged with supervision over the Egyptian soldiers managing his Jewish kin. Then, he makes a split-second decision in the heat of passion for his people that changes everything.

This is a forty-year journey filled with little-understood anxiety, as a boy became a man, and ending in the murder of an Egyptian at Moses' angry hand.

Scene 6

Just as in Jesus' early life, Moses' story is placed on fast-forward after his first breaths in the book of Exodus between chapter 2 verses 10 and 11. In verse 11, Moses is suddenly grown, perhaps thirty-five to forty years of age. We know very little about Moses as a child, teenager, or young man. We can assume, since he was taken in as a son of the Pharaoh's daughter, that he received the best Egyptian education possible. Because of his privileged status, he had a great deal of freedom to wander about the territory. These wanderings included the home area of his Jewish relatives and Egyptian worksites. At this point, it is not clear if his mother has said anything to him about his birth.

It would have been obvious by looking at his body, compared to Egyptians, that he was of the Hebrew race, those who didn't enjoy the riches and freedoms he now took for granted. One of the next incidents in this story, as recorded in Exodus, is an encounter with an Egyptian and one of Moses' kindred. The only detail we are given is that an Egyptian was beating a Hebrew person, *"One of his brethren."* Exodus 2:11 No discussion was recorded, just Moses' interceding and murdering the Egyptian. The last breath of this man coincided with the first breath of guilt and isolation felt by Moses. We are told he buried the dead man and went about his way.

In a parallel to the parable later told by Christ regarding the forgiving of a man his debt—who in turn refuses to forgive someone who owed him an even smaller debt—Moses goes out the next day to find two Hebrew men fighting. He determined who he felt was in the wrong and addressed him: *"Why are you striking your companion?"* Exodus 2:13

Now, Moses was no doubt wearing an Egyptian uniform or expensive clothing, as the accused Hebrew responded with something that struck fear in Moses, more fear than he felt even the day before. *"Who made you a prince and a judge over us? Do you intend to kill me as you killed the Egyptian?"* Exodus 2:14 So, who needed social media? Talk about a human grapevine! There were thousands of Hebrews in Egypt scattered over a vast area, yet in no more than twenty-four hours this man knew about Moses' previous day's activities!

With the phrase, *"Surely this thing is known!"* Exodus 2:14, not only did the previous life of Moses take a 180-degree turn, but the entire history of civilization changed. Moses fled Egypt to the land of Midian at the age of about forty. Although Moses' adoptive mother had dearly loved him as a son and would likely have attempted to protect him, he knew that Pharaoh would not be as forgiving: *"When Pharaoh heard of this matter, he sought to kill Moses."* Exodus 2:15

Breakout Session

Time to put Scripture on hold for a bit. The Holy Bible is truthful and inspired, yet its contents are biased in regard to world history. Scripture's agenda is to follow Christ's lineage from Adam to Bethlehem—to the cross—to the tomb—to the ascension—and to His third return to Earth on Judgment Day. There is absolutely nothing wrong with this agenda. Consider, however, that introducing someone to the concept of Jesus Christ will often come with resistance. The timeline of Scripture may be understandably rejected out of hand at first.

In light of this possibility, it might be helpful to acknowledge the fact that there were other things happening around the created Earth besides the genealogies that are prominent in Scripture. It would be beneficial to be informed as to what was going on in the rest of the world during this period. Sorry if that was a bit harsh. Please remember, this book is about the assumed conflict between faith and science over Creation and other faith-based traditions. We will get back to Moses and Scripture directly—but first, let's take a detour.

Following are a few things that were happening around God's created world about the time that Moses began hiding from Pharaoh in the Land of Midian, circa 1500 BC:

- Ever heard of the kingdom of Kush, a major military opponent of Egypt around the second millennium BC?
- How about the Punt civilization on the eastern coast of Africa about 2500 BC? [95]
- The Maya of South America in 1500 BC?
- The first Mesoamerican Civilization, the Olmec, 1600 to 1400 BC in southern North America? [96]
- The Harappan Civilization, flourishing around 2500 BC, in the western part of South Asia, what today is Pakistan and western India? [97]
- The Han dynasty of Ancient China, at 1500 BC in its death throes, with a troubled period in Chinese history dawning? [98]

Although what is about to happen in Egypt is a big Biblical deal to believers, in the eyes of secularists, scientists, and agnostics, it is not very significant. In light of this fact, perhaps the faith community should recognize that the coming of Christ did not take place in a vacuum, and resolve to make this event relevant within the broader context of history. This is an easy task—*if* you mastered the earlier explanation of the beginning of time and creation. Back to the story line.

Scene 7

After forty years of living in the lap of luxury and enjoying life as the stepson of the Pharaoh's daughter, playing with other children from the top echelons of Egyptian culture, eating the very best food, and receiving the highest level of military training, Moses found himself sitting by a well in a foreign country surrounded by total strangers. It isn't clear how far it was to the Land of Midian from the encounter with the Hebrew. One estimate is twenty to thirty-five miles, depending on where Moses was in Egypt when he left and where in Midian he arrived. Even so, the conditions were not great for an outdoor excursion… it could have taken him two or three days. He likely had nothing to eat or drink. Scripture says he *sat* down by the well. Truth is, he probably *flopped down* by the well.

As tired as he might have been, an impromptu incident developed around where he was sitting that got his adrenaline pumping. Moses was in very good physical condition. There is no reason to believe otherwise, since he had the best environment growing up any young man could have imagined. The incident involved some young ladies who were seemingly in trouble near the well. They were the seven daughters of a Midian priest who were tending their father's sheep. They were being harassed by shepherds over rights to the water in the well. The daughters apparently came every day to water their father's flock, and the shepherds were determined to stop them. Moses had been trained to protect helpless ladies in distress, so he did what a gentleman does, tired or not.

If Moses wrote the book of Exodus, he could have been the model for John Mark's style of authorship in the New Testament. Like Mark, the author of Exodus writes only basic detail—no frills. Consider in Exodus 2:15: Pharaoh declares he will kill Moses, and Moses winds up in the Land of Midian by a well. In verse 16, the daughters of the priest come to water his flock. In verse 17, the shepherds attack the ladies and Moses comes to their recue without being asked. We meet the daughter's father, Reuel, in verse 18, with Moses being declared a hero in verse 19. Verse 20 has Moses invited to dine with the father. In verse 21, the father gives Moses his daughter, Zipporah, to be his wife. Then in verse 22, they have a child whom they name Gershom meaning, *"I have been a stranger in a foreign land."* How's that for moving history along?

The end of the second chapter of Exodus closes with three clarifying pieces of information. First, the Pharaoh died. Second, the bondage of the Hebrew people had become unbearable, causing them to cry out to God. Third, God heard them and remembered His ancient covenant with Abraham, and (I love this phrase), *"… and God acknowledged them."* Exodus 2:25 In other words, he turned and stared, probably without a word, since Scripture chooses not to record any words spoken at such a pivotal point in the life of God's people.

Perhaps God said nothing because He doesn't react as humans do. Instead, at the beginning of Exodus chapter 3, we are introduced to the first of several events of biblical proportions that will take place over the next forty years or so. Enter the fire-resistant shrub—the famous burning bush. This isn't the first or last time God will choose to appear to humans without showing His face or revealing His identity. Remember the wrestling match Jacob had? Theologians will be quick to point out that God is perfect, and humans cannot look upon God's image without suffering dire consequences. We will get to see this happen in a bit, but for now, we need to stick to the story.

Moses is approaching eighty years of age when he encounters the burning bush, or a flame within a bush that will not catch on fire. He has been out for a period of time with his father-in-law's flock of sheep in the desert, looking for grazing pasture. Remember that I mentioned previously that God doesn't do things in the way that we totally rational humans do? Well, here you go. Compare these two encounters with God regarding His chosen people. In the first, thousands of Hebrews are living and working under the worst conditions possible in one of the most populated, urban areas in the known world. In the second, an eighty-year-old man is in the desert next to a mountain populated only by a flock of sheep. In the first encounter, God simply acknowledges His chosen peoples' cries without saying a word. In the second encounter, He speaks to an old man in the middle of nowhere through a burning bush.

Now ask yourself: How would a Doctor of Psychology or a scientist react to the burning bush scene? I am neither of these professionals… but let me take a stab at it. A doctor who believes and teaches "felt-feelings" would have probably said, *"If God really felt for the people He loved so dearly, He should have calmed their fears instead of simply acknowledging them."* In other words, did God really love these people? In the scientific view, *"Each action has an equal and opposite reaction."* If there was a fire, something should have turned to carbon, or perhaps someone poured that stuff on the plant used by stunt people in the movies when pretending to be on fire. And this will be just one of the smaller miracles and encounters with God over the next four decades.

Scene 8

Back to the bush—not just a burning bush, but a burning, talking bush. This conversation occurred on a mountain, in a desert, with the background sounds of bleating sheep echoing in the dry wind. If it were not such an important event in the life of the faith community, I would see a bit of humor in this scene. God calls out to Moses, but not in the loud, low-baritone, deep-chested voice of the God portrayed in the famous Ten Commandments movie. Instead, a softer, inviting voice. When Moses hears this voice, as if there is a crowd standing around, he replies "Here I am." I imagine that even God smiled at his reaction. I guess Moses' first thought was that the bush couldn't see.

This, and another comment a few verses later, highlights something most people assume about Moses' knowledge of his ancestry and Hebrew teaching. Moses was smart and knew a lot of things, but Jewish history and law were not a part of his background schooling. He was raised in Egypt by Egyptians. This is, after all, his first encounter with... God, *the* God.

There is a brief mention in this Scripture that is related to Creation that could easily be overlooked. I believe the order from God for Moses to remove his sandals has that connection, literally. I believe God wanted Moses to physically connect to His creation—skin to Earth, the very Earth God created, the very dust from which man was brought into this God-created world. I also believe the Holy Spirit, who participated in the Creation, was present here as well. Think forward to the day of Pentecost in the New Testament. There were what seemed to be flames of fire that did not consume anything or anyone, and people were understanding languages they did not speak. Those actions at Pentecost were believed to be the work of the Holy Spirit's breath. I believe that the Trinity was present during this interaction with Moses and the burning bush.

Three times—in verses 5, 15, and 16 of chapter 3 of Exodus—God recites Moses' lineage all the way back to Abraham. In verse 5, He says "father," meaning Moses' father, without using Levi's name. In the same manner, he uses the term "fathers" in verses 15 and 16, referring to all fathers of the children of Israel.

The other verse that indicates Moses' lack of Jewish tradition is in verse

13. He knows who the children of Israel are, but he has no idea what they call their God. He asks God, *"Indeed when I come to the children of Israel and say to them, 'The God of your fathers has sent me to you,' and they say to me, 'what is His name?' what shall I say to them?"* Here we come again to another breathless moment. What is God thinking? He is placing the success of His entire plan in the hands of a murderer and sending him back to the scene of the crime. Well, it has been forty years, people forget. But then, did I mention that Moses stuttered on occasion? Exodus 4:10

Before the bush returns to being, well… a bush, Moses exhibits another common characteristic of human nature. More than once, God tells Moses he is expected to lead the children of Israel to a land occupied by Canaanites, Hittites, Amorites, Perizzites, Hivites, and Jebusites. These are some of the most warring people alive. The existence of these communities was probably known to Moses, because it is certain that the Egyptians knew about them and their threat to Egypt. So, what is that human characteristic demonstrated by Moses? At the beginning of the first mention, and at the end of the second, God's list of war-like tribes summarizes with, "*…to a land flowing with milk and honey.*" All Moses heard was… milk and honey. He is all human!

Chapter 3 ends with God promising that He will do whatever it takes to convince the pharaoh to allow Moses to lead the oppressed nation of Israel away from Egypt. Unfortunately, a lot more children, and some adults, are about to die. This story will not end when the children of Israel reach the promised land. These people will never really be free until hundreds of years later, with the resurrection of Christ. It is at the resurrection that the central message of Jesus is intentionally sent out to the entire creation for the first time. The Jewish Nation may still have remained God's favorite people but… the messengers and their message would have a whole new audience.

The rest of the book of Exodus contains some of the most miraculous events ever recorded. These events have been discussed and studied by the scientific community for centuries. There are many, beginning with the ten plagues used by God to encourage Pharaoh to allow the Hebrews to leave Egypt. The last, the killing of all the firstborn sons of the Egyptians, had more than an emotional impact. The firstborn would have received the inheritance of a dying father. This act of God no doubt was an economic disaster when Egyptians tried to sort out the passing of wealth from one family member to another.

After being released from bondage, God parts a sea of water and the refugees walk through on dry ground. Later, they are starving, and God provides manna (white, bread-like wafers) every morning for food. When they are dying of thirst, the striking of Moses' rod on a rock produces water.

These events are followed by God giving Moses the Ten Commandments after calling him up to a high mountain. Regardless of how the commandments were written on two tablets—twice, by the way—God decides to begin explaining the ten rules by allowing them to be changed or expanded. Not by Him, but by religious priests, into 613 total rules or commandments. It is clear these were never intended to be suggestions.

The remainder of Exodus, and in fact the rest of the five books of the Old Testament, becomes a very precise instruction template as to how the Temple will be built and used for worship in the coming one thousand years. This template includes some very familiar items. God needed a place for the Ten Commandments to reside, so He gave the instructions to build an Ark of the Covenant. This created a need for a place to keep the Ark. That need was filled with the building of a tabernacle. This structure would contain curtains. When Christ exhaled His last breath on the cross, the tearing of the Temple curtain from top to bottom signified that no longer would His people be separated from direct access to God.

All of this created a need for a special group of men—priests—to direct the use of these very important devices, structures, and rules. Who better to head this up than the one in charge of that disgusting love fest out in the wilderness when Moses was busy on top of the mountain receiving God's hand-written ten commandments? Who better than the one who was the architect of the golden calf, worshiped by many of those same people freed from Egyptian slavery? Aaron, Moses' brother, became the first High-Priest of Israel, and his male descendants would follow him in this position after his death.

In the following years, after being led by judges and prophets from the days of Moses, the Jewish people decided that they needed a king. Moses had foreseen a day when his people would demand a king, and he wrote about it in the book of Deuteronomy prior to his death. Some 400 years later, the people complained to the prophet Samuel, insisting on the anointing of a king. Samuel reluctantly agreed. He appointed Saul, then following him, young David. David's son Solomon would build the first permanent temple to replace the mobile tabernacle. After Solomon, things began to deteriorate, and in about 400 BC, the Babylonians attacked and destroyed the Temple. [99]

Thus began a dark period of 400 years of Biblical silence, ultimately leading to a manger in Bethlehem. Around this period, a king of the Jews was appointed by Rome named Herod, later called Herod the Great. Herod was not from the previous kingly lineage. Although he ruled over the Jews in Israel in the time before Christ, he was not completely Jewish. He was born in Ashkelon, a southern Palestine seaport on the Mediterranean Sea in 73 BC to an Idumean man named Antipater and a woman named Cyprus, the daughter of an Arab sheik. [100]

The priesthood that fell during the Babylonian siege was not replaced from the tribe of Levi, but instead from the tribe of Seth. During Jesus' time, the high priest was Joseph Caiaphas, who was the son-in-law of Annas, a very influential former high priest. Annas likely ran Caiaphas' affairs from a distance. The high priest served at the pleasure of Rome, as did King Herod. This is the time Jesus chose to be born. These were a few of the people who would influence His time on Earth. Jesus' lineage remained intact, but everything else around him seemed to fall apart.

First Five Books Postscript

There has always been much debate about whether Moses wrote the first five books of the Old Testament and if he did, when. Many believe, as I do, based on evidence in the Scripture itself, that these books were written by Moses and they were compiled around 1400 BC. On the other hand, if more substantive proof can be brought to light, I am willing to listen with an open mind. Having not dug our heels into the ground on that point, what is the importance of these five books, and what actually are they? [101]

These books are often referred to as the Pentateuch. The five books of the Pentateuch are the first five books of the Christian Old Testament and the entire Jewish written Torah. The word *pentateuch* is a Greek term meaning "five scrolls" and refers to the five scrolls which comprise the Torah and which also comprise the first five books of the Christian Bible. It is unlikely that these five books were originally intended to be five books at all. Instead, they were probably considered all one work. The division into five separate volumes is believed to have been imposed by Greek translators. Jews today divide the text into fifty-four sections called *parshiot.* One of these sections is read each week of the year (with a couple of weeks doubled up). [102]

As I expressed earlier in this book, there are dubious things in the book of Genesis regarding the timeline. There is simply no way to bring everything together without a clear understanding of many missing or unrecorded events. Also, assuming Moses wrote it, much had already occurred prior to his birth. Anything he would have used to author Genesis must have come from oral tradition or manuscripts recorded by others, assuming such methods of writing even existed. The other source could have been divine inspiration, which I cannot deny.

What is certain in Exodus, if written by Moses, is the presence of a firsthand eyewitness. If you read the entire book of Genesis and then the first part of Exodus in one reading, there is a sense that the flow of writing changes. From chapter 1, verse 1 of Exodus, there seems to be, a certainty in recalling events which is unlike most of Genesis.

Regardless of these areas where clarity is needed, neither the scientific community nor anyone else has uncovered or discovered things that would disprove what has been written in the book of Genesis.

From the beginning of the Old Testament book of Joshua to the end of the Book of Malachi, there is a little over 1,000 years of births, deaths, extraordinary events, wars, prophecies, and a maturing of civilization. Between the end of Malachi at the end of the Old Testament and the beginning of the first book of the New Testament, Mathew, there are 400 years unaccounted for in Scripture. Toward the end of the last book of the Old Testament—assuming the books are in chronological order—empires begin to rise and fall, conquer and be conquered, leading up to the ruling presence of Rome in Judea. These moments in secular history seem to set the stage and complete the casting perfectly for the Book of Matthew and the coming of Jesus.

The Lead-up to the Book of Matthew

... from BC to AD

In 555 BC, the Persian king, Cyrus the Great, united the Persians and the Medes. Finally, in 539 BC, he took Babylon in a bloodless coup and established Persia as the dominant force in the Near East. This King permitted the return of the exiles of Judah to their homeland, with permission for them to rebuild the temple in Jerusalem after its destruction in years past. His son Cambyses conquered Egypt in 525 BC, making the Persian empire the largest the world had yet known.

During the years 499–400 BC, the Greek-speaking Ionians of Turkey revolted against Persia and received aid from the Greek city of Athens. This began 170 years of strife between the Persians and Greeks. Philip II of Macedonia, who united the Greeks, was assassinated in 337 BC. He was succeeded in 334 BC by his young son, Alexander. Later proclaimed Alexander the Great, he would go on to conquer an area even larger than the Persian empire he was fighting against. After dying in 323 BC, his vast kingdom was split into smaller entities. These small groups constantly fought against each other, and by the Maccabean rebellion of 66 – 63 BC, the Roman Pompey conquered much of the Near East.

Of all kingdoms that had come before, there would be no empire covering more territory than that of Rome. Rome was to rule for more than five hundred years after Pompey. Into this period came the Messiah with His message of peace for all the world, and it was this world that shaped the rise and fall of the early Church faith community. [103]

Welcome to the end of Creation's Winter. Spring is busting out all over...

SPRING

The birth of Christ through His ascension

CHAPTER 7
WHY BETHLEHEM

... Why a manger?
... The most important breath ever taken and exhaled in the epoch of history from beginning to end

And she brought forth her firstborn Son, and wrapped Him in swaddling clothes, and laid Him in a manger, because there was no room for them in the inn.

LUKE 2:7 NKJV

So, when Jesus had received the sour wine, He said, "It is finished!" And bowing His head, He gave up His spirit.

JOHN 19:30 NKJV

CREATION HAD BEEN yearning for eons to find an opportunity to burst out with a sigh of Spring splendor. It was a sigh of relief as well, because the opportunity had been held back for so long. It was full of splendor because it began with the birth of royalty, a tender baby due all the regal, ornate pomp and circumstance of the ruling class. When this child took His first breath, however, there was absolutely nothing regal or ornate… just a tender baby. At this time the mother, and especially the earthly father of the child, partially understood the significance of this birth. The humblest of births in a stable surrounded by His family and everyday individuals was the logical entry point for Christ, the Son of God… the entire basis for the faith community's position.

Being born into a wealthy, elite environment would have created unnecessary obstacles. His choice of social class for His human entry did not dictate to whom He would minister. Nor would His entry point isolate any social class. The choice of a manger was not because the wealthy are inherently evil or bad. Jesus was associated with and witnessed to several well-off individuals and public officials during His walk on Earth. Even rich dignitaries from afar visited Him as a newborn, some six months after His birth, bearing expensive gifts. It should not be overlooked that these same "*Wise Men*" were instrumental in making sure the murderous authorities did not locate Jesus and His family. It should also be noted that it was two rich members of the Jewish Court who boldly asked to remove His body from the cross, one of whom even donated his own tomb.

At Christmas time, we read the story of Jesus with the certainty that, although there was something strange about the science of the mother's conception, it started when the angel said to Mary, *"The Holy Spirit will come upon you…"* Luke 1:35 Unlike a young couple, wisely planning for the arrival of their first child after marriage, this baby's real Father had been planning this event for thousands of years… along *with* the thoughts of the baby to be born. Yes, it was complex. Does that surprise you?

Have you ever thought about what Jesus smelled in that first breath of air? Science tells us that a baby develops their sense of smell while still in the womb. By ten weeks, the receptors that a baby will use to detect smells have already formed. As a newborn, a baby has a highly developed sense of smell. The sense of smell is processed by a part of the brain that also controls memory. A baby will form strong associations between particular scents and the experiences attached to them. This means that years later, a scent may trigger a memory for a grown person and remind them of a time or feeling from their past.[104]

Jesus Christ's first breath must have been overwhelming. He had never smelled creation as a human, just as a God of the Trinity. Conversely, no one on Earth would know what "Holy" smelled like. Remembering where Jesus was born, have you ever been in a stable or a barn? The odor is indeed overwhelming. Not necessarily repulsive, depending on the type structure you are in, but at a minimum... distinctive. In most cases, especially a stable where livestock is housed and fed, a predominant odor is hay and straw. Why is that important?

Everything eventually gets back to the Creation; what started this adventure. Like Moses removing his sandals on the mountain while receiving the Ten Commandments, it was important for Jesus Christ to experience His creation in the first intake of breath. God wanted Moses' bare feet to be in direct contact with His Earth with nothing man-made between. And as science has proven, later in life, when Jesus encountered the same smell He did at birth, He would be reminded of His first entry into creation and the gentle life that was all around Him in that humble stable.

In comparison, the last breath Jesus would draw and then exhale as a human would not remind Him of anything associated with His birth. The only common thread would be His memory of seeing His mother for the first time through the blood created by the birthing process. As then, He was now seeing her through His blood flowing from open wounds created by a crown of thorns. Just as when He was a baby, He could not take care of His mother any longer. He was moved to request that one of His best friends, John, take care of her. The Spring we are about to encounter lasts about thirty to thirty-five years—decimal dust compared to the Winter of Creation just left behind. But the events following this birth will be the most important span of eternity's timeline to the faith community. And, unlike the Winter of Creation, these years can be measured with a certain amount of accuracy.

So, we have discussed the reason for a manger. But why was Bethlehem important at this time in history? The answer is fairly simple: prophecy, a humble birth, a census, and taxes. This appearance in history can be described as strategic. In summary, nothing about this occurrence was an afterthought or spur-of-the-moment decision. Also, none of this has to do with common sense, because God doesn't have common sense... He is God. That is why much of what He does makes no sense to us… the common people.

It is rumored that Abraham Lincoln said, "God must have loved the common people… because He made so many of them." This quote didn't appear until twenty-years after Lincoln's death, but Scripture proves God's mind cannot be duplicated by a human—theologian or scientist. We will never think like Him. So, guys, your wife is correct… you are not perfect.

The question as to why Bethlehem and why that time in history is interwoven. While there is prophecy being fulfilled with this place and time, including lineage, we can look back with 20-20 hindsight and understand God's plan for the arrival of His Son. It is important to remember that the world didn't stop the moment Christ was born. For most, He was just another baby. Truth is, within the magnitude of world events, His birth was entirely insignificant… at the time. It would continue to mean nothing to that part of the world He lived in for another thirty years or so.

Just as the seemingly unrelated four hundred years of world history leading up to this birth were nothing more than continuing wars and power struggles, God was slowing setting things in motion during Jesus' first thirty years of silence. Looking back, this period of history was the perfect time for Christ's entrance onto the scene. Just as God was working through the world kingdom buildup leading to the domination of Rome, He was also working in the lives of thirteen specific men to prepare them, physically and mentally, for a history-altering, three-year journey with God's Son. This would be followed by years spent inoculating the world with a message-vaccination, against Satan and the evil he had brought into God's creation from the very first breath taken by a human.

I will elaborate later, but let me list those thirteen names here: (Matthew 10:2-4 – paraphrased for this example)

- Jesus' cousin, John the Baptist
- Simon, who is called Peter
- Andrew, Peter's brother
- James, the son of Zebedee
- John, the son of Zebedee

- Philip
- Bartholomew
- Thomas
- Matthew, the tax collector
- James, the son of Alphaeus
- Lebbaeus, whose surname was Thaddaeus
- Simon, the Canaanite
- Judas Iscariot, who also betrayed Him

Jesus attracted and called many disciples, but these twelve, along with His cousin John, were eventually selected for the final push. The timing of Jesus' birth was much more than just prophecy. In many ways it was as complex as creation itself. Everything had to be in place over the next thirty years or so.

- Nations, rulers, emperors, kings, priests, courts, judges, and temple leaders
- The Jewish Nation
- The non-believing half-brother of Jesus: James
- The economy
- Peace among warring nations
- Literacy
- Class-defined society
- The thirteen men listed earlier
- A band of heavenly star astrologers from the East who show up about six months after the birth
- And not the least of these, a young virgin who would give birth to the Messiah without the assistance of a man.

In my life as a Christian—most of my seventy-two years—I have heard an untold number of sermons, Bible lessons and discussions. It seems most of these regarding Jesus' mother omit the term "*virgin.*" Why is that? There are two good reasons I can think of: If children are present, it eliminates an uncomfortable discussion at Sunday lunch when the child asks what the word means; and secondly, you don't have to get into a debate with the science community. I really think both of these are inexcusable. Avoiding these discussions with children is to ignore their capacity. In the case of the science community, it throws away a terrific opportunity to have a substantive discussion with someone who may know science, but needs to begin a relationship with God through His Son, Jesus Christ.

The discussion with your children is a personal matter that parents just need to work through. Every child is different. Don't let that prevent you from telling your child the truth instead of them hearing it from someone you may not entirely trust or even know.

The discussion with the scientific community takes a little thought on your part. Let's open the lid and peek inside.

We will begin with the basics, from a scientific viewpoint—from my research, of course. "Human development begins after the union of male and female gametes or germ cells during a process known as fertilization (conception). Fertilization is a sequence of events that begins with the contact of a sperm (spermatozoon) with a secondary oocyte (ovum) and ends with the fusion of their pronuclei (the haploid nuclei of the sperm and ovum) and the mingling of their chromosomes to form a new cell. This fertilized ovum, known as a zygote, is a large diploid cell that is the beginning, or primordium, of a human being."[105] This new cell, this fertilized ovum, works its way into the woman's womb where it will grow until birth. Here's my unscientific two cents: In the Virgin Mary's case, we are missing the, *"after the union of male and female."*

In our modern age of medicine, we now know it is possible to skip the "union" part of the equation and artificially implant the male's contribution. Even still, the male contribution must come from another human, a man. A woman cannot do this by herself. The faith community should not be uncomfortable explaining the miracle that took place during this event. Without this miracle, how can anyone accept the entire miracle of creation? This is not the first time a human was born without the scientific reaction we depend on today for a normal delivery of a child. Back to creation.

Adam, the first born of the Earth, was formed from the dust of the earth followed by a breath of Life from God Almighty. Eve, the second human, was born when God removed a rib from the first human and designed it into a woman, which in Scripture means she came from a man. Genesis 2:23 Science is today rapidly learning how to grow body parts from cultured cells of living beings. [106] [107] How large of a step would it have been for God—creator of Adam, Eve, all living things, the Earth and the heavens—to be more advanced than science in 2021? Was a virgin birth possible when Jesus was born? Yes. Now for something a little deeper in the discussion: DNA.

When a man and a woman have a baby, and friends and family gather around to see the newborn, one of the anticipated comments will be, "She has her mother's eyes" or "Thank goodness he doesn't have his father's nose!" This happens because of the combining of DNA at conception. In *this* miracle of Jesus' birth, whose DNA was combined? Was the miracle that the only DNA used was Mary's? Or, somehow, was God's DNA, through the Holy Spirit, combined with Mary's to form a human version of the third part of the Trinity: Christ? Though some would consider this a silly idea to explore, it is at the core of a non-believing member of the scientific community. It is a valid question. Who *did* Jesus look like? One thing for certain: It wasn't Joseph.

The darker side of this debate includes several theories by ancient and contemporary Gnostics and fiction authors. Some include Joseph as the father, or another past sexual relationship. Some writers have even included the possibility of a relationship between Jesus and Mary Magdalene, who then had children, thus passing Christ's DNA though future generations.

Since none of the above can be proved, this will remain something the Faith and scientific communities will have to either agree to disagree or accept each other's facts and reasons for existence. At the end of the day, as stated throughout this writing, both need each other.

God created the heavens and the Earth and all living things. Those living things include curious human beings who thirsted for knowledge of the world around them and developed the discipline of science to better explore it. Yet the more science discovers, the more it appears God is real and science is drawn closer to accepting the same truths as their fact-based faith when boarding a multi-ton airplane and flying at 35,000 feet on invisible currents of air molecules. Like the faith community, they know it is there, they know how it works, but at some point, they have to rely on their faith in the process used to discover those currents. They have to get on the plane.

The faith community walks out into the light of the sun and enjoys God's promise. The scientific community walks out into the sun and hopes the fragile distance between the Earth and that distant star doesn't vary by a relative fraction, lest the earth burns to a crisp. With every discovery, it would appear that it gets harder and harder for science to hold onto the notion that Earth's existence is a mere accident. Likewise, it becomes more difficult to explain a six, twenty-four-hour day, Creation.

CHAPTER 8
ABRAHAM… TAKE TWO

… God must just love a challenge
… but what better way to get everyone's attention?

There was in the days of Herod, the king of Judea, a certain priest named Zacharias, of the division of Abijah. His wife was of the daughters of Aaron, and her name was Elizabeth.

Luke 1:5 NKJV

But they had no child, because Elizabeth was barren, and they were both well advanced in years.

Luke 1:7 NKJV

WELL, HERE WE go again. But unlike Abraham, this priest named Zacharias paid attention. He would be the uncle of the Messiah, not a distant relative as Abraham had been. At the time, we are not told the exact age of Zacharias or his wife Elizabeth. We are simply told that they were both advanced in years. They also had no children, as Elizabeth was barren. The opening scene finds Zacharias tending the Temple. We later discover his assignment required him to be there for several days at a time. While he was on his shift, one of his duties was to light incense at certain times inside the Temple while people, who were not allowed in, prayed outside.

I find it interesting that an angel appeared to Zacharias *inside* the Temple near the Holy Curtain through which only the high priest could enter into the Holy of Holies. This is the same curtain that would split from top to bottom as Christ died on the cross. God does have a sense of humor. That sense of humor and understanding of human nature continues as the angel tells Zacharias how things are going to happen. Upon hearing the angel's proclamation, Zacharias has a moment of unbelief. Surely, he knew the story of Abraham. Well, so do I, but if this were to happen to any one of us, disbelief would be showing all over our faces.

Still, his unbelief is temporarily punished. It was after Zacharias' second, telling question that the angel becomes irritated. So much so that for the first time, he lets Zacharias know he is more than an angel. In verse 19 he states, "*I am Gabriel who stands in the presence of God, and was sent to speak to you and bring you these glad tidings.*" The use of the word *glad tidings* here is significant as it relates to the birth of Zacharias' coming son. These same words will be used again in Luke 2:10 by an angel speaking to the shepherds announcing the birth of Christ the Lord!

I love the imagery in the following verses. Zacharias is struck mute by the angel so that he will not tell anyone until the announced events have taken place. There is a brief line in verse 22 of chapter 1 where the people crowd in on Zacharias, after he finally leaves the Temple, wanting to know what took so long. Since he cannot speak, I see him waving his arms and hands, first pointing to the sky, then the Temple. Perhaps he is even trying to form words with his hands or cupping his arms as if holding and rocking a baby! What would your friends and those who work with you have said about you, if this had happened to you at work?

When speaking about Jesus' ministry, the term *twelve* is very often used to emphasize His special group of hand-chosen men that were there during His three-year ministry. They would ultimately be the nucleus of the Christian movement. I prefer the number thirteen instead of twelve, however, because this story about the birth of Jesus' first cousin is equally important to the establishment of Jesus' future message and ministry. If God had not allowed John the Baptist's early demise by a tragic beheading,

I am confident his legacy would have been just as instrumental in the years following Jesus' death, burial, and resurrection as the remaining eleven of the original twelve disciples.

The incident beginning in Luke 1:39 is one of the greatest demonstrations of the Holy Spirit's presence during this time of Christ's transfer to human form. During the sixth month of Elizabeth's pregnancy, Mary traveled to Zacharias and Elizabeth's home. In verse 41, the Scripture says that when Elizabeth heard Mary enter their home, the babe inside her leaped. Not to correct Scripture, but it was the babe who heard or sensed the very presence of his cousin, the coming Savior of the world. The babe, soon to be called John the Baptist, was moved by the Holy Spirit and leaped inside the womb like a Pentecostal preacher coming around the pulpit to emphasize his point. This was the first instance of John the Baptist's proclamation of the coming Messiah. He may have leaped in the womb, but I want to believe that his almost-born lips were forming the word, "Messiah!"

We are missing a great deal of the early years of Jesus' life. I can't help but believe that growing up, there were several times when these two miracle cousins played together as boys. The movement by John in his mother's womb, when sensing the presence of Jesus' mother, would surely suggest such a possibility.

Back to poor old Zacharias' speech problem. It just goes to show you—don't upset Gabriel. It has been over nine months since Gabriel zipped Zacharias' lips. Elizabeth gave birth to a healthy, active boy. Eight days later, as we discussed earlier in another chapter, it was time for the baby to be circumcised. The ceremony required calling out the baby's formal name. Tradition would have dictated the boy be named after his father, Zacharias. But before the proclamation, Elizabeth stopped them and announced that the baby would be called John. The group was baffled by this break with tradition, particularly in light of the fact that no one in the family went by that name. The crowd turned to the father to see if he objected. They were making gestures to see whether he approved, and he was trying to communicate, again with hand and arm motions. In a final act of frustration, Zacharias pointed to a writing tablet. He wrote these words, "His name is John." They all marveled. But, happy wife, happy life. Another benefit of having a baby at an old age—the man had figured out his place.

And then it happened! He got his voice back!! After the events of the past nine months, the crowd wondered, "What kind of child will this be?" This was followed by a Holy Spirit-filled sermon that Zacharias had been putting together in his head for the past nine months. After the introduction, his sermon turned to prophecy concerning his newborn son's future ministry. He proclaimed that John would be called, *"the prophet of the Highest; for you will go before the face of the Lord to prepare His ways…"*

John was an interesting personality with a peculiar diet and wardrobe. Prior to his introduction to Israel, he had been living in the desert. The next time we encounter John, he will be witnessing the Holy Spirit appearing with the gentleness of a dove.

CHAPTER 9

HERODS... ROME'S KING OF THE JEWS FAMILY

... a descendant of Esau
... a political appointee
... and two first cousins

When Herod the king heard this, he was troubled, and all Jerusalem with him.

MATTHEW 2:3 NKJV

Then Herod, when he saw that he was deceived by the wise men, was exceedingly angry; and he sent forth and put to death all the male children who were in Bethlehem and in all its districts, from two years old and under, according to the time which he had determined from the wise men.

MATTHEW 2:16 NKJV

HEROD "THE GREAT," as he is popularly known, was appointed by Rome as the king over the Jewish Nation after the death of Julius Caesar. This followed a series of military conquests by Rome, including the capture of Jerusalem in 37 BC. Herod reigned until his death in 1 AD or 4 AD, depending on which lunar eclipse you choose. His one major accomplishment was the initiation of the rebuilding of the Jerusalem Temple, which was finally completed sixty years after his death. Several years later, it was totally destroyed: "*But when the king heard about it, he was furious. And he sent out his armies, destroyed those murderers, and burned up their city.*" Matthew 22:7 [108]

Herod the Great is also known for another infamous act. He had heard about scholars from the East seeking a king, and he invited them to his palace. Upon learning from them the approximate time of the foretold coming of the King of the Jews, he ordered the murder of all Jewish-born males, two years old and younger. History also tells us that he had three sons. All of them were crazy, but one became just as notorious as his father. This eldest son, Archelaus, assumed the throne upon his father's death. It is thought that Herod the Great died of chronic kidney failure. The disease, without the treatment that was later developed, ravaged his body with sores and putrefied puss in his extremities. He died a very drawn-out, painful death.

The Roman Emperor, Augustus, would not confirm Archelaus until he proved himself. This never happened, since Archelaus slaughtered 3,000 prominent citizens soon after taking over. Not only did the Emperor not confirm him, he removed him two years later and stripped the title from the entire Herod family. Later, a less radical but still very disturbed member of the Herod family would reign during the ministry of Jesus: Herod Antipas.[109]

Herod Antipas, ruler of Galilee and Perea from 4 BC to 39 AD, had John the Baptist beheaded to keep a promise to Salome, daughter of Herodias. This would put an end to John's prophetic ministry, which played a crucial role in the development of the most profound of all religious faiths in the history of the world. God did not allow the cruel beheading of John until he had set the stage for Christ's coming ministry. It all began on a seemingly normal day with an encounter between John and his cousin, Jesus.

Before reviewing this special encounter, let's take a look at a non-Biblical account that addressed these events. Josephus lived within three decades of Jesus' death, burial, and resurrection. He is recognized by Jewish and Christian theologists as a defining resource concerning the historical accuracy of New Testament Scripture. He records in *Book 18, Antiquities of the Jews:*

> *Now many people came in crowds to him, for they were greatly moved by his words. Herod, who feared that the great influence John had over the masses might put them into his power and enable him to raise a rebellion (for they seemed ready to do anything he should advise), thought it best to put him to death. In this way, he might prevent any mischief John might cause, and not bring himself into difficulties by sparing a man who might make him repent of it when it would be too late. Accordingly John was sent as a prisoner, out of Herod's suspicious temper, to Macherus, the castle I already mentioned, and was put to death.*[110]

Just as there is a variance as to the time of the death of Herod the Great, there is a difference of about three years between Josephus and Scripture regarding these events. However, looking at a time span of over 2,100 years, the margin of error is small. I trust that will be acceptable to the science community. Back to that encounter between the cousins.

On a seemingly normal day, John's established ministry and the early stages of Jesus' ministry converged in what appeared to be a chance encounter. It occurred not with a bang, but rather with a calm whisper. In addition to the multitudes that gathered to hear John, there were other characters present. These characters included the Trinity... yes, all three of them. The Son appearing in the form of a man, the voice of God the Father announcing He is pleased with His Son, and the Holy Spirit appearing as gentle as a dove. This is an extremely rare event in Scripture. The scene is surrounded by creation and those Who created everything. After being placed under the water by John the Baptist, Jesus Christ would emerge a baptized individual, gasping a breath of the air He helped create long ago in Genesis.

At that moment, with Jesus at the beginning of His ministry and John toward the end of his, the first sacrament, baptism, came into being. Three or so short years later, at the last supper, Jesus would institute the second Sacrament: Holy Communion. These sacraments of baptism and communion, along with Jesus' resurrection, would forever separate Christianity from every other recognized religion.

Of interest, but not for discussion here, is the following question: We know that Jesus Himself performed the first Holy Communion, but did He ever perform the sacrament of baptism by baptizing anyone?[111] In Scripture, John 3:22-23 seems to indicate that He did; however, in John 4:1-2 it clearly states that it was the disciples who were doing the baptizing. If Jesus did baptize as John, this is in sharp contrast to a quote from John the Baptist in Matthew 3:11: "*I indeed baptize you with water unto repentance, but He who is coming after me is mightier than I, whose sandals I am not worthy to carry. He will baptize you with the Holy Spirit and fire.*" NKJV

The object of the cross, whether displayed with or without Christ's image, is the most recognized symbol of Christianity. This cruel death of Christ on the cross by crucifixion, prophesied over 1,000 years earlier, was necessary for the resurrection to take place. Without that death, the resurrection is meaningless. Most often, dying is never associated with a miracle. Now, raising oneself from the dead… *that* is a headline.

The closest contemporary non-fiction literature paralleling the acclaimed 1955 novel, play, and movie *Inherit the Wind,* written by Jerome Lawrence, is a non-fiction book titled *The Case for Christ,* written by Lee Strobel in 1998 and made into a documentary film in 2017 by the same name. While these two works approach the battle between the science and faith communities from different points of view, they are both excellent examples of how and why these two communities need to talk to each other instead of arguing over who is right.

Lee Strobel documents in his book, *The Case for Christ,* his rebellion against his wife's attempt to convert him. He begins the book from the dogmatic position of an atheist and ends as a born-again Christian. After a period of painful encounters with his wife over the subject, he decides to begin an objective investigation into the subject without informing his wife. He initially sees it as an attempt by misguided followers to convert others to a belief in the seemingly fantastical notion of the birth, death, and resurrection of Jesus. Early in his research, he comes to understand that the whole key to the Jesus thing is… the resurrection—something that dies coming back to life on its own.

He begins by trying to prove that Jesus never died in the first place, but simply was in a coma and later woke up in the tomb. The evidence, however, proved otherwise. Strobel's writing is overwhelming and his sources are credible. I encourage you to read this book. I would also encourage atheists, agnostics, and your Christian friends to seriously research other reliable sources who wrote close to the time of Christ, and who reflect eyewitness accounts of these events. And please… with open minds.

Returning to that famous encounter between two first-cousins—the closing scene of the chapter is both touching and endearing:

> 13 *Then Jesus came from Galilee to John at the Jordan to be baptized by him.* 14 *And John tried to prevent Him, saying, "I need to be baptized by You, and are You coming to me?"* 15 *But Jesus answered and said to him, "Permit it to be so now, for thus it is fitting for us to fulfill all righteousness." Then he allowed Him.* 16 *When He had been baptized, Jesus came up immediately from the water; and behold, the heavens were opened to Him, and He saw the Spirit of God descending like a dove and alighting upon Him.* 17 *And suddenly a voice came from heaven, saying, "This is My beloved Son, in whom I am well pleased."* Matthew 3:13-17 NKJV

Following this brief encounter, Jesus Christ travels alone into the wilderness and fasts for forty days and nights. There He encounters Satan, who finds out that the battle ahead is not going to be another Victory Garden event. Some have rightly asked why Christ, who was without sin, would need to be tested. Personally, I believe the interaction was for the benefit of Jesus. While the Son was familiar with Satan's background and his devious ways, the man Jesus was not. This was an excellent opportunity for Jesus to see and hear this fallen angel. Also, Jesus would learn to depend upon the Son's control over nature while not starving during this forty day and forty night fast.

Following this Satanic trial, Jesus is cared for by angels. Upon returning to civilization in Nazareth, Jesus first hears that His cousin is now in prison. Fulfilling the prophecy of Isaiah, He leaves Nazareth, which is in Galilee, and travels to Capernaum in the regions of Zebulun and Naphtali. Beginning there in Capernaum, Jesus begins to repeat the words of John the Baptist, "*Repent, for the Kingdom of Heaven is at hand.*" Matthew 3:2 NKJV

It was in this region, by the Sea of Galilee, that Jesus called His first four disciples. Three of these men would become His closest friends - Peter, James, and John. The fourth, Andrew, the older brother of Peter, would become the best example of Jesus' teaching on leadership ministry.

... and so, the ministry begins.

CHAPTER 10
LEADERSHIP TRAINING DONE RIGHT

... teaching shepherds to tend flocks, not herds
... nothing about administration
... everything about introducing
others to Jesus Christ

19 Then He said to them, "Follow Me, and I will make you fishers of men." 20 They immediately left their nets and followed Him.

MATTHEW 4:19-20 NKJV

22 And immediately they left the boat and their father, and followed Him.

MATTHEW 4:22 NKJV

ANY CHRISTIAN CAN teach a leadership class on servant-leaders. The only background requirement is that they can read. Even then, if they cannot, they ask someone else in the class to read. To read what? The 10th chapter of Matthew! This chapter is written, first about the completion of Jesus' recruiting efforts and second, a recounting of the most effective servant-leadership training techniques ever taught. Whatever you do, don't improvise on any of Jesus' outline! More on this later—but first, let's take a look at what happens in Matthew chapter 11 after class is dismissed.

> 1 *Now it came to pass, when Jesus finished commanding His twelve*
> *disciples, that He departed from there to teach and to preach in their*
> *cities.* 2 *And when John had heard in prison about the works of*
> *Christ, he sent two of his disciples* 3 *and said to Him, "Are You the*
> *Coming One, or do we look for another?"* 4 *Jesus answered and said*
> *to them, "Go and tell John the things which you hear and see:* 5 *The*
> *blind see and the lame walk; the lepers are cleansed and the deaf hear;*
> *the dead are raised up and the poor have the gospel preached to them.*
> 6 *And blessed is he who is not offended because of Me."*
>
> 7 *As they departed, Jesus began to say to the multitudes concerning*
> *John: "What did you go out into the wilderness to see? A reed shaken*
> *by the wind?* 8 *But what did you go out to see? A man clothed in soft*
> *garments? Indeed, those who wear soft clothing are in kings' houses.*
> 9 *But what did you go out to see? A prophet? Yes, I say to you, and*
> *more than a prophet.* 10 *For this is he of whom it is written: 'Behold,*
> *I send My messenger before Your face, Who will prepare Your way*
> *before You.'* 11 *Assuredly, I say to you, among those born of women*
> *there has not risen one greater than John the Baptist; but he who is*
> *least in the kingdom of heaven is greater than he.* 12 *And from the*
> *days of John the Baptist until now the kingdom of heaven suffers vio-*
> *lence, and the violent take it by force.*
>
> 13 *For all the prophets and the law prophesied until John.* 14 *And if*
> *you are willing to receive it, he is Elijah who is to come.* 15 *He who*
> *has ears to hear, let him hear!* 16 *But to what shall I liken this genera-*
> *tion? It is like children sitting in the marketplaces and calling to their*
> *companions,* 17 *and saying: 'We played the flute for you, And you did*
> *not dance; We mourned to you, And you did not lament.'* 18 *For John*
> *came neither eating nor drinking, and they say, 'He has a demon.'*
> 19 *The Son of Man came eating and drinking, and they say, 'Look,*

a glutton and a winebibber, a friend of tax collectors and sinners!' But wisdom is justified by her children." Matthew 11:1-19 NKJV

What happens between verses 1 and 2 of Matthew 11 is similar to watching a TV drama when there is a sudden change from one scene to another, to cover another aspect of the story line. Your mind remembers what just happened and tries to relate what you just saw to what is currently in front of you. It is important to remember what happened in verse 1 as you continue to read.

Verse 1 tells us that Jesus' servant leadership class (one session around the table, not a series of endless power-point slides with examples) has ended. Nothing was said about who was in charge of what, budget concerns, the correct order of worship, appropriate music to be used, innuendo, organizational movement, or micromanagement. Now it was time to apply what Jesus had taught to real people in their real-life situations. Verse 2 is a sudden scene change to a dingy, Roman prison cell where we see an emaciated, half-alive human covered in filth. It is John, formerly called the Baptist.

He has been lingering there for some time. His misery, not his soul, has brought on doubts about the claims he made earlier regarding his cousin's prophetic coming as the Messiah of the Children of Israel. He sends two of his disciples to inquire if he had mistaken Jesus as the real Messiah, and if so, must he start over. John surely hoped he knew what the answer would be for two reasons. First, he really wanted it to be true and second, if he had been wrong, he had probably missed his opportunity for a redo; his current situation could end only one way. There would not be a second chance.

Jesus first addresses John's disciples, reassuring them He is who John proclaimed Him to be. Then He turned to the crowd with a bit of sarcasm toward them and high praise for His incarcerated cousin. There was a third group present, His disciples. This image should burn into the minds of spiritual leaders and pastors in the present age. The disciples He just taught witnessed their leader address the doubts of the faithful and the distrust of the unbelieving, all in one uninterrupted motion. He spoke about love and concern for His cousin, clarity of who He was, what those listening should believe, and what to do about it.

What should the 21st-century pastor and church leader do about it? First, a short list of problems, followed by a longer list of solutions. Let's begin by contrasting the implied scale in the last sentence.

Problems for 21st-century pastors and lay-leaders:

- Christian churches, of all denominations and brands, are dying
- Generational diversity
- Growth of undefined Gnosticism

Solutions for 21st-century pastors and lay-leaders:

- Pastors should stop tending their herds and tend their flocks.
- Churches should hire pastors for their various ministries with the expectation that they will train others to minister, not to do the ministry for them.
- Pastors should visit the sick and the well, after their trained lay-leadership have made the first visit and advised the pastor of special needs.
- Lay-leaders should draw their priorities from Matthew chapter 10, not their board or committee assignments.
- The majority of all ministry is outside the walls of the church building, applied from within the walls of the lay minister's heart.
- Programs, spiritual and physical, should be developed for the community, utilizing Christians within the church who have a multitude of gifts and talents, with identified needs.
- Programs, spiritual and physical, should be developed for the community, matching unused facility space on days and nights other than Sunday or Wednesday evening.
- Churches must stop referring to growth as a "numbers game" but

rather as the Great Commission of Christ given by Him as He ascended into heaven at the end of His earthly ministry.

- Churches should stop rephrasing statements concerning growth to emphasize spiritual growth as much more important than physical growth—they are equally important.
- Churches should realize spiritual growth is expected, physical growth is commanded and measured, and the two are intertwined, meaning that one without the other is a failure.
- Treat stewardship as a blessing and a privilege, instead of something to avoid.
- Stop treating the stewardship of money as an afterthought because it seems more religious to talk about the stewardship of time and talent.
- Churches should stop idolizing the Red-Letter editions of the Bible and emphasize what Jesus did more than what he said.
- Churches should develop ministry around the gifts and talents of the people God brings to the congregation, not attempt to change people to fit popular programs the church would like to develop, or worse… continuing unproductive programs.
- Accept that the first vital step to revive a failing church is to admit your church is dying… out loud… to God and…. to each other.
- If a church is dead… be kind… bury it and distribute its assets to a recovering ministry… not another terminally ill church.

In an earlier book I wrote in 2008, *The Little Brown Church in the Vale… when did the lights go out?,* I go into some of the above issues within the US denominational crisis. It continues to be a serious and debilitating problem. Since 2008, the issues have only grown more pressing, and churches continue to close. This leaves a void in communities in some cases, while in others, it has been long forgotten that a church even existed.

The latter was the problem when Christ gasped for His first breath of air in that Bethlehem stable. Actually, a church didn't even exist, as we know it today. There was the Temple, an ornate structure designed and run by centuries-old laws originally set down by Moses. Over time, the religious rulers and teachers constantly added to and changed the meaning of those laws to fit their contemporary needs. At the time of Christ's birth, the Temple was a corrupt organization held in place by a Roman Government that was singularly interested in keeping the Jews calm and compliant.

The leaders of the Jews and their temple were no longer from Israel's lineage, but appointees of Rome with just enough blood line to be acceptable. Below that were local synagogues. These were the meeting places of common people where they were taught Jewish law and traditions. More than the corruption of the organization and the polluted lineage of its leadership, this place of worship had no direct access to God by a common person. That was the privilege of a chosen few. Since the fall of humankind in the beginning of Genesis, the Trinity had waited patiently for this exact time and place to finally give humankind access to the God who breathed life into their very being. At the end of that very long day in the human life of Jesus' suffering, His Father had only one objective. That objective would be carried out by the mover and shaker of the Trinity: the Holy Spirit. The objective? Split the curtain in the Temple from top to bottom. Humankind would finally be able to again commune directly with their Maker, and one day, live with Him in His Kingdom. But first, Jesus Christ had to train a few, very special, common people: the twelve.

We have discussed God's brain several times, so His method of choosing the twelve should not surprise us. First, the Jewish Nation was expecting a mighty king to come and save them. Why not? The honored example they had was their delivery by Moses and Joshua from Egyptian oppression and bondage to the promised land! In contrast, what did they get this time? A Nazarite, a son of a common carpenter. He was nothing like Moses or Joshua. He was… well… common, and for the most part, soft-spoken and a pacifist. War and anarchy were not on His lips. Maybe, they thought… He would choose followers with military credentials? No, He chose fishermen, tax collectors, traitors, and doubters, among other professions. Maybe He would train them and have them form a mighty army?… didn't happen.

Today, we look up to and respect the opinions and interpretation of theology from those who spend years in school, college, universities, and seminary. God's plan, unlike ours, was to skip all of that education and choose men who knew common life truths. They would get an education centered on observation and explanation around campfires, on long walks, and while sitting in fishing boats. Unlike today, their practical education classes lasted about two and a half years, not seven to twelve with a required thesis and post-graduate work.

Whether it was Jesus' ability to look into the hearts of these twelve men and see honest sincerity, or perhaps their understanding of Jesus' call on their lives, or both—in the end, all of them but one finished the race. And even the one who betrayed Him repented. Nowhere in Scripture does a disciple reply to Jesus, "I know You said you would make us fishers of men, but I already know more about fishing than you do." Or Matthew interrupting Jesus with, "You know I'm a tax collector, and I'm a smart guy, so let's try it my way." What they did do and say included betrayal, unbelief, and doubt. No, in about two and a half years, many still didn't fully get it, but they would... all of them. And with the exception of John, who lived to write the Book of Revelation—and Judas who took his own life, racked with guilt and depression—they all followed Jesus in persecuted deaths for the glory of Christ.

Jesus performed a lot of miracles while on Earth, but He is given very little credit for His greatest achievement: choosing and training these twelve men. They, along with many others, would be the catalyst of the greatest movement in history. The New Testament they authored by their life's work would be heard and read by more people than any other document that has appeared before or since in the history of the world.

The irony is that learned people throughout history had to study the books written by these uneducated men to earn their multi-year educations and degrees. The other takeaway from these stories is the missed opportunities of countless pastors who fail to stop and sense Christ's direction for their ministries, specifically as it relates to their comparative disciples: church leaders, elders, deacons, board and committee chairs, and anyone else in a ministry who shows promising abilities to be a part of a ministry outreach introducing others to Jesus Christ.

Most pastors I have known and worked with, understandably, get distracted by the day-to-day grind of administration and tending the herd. I have used this term *herd* several times. Let me explain what I mean by this, as I mean nothing derogatory. For illustration purposes only, I use this term to separate the congregation, or the herd, from church leaders, or the flock. It is a matter of scale for someone trying to manage, or in this case minister to, a large group of people. If a pastor is literally sucked into doing all of the personal contact, common sense will dictate something else has to be sacrificed. If a pastor spends most of their time visiting the sick and meeting with only those in the congregation who have issues with the direction of the church, those who are not sick and are trying to offer solutions as opposed to problems will suffer. Where is the time for a pastor tending the herd to connect with new people who could become dynamic fuel to a declining ministry? Gone, gone, gone…

The best time to set these parameters is when a pastor first begins a new ministry. Often, many dying churches and ministries have no current pastor, or are using interim supply pastors not fully dedicated to a specific ministry. There are also many pastors who have been in a church or ministry so long they are just trying to get to retirement before the church dies. This has been one of the reasons for past success by so-called non-denominational churches. Many of these have grown because from the outset, the non-denominational pastor's only assignment is to train others to minister and introduce people to Jesus Christ. As a result, they do not become mired in the herd.

Unfortunately, there are other reasons that non-denominational churches initially flourished, and some of these reasons had to do with the number of dying *denominational* churches within that community. Members of established churches became disenchanted with their established church's lack of new activity and prayerless direction. Many of these members just stopped going to church. Some just left their church after being attracted to a new, vibrant, community-serving ministry. And the other reason? A younger generation that saw church in a completely different way than their parents. A new generation whose desire is to be the church, not sit in it. Some of those non-denominational churches grew quickly into very large congregations.

Their size created large facility and personnel demands that placed them at risk for any small reduction in size and income. Some have been in existence now long enough for their facilities to begin wearing out. It is one thing to replace or repair a church building that holds 200, but quite another task for a facility seating 3,000 to 5,000. These new ministries have also lasted long enough to run into the same issues that helped them grow: people looking for yet another new way to minister or be ministered to.

Non-denominational churches had another weakness that didn't become visible until they began running into the aforementioned problems: They were on their own. There wasn't an organization that could stand by them until they solved what should have been a temporary situation. The bottom line is that every ministry has issues, and many of those issues all come back to a disregard for the central purpose of the church, the Great Commission. Anything that gets in the way of that commandment to go into all the world and proclaim the name of Jesus Christ, and the examples of Christ's leadership training outline in Matthew 10, is a recipe for disaster and replacement of an open church door with a for sale sign.

Although it is tempting to draw a parallel today of our struggle between denominational and non-denominational 21st-century congregations, and the 1st century struggle between the established church and the developing New Testament Church in Acts, there is little to compare. The established church in Jerusalem was the end result of at least 2,000 years of specific family, or tribe domination and attached tradition. The established church in the US and consequently around the world has its roots in the late 1700s and early 1800s. I am referring to the outbreak of denominationalism during the Second Great Awakening, starting in the newly developed areas around Kentucky and Tennessee and spreading east to the Atlantic.

From this outbreak of Holy Spirit-injected revivals and camp meetings, thousands were converted to Christianity, forming what we recognize today as massive denominations. Among their number are Cumberland Presbyterian, Baptist (all varieties), Methodist, Nazarene, Church of Christ, Church of God, and many other smaller groups. Until that time, the Church of England, Anglican, Lutheran, and the Catholic Church dominated much of Europe, England, Scandinavia, and the American colonies. With this

Second Great Awakening and the end of the War of 1812, the United States eventually became a stronghold of denominational might.

Although not a parallel, a common thread between 1st- and 21st -century church development was and is a yearning for a closer relationship between mankind and God. In the 1st century, that yearning was for the fulfillment of a 2,000-year-old promise of a coming Messiah. In the 21st century, the yearning doesn't seem to be so much for the second or third coming of Christ—depending on how you count Christ' presence at the Creation in Genesis—but a need for connecting to God, not connecting to other churches. Both involved dynamic and sociologic change. Both were and are extremely disruptive to the existing established church. Comparing the birth pangs of these two movements would be a mistake. They are two distinct reactions to two totally different motivations.

1st -century motivation included establishing direct access to God without the need of a priestly intermediary. 21st -century motivation seems to include disconnection from the church formed by 1st -century motivation. The nemesis of both movements was, and is, organizational structure. As things grow, there seems to always be a nagging movement toward organization. Organization leads to structure, structure to management, management to micro management, micro management to power, power to absolute power, absolute power to flat-lining, and flat-lining to digression, or a death spiral. In other words, the current condition of the 21st -century church as a whole.

It is important however, that the 21st -century church understand the total circumstances of the early New Testament church movement. While there was much energy and forward movement in the first churches, there was also equal amounts of division fueled by unrelenting tradition and control issues. The early church was devastated, time and again, over the issues of circumcision and baptism. The act of taking the gospel to the world included placing Jews and Gentiles in the same building, eating and worshiping together. These bitter disagreements were not confined to what we call congregations today, but also to very top leadership positions. These issues followed the Apostle Paul from day one of his first Missionary Journey until the day he was executed. A lot of things change over time, but human nature does not.

CHAPTER 11
WHY THEN?

... Jesus' introduction, appearing
on a late-night talk show...
... Breaking news, a Man calling himself
Jesus, the Son of God performed...
... The latest Tweet from Jesus seems to
downplay the position of the Pope...

And I will put enmity between you and the woman, and between your seed and her seed; He shall bruise your head, and you shall bruise His heel.

Genesis 3:15 NKJV

Therefore the Lord Himself will give you a sign: Behold, the virgin shall conceive and bear a Son, and shall call His name Immanuel.

Isaiah 7:14 NKJV

N.T. WRIGHT AND Michael F. Bird co-wrote a book in 2018, *The New Testament You Never Knew*, after research and observation about the people and locations of many of the events recorded and referenced in the New

Testament. This is an excellent book and study series for any individual or group desiring a better understanding of the background surrounding the collection of books, letters, and epistles called the New Testament.[112]

One of the sections of this book asks the questions, why was Jesus born, when He was born, and why was He born where He was born? These are excellent questions. Understanding both can give a Christian great insight into the heart and mind of the Trinity. In support of the New Testament, the earlier writings in the Old Testament contain over 300 prophecies concerning the coming of a Messiah. In these prophecies lie specific information regarding where and when this amazing event will take place. Two specific references, one at the beginning of the Old Testament in Genesis, and another at the beginning of the Nativity Story in Matthew, contain chilling insight about the why and where of Jesus' arrival.

The reference in Genesis 3:15 must have been hard to comprehend for those reading these words in Moses' time, when Genesis is believed to be first drafted: *"And I will put enmity between you and the woman, and between your seed and her Seed; He shall bruise your head, and you shall bruise His heel."* Genesis 3:15 NKJV

For context, the words in this passage of Scripture take place right after an encounter between Satan and Eve in the Garden of Eden. I discussed this Old Testament chapter earlier in the Winter of Creation section of this book. Even after settling how and why Satan was in the Garden, this passage raises multiple other questions. Let's take a peek into what I believe to be one of the most important prophecies in the entire Bible.

Here is the sequence of events as told in the third chapter of Genesis: Satan just shows up and confronts Eve. This is our first introduction to Satan in the Bible, except for a vague reference in Genesis 1:2, "*...and darkness was upon the face of the deep.*" Adam seems to be present during this conversation between Satan and Eve. Satan dares Eve to eat from the tree that God specifically told her and Adam to leave alone. The creature makes a convincing argument for disobeying God, and Eve relents. Adam, after hearing this discussion between his wife and this creature, is offered a bite by Eve, and without question eats the offered fruit. Scripture says in verse 7, "*Then the eyes of both of them were opened*" NKJV, and they left to make clothes to cover their nakedness.

The couple later heard God walking in the Garden and ran to hide. When asked by God why they were hiding, Adam confesses what they did, and that they hid because they were afraid that God would see they were naked. God asks a question in verse 11, much in the way that a parent would investigate the innocence or guilt of their child: "*Who told you that you were naked? Have you eaten from the tree of which I commanded you that you should not eat?*" NKJV The man answered, initiating an age-old tradition for all men who would come after him: he blames someone else, in this instance, his wife. Following natural instincts, Eve turns and accuses Satan. Or, as comedian Flip Wilson used to say, "The Devil made me do it!" It was funny when Flip said it, but this was dead serious… literally. So serious, that Adam and Eve and every human born after that day, including the prophesied Jesus, would hereafter exhale their last breath and die. All but Jesus, and a few other saints whom God just took, would return to the dust from which they were created.

Now, within that context, we find the first reference in Scripture in Genesis 3:15 containing the foretelling of the birth of Christ. After Eve points out the involvement of Satan, God turns to the creature and says, "*And I will put enmity between you and the woman, and between your seed and her Seed; He shall bruise your head, and you shall bruise His heel.*" What is enmity, and why just between Satan and the woman? Does Satan have seed, can he be part of a male and female union that will create a pregnancy? Or was God saying Satan could no longer impregnate a human woman?… and just when will this heavy welter-weight fight take place, between whom, and where? Whoever and whenever, I am betting on the one who only winds up with a bruised heel! The other person is going to have a serious concussion. For future consideration it is important to notice it does not say the bruised heel will cause the death of the one whose head is bruised. More on this in the Fall Season of Creation.

Does this help when considering how someone reading or hearing this for the first time around the parting of the Red Sea might look a little lost in their thoughts? As before, that has been the norm when something involves one, or all three, of the Trinity. Humans just don't think like their Creators and probably never will.

On the other hand, what does the science community think about all the above questions related to this prophecy? It might surprise you. Remember that it was this scientific community that came up with the theory of the birth of humanity happening by accident, either from billions of years of something not living, evolving in yet unnamed chemical elements, passing through that instant when the nonliving became alive, or, life arrived on Earth as a single, living cell riding on an asteroid or comet that banged into the Earth. So ask yourself, which one of these stories is more believable—those of faith or those of science?

While neither the faith nor science community has any empirical proof, there is one staggering difference. The faith community's assumptions and beliefs are based on events that take place over thousands of years, while the science community insists on the events occurring in an instant, but, developing over billions of years. This might be a great starting point for a conversation. Going into such a discussion, the faith community should remember that they have yet to give a serious explanation concerning the dinosaur's existence and when they would have appeared on their timeline.

After that little detour, let's get back to Genesis 3:15 as one of the most important prophecies in the entire Bible. If this prophecy in Genesis 3 does not come true, nothing else matters. At the time and place Jesus was born, the future history of the Earth was again in serious danger of drifting into another situation that would have angered God. Looking back to biblical events—the Tower of Babel, Sodom and Gomorrah, the flood of Noah, disobedient Children of Israel on their march around Mt. Sinai, and others—a devastating reaction from God would not have been out of the ordinary. However, considering this prophecy in Genesis, God must have been planning some event even before the previously mentioned disasters. Apparently, those earlier, angry reactions of God took place because the time had not yet been right for His Son to enter the human race. That once-in-a-creation opportunity was now at hand.

Before Genesis 1:1 there was time, in some form. There was God; the Trinity. They must have talked... a lot. They must have planned... in detail. Part of that planning, before Creation, likely included a rupture in the human timeline open long enough for Christ, God's Son, to walk into our lives disguised as a birth. This would be followed by two more ruptures in time: the rolling away of the stone representing a dead human's rise to life—remaining in human form— and a third incident when time would spread apart as Jesus Christ rose into the clouds and disappeared back to His place with the Father.

It could be argued that there was a fourth as well. Jesus said that although He was leaving to prepare a place for us, He would send a comforter, a figure we believe to be the Holy Spirit. Although there is record of the Holy Spirit being present before this time, Jesus' words appear to say He is *sending* Him, meaning He wasn't there at that time for some undisclosed reason. We know from the scriptural account of Jesus' baptism the Holy Spirit was present at the beginning of Jesus Christ's adult ministry. This could start an interesting conversation most folks have never had about who and what the Holy Spirit is and does. Gather a few friends around a campfire and start the fireworks!

Just to flesh out why Christ chose to be born as a human at this time, (no pun intended), let's compare the conditions that existed at the time of the arrival of Christ to a few other events on the timeline of history. Let's begin with the 18th century. Why not, it seems it is always about us, so why not consider the period of history involving the United States' struggle for release from their oppressors, leading to the birth of a new nation. Realistically, history is not just about we who call ourselves Americans, or North Americans. When we use these terms, we overlook our human neighbors who are also Americans, who live in South America. North America is also not descriptive, since Canada also has claim to this continent. The US conflict in the 18th century is not the only thing going on that would have made it more difficult for Christ to have arrived with the same 1st-century impact.

During the 18th century, elements of Enlightenment thinking culminated in the American, French, and Haitian revolutions. The age saw violent slave trading and human trafficking on a global scale. The reactions against monarchical and aristocratic power helped fuel the revolutionary responses against it throughout the century.[113]

The 19th century saw large amounts of social change. Slavery began to be abolished, and the First and Second Industrial Revolutions led to massive urbanization and much higher levels of productivity, profit, and prosperity. This paved the way for the growing influence of several nations—the British Empire, the Russian Empire, the United States, the

German Empire, the Second French Empire, the Kingdom of Italy, and Meiji Japan—with the British boasting unchallenged dominance after 1815. The British Empire grew rapidly in the first half of the century, especially with the expansion of vast territories in Canada, Australia, South Africa, and heavily populated India, and finally, in the last two decades of the century, the rest of continental Africa. By the end of the 19th century, the British Empire controlled a fifth of the world's land and one-quarter of the world's population.

By the end of the 18th century and the beginning of the 19th century in the US and parts of the world, Christianity was already in need of a Second Great Awakening.[114] Compared to the reign of the Roman Empire, imagine the shape of the world had Christ chosen to be born in either the 18th, 19th, or 20th centuries.

Just for fun, let's quickly review peace opportunities in the 20th century:

- 1898—1901 Boxer Rebellion
- 1899—1902 Boer War
- 1904—1905 Russo-Japanese War
- 1910—1920 Mexican Revolution
- 1912—1913 First and Second Balkan Wars
- 1914—1918 World War I
- 1915—1918 Armenian Genocide
- 1917—Russian Revolution
- 1918—1921 Russian Civil War
- 1919—1921 Irish War of Independence
- 1927—1937 Chinese Civil War
- 1933—1945 Holocaust
- 1935—1936 Second Italo-Abyssinian War (also known as the Second Italo-Ethiopian War or the Abyssinian War)
- 1936—1939 Spanish Civil War

- 1939—1945 World War II
- 1945—1990 Cold War
- 1946—1949 Chinese Civil War resumes
- 1946—1954 First Indochina War (also known as the French Indochina War)
- 1948—Israel War of Independence (also known as the Arab-Israeli War)
- 1950—1953 Korean War
- 1954—1962 French-Algerian War
- 1955—1972 First Sudanese Civil War
- 1956—Suez Crisis
- 1959—Cuban Revolution
- 1959—1975 Vietnam War
- 1965—Civil rights in the US turns violent on March 7th in Selma, Alabama
- 1967—Six-Day War
- 1979—1989 Soviet-Afghan War
- 1980—1988 Iran-Iraq War
- 1990—1991 Persian Gulf War
- 1991—1995 Third Balkan War
- 1994—Rwandan Genocide[115]
- 1994—1997-Iraqi Kurdish Civil War
- 1994—1996-First Chechen War
- 1995—2019- Insurgency in Ogaden
- 1996—2006- Nepalese Civil War
- 1996—2001- Civil war in Afghanistan
- 1997—1999-Republic of the Congo Civil War

- 1997—Clashes in Cambodia
- 1998—Ongoing- Al-Qaeda insurgency in Yemen
- 1999—2003- Second Liberian Civil War
- 1999—2009- Second Chechen War
- 2001—September 11th bombing of the Twin Towers in New York city
- 2001—2003- Kurdistan Islamist conflict
- 2001—Ongoing- War on Terror
- 2002—2015- Operation Enduring Freedom – Philippines[116] [117]
- 2003—Ongoing- War in Darfur
- 2004—2017- War in Northwest Pakistan
- 2004—2015- Houthi insurgency in Yemen
- 2006—Ongoing- Fatah–Hamas conflict
- 2006—Ongoing- Mexican Drug War
- 2006—2009- War in Somalia
- 2007—Lebanon Conflict and Hamas' takeover of Gaza
- 2009—Ongoing—Sudanese nomadic conflicts
- 2011—Ongoing—Sinai insurgency
- 2011—Ongoing—Syrian Civil War
- 2014—2017-Iraqi Civil War
- 2014—Ongoing—Libyan Civil War
- 2014—Ongoing—International military intervention against ISIL involving over fifty countries
- 2017—Ongoing—Islamist insurgency in Mozambique
- 2019—India/Pakistan standoff
- 2019—2020—Persian Gulf crisis[118]

Just a few more comments in addition to the above short list: There was a Cold War between the US and the now-defunct Soviet Union between 1947 and 1991. This period followed the US bombing of Japan with nuclear weapons on August 6 and 9 in 1945 to end World War II. These bomb strikes, by introducing nuclear weapons, set in motion a period some will remember as a time for worthless nuclear bomb drills in public schools, followed by the world's closest brush with world annihilation during the Cuban missile crisis on October 16-28, 1962.

In October 2001, after the Twin Tower bombing in New York City on September 11, 2001, the US declared war on Afghanistan and began strikes on the Taliban. That war worked its way through multiple phases and presidents until its first formal end in December, 2014.[119] However, US forces still remained in Afghanistan for another seven years. As of this writing, all US military forces were removed by October 31, 2021. At that time, it remained unclear just who was still on the ground in Afghanistan regarding US citizens who wished to leave and Afghan individuals who were in danger because of their past contribution to US and Allied Forces during the extended war.

The above list is not exhaustive and as you will note, many of these wars and conflicts are either ongoing or start and stop for various reasons; many overlapping. A logical summary of the 20th century would be that it was a terrible time to introduce a new wave called Christianity. Instead, in the Trinity's timing, the 20th century needed an already fully developed Christ-based, worldwide church to minister to the chaos it would produce. Without the presence of Christ in the world over the last few centuries, the end would surely have already occurred.

And, what about the 21st century? Could I get a ditto to the 20th?

Starting Christianity at any other time in history than when it occurred would have included competing with global wars and other belief systems, because Christianity would not yet have been formulated. Christ arriving in our world today, or any other time in history, would by necessity have had a different application and quite possibly a different result. Let's review the timeline leading up to the manger in Bethlehem.

Preceding Jesus' birthday, whatever exact date and time that was, the records in the Old Testament cease with the writing of the last books to be included, presumably 2 Chronicles. This left a gap in the biblical record of some 350 to 400 years before the writings of the New Testament began to be written. Alexander the Great conquered the ancient Near East and Judea in 322 BC, ending decades of conflict between warring empires. Following that, in 167 BC, Antiochus Epiphanes desecrated the Temple in Jerusalem. Seven years later, in 160 BC, the end of the Maccabean wars led to almost a century of Judean rule. Pompey entered Jerusalem in 63 BC, followed in 27 BC by Octavian, who became the sole ruler of the Roman Empire, which included Judea.

The centuries-long rule of the Roman Empire, as oppressive as it was for the Jews and countless others, created a lasting world peace that had not been enjoyed before or since in that area of Judea and the Mediterranean Basin. Roman rule began close to the time of Christ and lasted many years after. This period of relative peace—or domination, if you prefer—in the known world was a perfect setting for Christ to enter history, establish a ministry, and allow sufficient time for Christianity to grow and take root. Looking at the entire timeline, whether it is thousands of years or billions of years, there simply was no better time for Christianity to be born and develop. This was that moment on the timeline overseen and developed before Creation by the Trinity, when the only person in the history of the universe had the ability to choose both when they would be born and where they would be born. This is amazingly awesome!

CHAPTER 12

BUT… HE'S JUST A BOY…

… Mary, listen! He said "Papa!"
… Don't worry. He'll show up. You know how He wanders off.
… Wouldn't you know it? His first miracle would involve alcohol at a party.

42 And when He was twelve years old, they went up to Jerusalem according to the custom of the feast. 43 When they had finished the days, as they returned, the Boy Jesus lingered behind in Jerusalem. And Joseph and His mother did not know it.

Luke 2:42-43 NKJV

1 On the third day there was a wedding in Cana of Galilee, and the mother of Jesus was there. 2 Now both Jesus and His disciples were invited to the wedding. 3 And when they ran out of wine, the mother of Jesus said to Him, "They have no wine."

John 2:1-3 NKJV

WE KNOW VERY little about Jesus' early years after His birth. In Scripture, we are told that after Mary's purification—believed to have been forty days after birth, according to the Law of Moses—they went to Jerusalem to present the child to the Lord and offer a sacrifice of two turtle doves and two young pigeons.[120] Here we find the story of two people associated with the Temple, Simeon and Anna. Simeon was a just and devout man, and the Holy Spirit was upon him. The Spirit had revealed to Simeon that he would not see death until he had seen the Lord's Christ. On the day Mary and Joseph brought Jesus to the Temple, the Holy Spirit led Simeon to the same location. When Simeon saw the child, he took Him in his arms and blessed God saying,

> 29 *Lord, now You are letting Your servant depart in peace, According to Your word;* 30 *For my eyes have seen Your salvation* 31 *Which You have prepared before the face of all peoples,* 32 *A light to bring revelation to the Gentiles, And the glory of Your people Israel.* Luke 2:29-32 NKJV

The Child's parents were struck by Simeon's words. Why? First, he was a total stranger who asked to hold their one-and-half-month-old child, but more importantly, he said Jesus would have a profound impact on… Gentiles!

> 34 *Then Simeon blessed them, and said to Mary His mother, "Behold, this Child is destined for the fall and rising of many in Israel, and for a sign which will be spoken against* 35 *(yes, a sword will pierce through your own soul also), that the thoughts of many hearts may be revealed.* Luke 2:34-35 NKJV

Then, a second encounter occurred.

While still within the Temple grounds, the parents encounter a very old prophetess from the Tribe of Asher. Her name was Anna, a widow of eighty-four years who served the Lord day and night through fasting and prayer. Scripture tells us that at the instant Simeon gave his blessing over Jesus, Anna entered the area and witnessed what took place. After hearing what was said, she gave thanks, and from that day on told of Jesus' coming to all in Jerusalem who were in need of redemption. Luke 2:36-38 NKJV

After this amazing encounter, the parents with their child return to their home in Nazareth. One verse sums up everything we will hear about this child for the next twelve years,

And the Child grew and became strong in spirit, filled with wisdom; and the grace of God was upon Him. Luke 2:40 NKJV

It seems much of Jesus' early life involved Roman or Temple laws and traditions. Jesus was born in Bethlehem during a governmental census. He was taken to the Temple in Jerusalem at the age of eight days for circumcision and to offer a sacrifice at forty days old, according to the laws of God and Moses. We find in Luke 2:41 that this was a tradition that Mary and Joseph, along with their children and many friends and relatives, would take annually to Jerusalem for the Feast of the Passover. What is often overlooked is that this annual family vacation gave Jesus several opportunities to become familiar with this very large, urban metropolis of ancient times.

As he grew older, close to his teenage years, he and his friends likely roamed the city, encountering its wonders and cruelty. Their young eyes observed how Rome reigned over all people. Their ears heard the priests' proclamation of God's Law, and their interpretations of those laws bent toward securing their place of prominence within Jewish life. In this setting, the words in Luke 2:40 permeate this time in Jesus' life: "*And the Child grew and became strong in spirit, filled with wisdom…*"

We are brought back into the young life of Jesus when He was twelve years old, a very specific age recorded in Scripture, Luke 2:42. It was not the first time Jesus had made this trip, but a traditional, annual event He attended with His family. We are given very little information about the visit, suggesting it was routine. We are simply told that the family traveled to Jerusalem; it was implied that they participated in the Feast and then left for home with friends and family. The story is not unlike today, when a twelve-year-old wanders away at the mall. The parents are not initially concerned when they discover their child is missing, assuming He is just playing with friends and relatives in some other area of the caravan. They know He will come back to them when He gets hungry. After all… He is just an average twelve-year-old. Well, as it turns out, He was not average, and He doesn't show up.

After He had been missing for a full day, the parents are very alarmed and start running from group to group asking if anyone had seen their boy. No one had seen Him since they left the city. Mary and Joseph quickly returned to the city to look for their lost child. It took three days of searching and asking strangers before they found Him. What they discovered was another one of those Simeon moments. Jesus had been missing now for five days; one day before they discovered Him missing, one day to return to the city, and three days looking for Him. Was He okay? Had some notorious group kidnapped Him to sell Him into slavery? Where did he sleep, what did He eat, how could such a young boy in such a large city survive?

And where did they find Him? Where else… in the Temple courtyard, sitting in the middle of the teachers, both listening and asking questions. The parents, out of respect for the learned Temple instructors, probably did not interrupt, but sat and listened to the conversation. They were amazed, as were the teachers, that this young lad had such an in-depth knowledge and understanding. They also probably realized for the first time that, although Jesus was always acting like a child during lessons in the Synagogue, He was hearing everything the rabbi was saying. Sound familiar, parents?

After scolding Him for not telling them where He was going, they are bewildered by His response in Luke 2:49: *"And He said to them, 'Why did you seek Me? Did you not know that I must be about My Father's business?'"* This would not be the last time Mary, and later Jesus' step-brothers and step-sisters, would hear Him say unusual things about their family relationship.

We lose track of Joseph in Jesus' later years, but surely Mary will remember this event when Jesus heads into the wilderness in preparation for His coming ministry. A mother's worry over her children never changes, regardless of the child's age.

To emphasize its importance and to prepare us for another long silence concerning Jesus' early life, Scripture repeats itself. As in Luke 2:40, we find these words in Luke 2:52 to close out this chapter: *"And Jesus increased in wisdom and stature, and in favor with God and men."* We will not hear from Jesus for almost two decades, when He is around thirty years old. He will be at a wedding, with His mother, and... His disciples.

Since God, and the Trinity, doesn't do things without a purpose, why are we left, as with the 400 years of silence between the Old and New Testaments, with these huge gaps in the early life of Jesus? It could be that His life was just that of a routine little boy and later young teenager, and there was nothing that happened to add to the moving timeline. Could it simply be nothing was recorded? Actually, given God or the Trinity never thinks the way we do, there are some logical reasons to think otherwise.

Jesus was by no means... normal. There is no reason to believe that Jesus would have just returned to being a normal child after the incident in the Temple at age twelve. There are several ancient books, not included in the Bible we know today, that were either written by various people or labeled as written by various people in the second century or earlier. One of these in particular focuses on Jesus as a child.

This book is called "The Infancy Gospel of Thomas."[121] The book is thought to be Gnostic in origin. F.F. Bruce writes (in "Jesus and Christian Origins Outside the New Testament," p. 87):

> *Then there is the Infancy Gospel of Thomas, which purports to describe the doings of Jesus in his boyhood. Jesus proves to be an infant prodigy at school, instructing his teachers in the unsuspected mysteries of the alphabet; he astounds his family and playmates by the miracles which he performs. This is the document which tells for the first time the familiar tale of the twelve sparrows which Jesus, at the age of five, fashioned from clay on the sabbath day."*[122]

Trying to fill the silence intentionally injected into our timeline by those who controlled the assembly of the Bible we use today, while frustrating to some, was well-accepted in the past by those who would never question the origins of the Holy Bible. To be clear, I am not questioning anything in Scripture. I am only frustrated by what I don't know and an absence of creditable records to find answers. What questions? Here are a few:

- Since we call Jesus a carpenter, what was His relationship with His father, Joseph, and did He actually pursue this vocation or even work in Joseph's workshop?
- What happened to Jesus' step-father, Joseph?
- If Jesus was part man and part God, how did God make it as a teenager?
- What happened after Jesus and His family returned to their home following the incident at the Temple when He was twelve?
- Was turning water to wine Jesus' first miracle or did He discover this ability younger in life?
- Since Jesus asked His best friend, John, to care for His mother while He was dying on the cross, where were Jesus' other half-brothers and half-sisters?
- Why, after living with Jesus for years, did His half-brother, James, not believe Him to be who He said He was until after Jesus' death and resurrection?

This is understandably a short chapter in this book. In my research, it was the opinion of many scholars that information and stories contained in the Old and New Testament are there because they all point to and add to the story of salvation. In other words, the decisions made by those assembling the Bible over centuries were driven by information that was specific to this purpose of redemption. Much of Jesus Christ's early childhood, teenager years, and then in His twenties were just normal (for the Son of God). There was also a belief that His parents were protective of Him based on Herod's early efforts to find and kill him. The scenes we are given are all connected in various ways to the rest of the Jewish story.

The science community takes pride in its insistence on supporting replicable data when discussing or writing about anything they are, or have been, involved in researching. Perhaps the faith community should not dismiss those who have doubts when they point to gaps in our data. They are only doing what they do best... questioning everything. As a part of the faith community, I have learned over the years to stop questioning something only when I have reached a point where it just makes sense that there is no answer. Until that point, I should be just like my nonbelieving friends and associates. In fact, as I recall, our savior was really good at asking questions. And this is a great example to prove my point. Here is a question to which there is no answer: "Why would a divine being, Who knew everything, need to ask any questions?"

CHAPTER 13

THE FOURSOME

... one game plan, or two?

... "I don't seem to be relevant anymore to this younger generation," said the Father.
... "Let me become like one of them. I can relate," said the Son.
... To which the Father replied, "You're only as good as those you hang around!"
... To all of whom the Spirited Friend next door said, "Finally, we have a foursome! I'll make the tee-time."

13 When Jesus came into the region of Caesarea Philippi, He asked His disciples, saying, "Who do men say that I, the Son of Man, am?" 14 So they said, "Some say John the Baptist, some Elijah, and others Jeremiah or one of the prophets." 15 He said to them, "But who do you say that I am?" 16 Simon Peter answered and said, "You are the Christ, the Son of the living God."

MATTHEW 16:13-16 NKJV

I HAVE ONE gigantic question: Was Jesus two persons—Jesus the man as well as Christ, the Son of God—or just one person? Christ existed long before there was a man called Jesus. Christ existed after Jesus died, but in a transfigured state. Simply asked, was Christ residing in the body of a human called Jesus? To look at this question from both the viewpoints of faith and science, let's begin with some scientific views.

In an article by Natalie Wolchover published by livescience.com on July 8, 2013, we find an interesting twist from the scientific community. The article is titled, "Proof of Jesus Christ? 7 Pieces of Evidence Debated." In her article, Ms. Wolchover addresses seven of the most common items associated with the crucifixion: the Shroud of Turin, the cross, nails, various non-biblical books, the Dead Sea Scrolls, and Christ's crown. All of her references, in one form or another, use carbon dating as a major piece of evidence in denouncing the authenticity of these Christ-related artifacts. As you are aware, scientists use this same method to validate their claims of human and animal life dating back millions of years. Remember dinosaurs?

This article does say some very interesting things about science and faith learning to coexist. Consider the following excerpts:

> *Most theological historians, Christian and non-Christian alike, believe that Jesus really did walk the Earth. They draw that conclusion from textual evidence in the Bible, however, rather than from the odd assortment of relics parading as physical evidence in churches all over Europe.*
>
> *"We do know some things about the historical Jesus—less than some Christians think, but more than some skeptics think," said Marcus Borg, a preeminent Biblical scholar, author and retired professor of religion and culture at Oregon State University. "Though a few books have recently argued that Jesus never existed, the evidence that he did is persuasive to the vast majority of scholars, whether Christian or non-Christian."*[123]

Somewhere between science and faith lies a vast wasteland of unanswered questions, understanding, misunderstanding, and hopeful doubt.

Cullen Murphy, in an *theAtlantic.com* article on December 1, 1986, writes an excellent narrative which begins with a quote from Scripture of Jesus posing very pointed questions to His disciple followers: "Who do men say that I, the Son of Man, am?" Matthew 16:13 NKJV and in verse 15, "*But who do you say that I am?" Then, in verse 20 we find a very curious closing response from Jesus regarding this brief group discussion: "Then He commanded His disciples that they should tell no one that He was Jesus the Christ."*

The writer continues, citing what he calls "The First Quest": "Consider one recent (sic—recent in terms of 1986) but basic Catholic text, *A Christological Catechism*, by Joseph A. Fitzmyer, a leading biblical scholar. Did Jesus claim to be God? Fitzmyer replies: 'The Gospels have not so presented that claim... . It is impossible to imagine how such a statement would have been understood.' "

More telling are Fitzmyer's following assumptions: "What about the Resurrection of Jesus? 'The New Testament never presents the resurrection of Jesus as a resuscitation, i.e., a return to his former mode of terrestrial existence.' The so-called virgin birth: 'New Testament data for this question are not unambiguous.' The historical reliability of the Gospels? 'The only answer which can be given to this question is meager indeed.'"[124]

It seems, beginning at the far edge of scientific understanding and moving toward the center of gnostic revelation, hopeful doubt grows. That hopeful doubt, somewhere in the middle, is met by an opposing, repulsive magnetic force generated by a faith community bent on tradition and documents they proclaim as divine and inerrant. So, does someone have to be wrong? Can science ever admit there is a divine being... a creator? Can the faith community ever admit, while holding on to the basic tenets of their faith, dinosaurs and other living things did exist on earth more than 10,000 years ago? Let's take a look at several faiths and what they believe.

The Catholic Church

We will begin with one of the first pillars of faith and belief, the Catholic Church, and more to the point, the Catechism, or confession of their faith and belief. In Part One, Section Two, "The Profession of the Christian Faith," Chapter 2 states, *"I believe in Jesus Christ, the only Son of God."* It is here, in Article 3, where the ageless battle between science and religion has one of its first disagreements: *"He was conceived by the power of the Holy Spirit, and born of the virgin Mary."*

The Catholic Church simply says it happened based on what someone said and wrote down. Science often suggests (paraphrased), *"Show me first the proof, that there is a God, a Holy Spirit, and that a woman can become pregnant and bear a child without the assistance, in some form, of a man."* With this proclamation, the Catholic Church fought holy wars, dominated politics in many forms of government, and became rich beyond measure, and are today recognized as their own sovereign country inside Italy.

The Nicene Creed answers the question, *"Why did the Word become flesh?"* by confessing, *"For us men and for our salvation he came down from heaven; by the power of the Holy Spirit, he became incarnate of the Virgin Mary, and was made man."* [125] Apparently, Christ wasn't always part man and part God, according to the Catholic Church's own confession. Like all humans, man and woman, He had a beginning and an end. It isn't just science that struggles with this statement; so do many Christians within the faith community.

Before leaving the Catholic view of this question, let's consider the thoughts of a group called the Order of Friars Minor or OFM. Pat McCloskey, OFM, writing for *franciscanmedia.org*, posted comments on a couple of related questions posed to him. Those two questions were: "*When Did Jesus become a Catholic?"* asked by a Catholic lady and, by a Jewish friend, *"When did Jesus become a Christian?"* His observations are telling concerning the struggle by many to answer the question about the internal relationship between Christ and Jesus.

Friar McCloskey points out, *"At least since the sixth century before Christ, the bedrock of Judaism has been monotheism, belief in one God. God's self-revelation in the Scriptures progressively insisted on monotheism."* He summarizes, *"From the very beginning, the followers of Jesus asserted that they were monotheists, that the Father, Son, and Holy Spirit were not three gods in the same sense that pagans, for example, considered Jupiter and Apollo as gods."*[126]

The original question had to do with one or two persons. The previous remarks make it about a choice between one person, God, or four persons; God, Christ, Holy Spirit, and Jesus.

Having this struggle doesn't lessen the position of Christ within the Trinity, or as the Son of God. It simply enforces the other question no one seems to want to honestly address: Was Jesus Christ the Messiah, or was Jesus, the created man who would temporarily house the Word, the Son of the living God? This position is also called "Adoptionism."[127] I am sure you might be asking yourself, "Why is this important?" Because it is important to those who need to be introduced to Him—introduced in a way that makes sense, even if for just a while, until they can learn to trust Him.

In the United States of America, the struggle for church identity is as deeply rooted in their heritage as their struggle for independence from their motherland, England. In fact, you could say both struggles were inertly combined. Many historians will assert the Colonies didn't as much win the Revolutionary War of Independence, but that England became so in debt from attempted conquests all over the globe that they simply went home. To balance their budget, regardless of the reason, their soldiers and guns went home. However, the church, *the* Church of England, remained and was a considerable influence in colonial affairs until after 1815.

The Presbyterian Church

By the late 1700s and early 1800s the Presbyterian Church had taken some foothold in the shadow of the Church of England. The Church of England is sometimes referred to as the "Anglican Church." The Church of England allowed the Presbyterian Church, a Protestant faith originating within Swiss and French Calvinistic churches and eventually founded by John Knox in Scotland in 1557, to exist. The Presbyterian Church was so similar in style and foundation that it was never considered a threat.

There was, however, a threat blowing in from newly developed territory in the US Cumberland Valley in and around Kentucky and Tennessee. The wind of change blowing in would eventually be called, *The Second Great Awakening.*

This awakening would become the grass roots for much of what the protestant faith would become. Martin Luther must have been smiling from Heaven. Many of the major denominations of today had their birth during this period: Methodist, Church of Christ, Nazarene, Cumberland Presbyterian, Church of God, and several different Baptist congregations. The need for these churches was the result of an explosion of growth west of the Atlantic Seaboard as people flocked to these areas with promise of free homesteads and opportunity.

This growth created a ripe market for commerce, industry, saloons, and a savory group of fortune hunters. Churches did spring up to offset bad with good, but there was a problem—very few pastors or ministers. Why? The Church of England, and even traditional Presbyterian churches, required a formal religious education—preferably in England—before the minister could be ordained and accept a position.

Necessity creates invention; however, the formal church would have no part of untrained leadership being replaced by ordinary people who felt a call from God to preach. The result was a significant break from traditional churches and the birth of a new faith community. The foundation of this movement would become the ballast needed to ultimately win the Revolutionary War, Part 2—the War of 1812. Not only did that war finally release the Colonies from the oppression and taxation of England, but it also signaled the end of the domination of the Church of England. Its head was the King of England, ever since King Henry the VIII decided to anoint himself after bitter disagreement with the Pope over divorce.

The Baptist Church

A branch of the new Baptist congregation eventually grew to be one of the largest denominations in the world. These folks would become known as "Southern Baptists." While many of these different denominations share a few common beliefs, there are differences which cause people to choose one over the other. Perhaps they are part of a particular denomination because their parents belonged to that group, and therefore raised their children to have the same beliefs. Sometimes these differences can cause a lot of stress. One such stress point has to do with the question I asked earlier: *"Was Jesus two persons—Jesus the man and Christ the Son of God—or just one person?"*

One or Two Wills

There is an interesting article written by John E. McKinley in the Southern Baptist Journal of Theology in the Spring of 2015 titled, "A Model of Jesus Christ's Two Wills in View of Theology Proper and Anthropology." The article is very thorough and well-documented.[128] Dr. McKinley, Associate Professor of Theology at Biola University with a PhD from the Southern Baptist Theological Seminary, takes a remarkable journey into the question I pose. The article is detailed and quite lengthy. I will attempt to summarize, but still encourage you to read the entire article online.

My question raises a lot of issues rarely spoken of in sermons or Bible Study Groups: *"Was Jesus two persons - Jesus the man and Christ the Son of God - or just one person?"* suggests another question. If Christ just occupied Jesus' body, could He have left that body like the demons Jesus commanded to come out of other humans? Until I began researching this topic, I had no idea that it went much, much deeper and has in fact been a significant topic of discussion since the late 7th century. Dr. McKinley probes this topic under the doctrine of Two Wills.

As is often the case in academia, everything has a stuffy name drawn from some ancient language. Two Wills is no different. Someone who believes that there were two independent wills possessed by Christ the God and Jesus the man is known to hold to "Dyothelitism." Conversely, the idea of those who profess that Christ and Jesus shared a common will is known as "Monothelitism." According to Dr. McKinley, the issue was actually settled for all Christians in 680-681 AD. Here is a quote from the summary of this event:

Further cause to dismiss consideration is the shadow of political intrigues and failed ecumenical strategies that surrounded the ancient discussion of Christ's wills in 680-81. Nevertheless, few questions in early theology received such focused attention as this one when the Sixth Ecumenical Council weighed arguments and rendered a verdict at Constantinople in 681. Brilliant theologians of that time understood the great importance to theology that Jesus possesses two wills, one divine and one human, since he is truly God and truly human. [129]

Taking this position does create a series of questions that one must address to support his position. For instance, if Christ was truly God and truly human, how do these two opposing wills agree when the truly man is tempted? Clearly, God cannot sin—but can He be tempted? Obviously, yes. He was tempted by Satan when in the wilderness, as recorded in Luke chapter 4. We have no record of Jesus giving in to sin after being tested. What happened to the fallen will of man, if the man Jesus also had a will?

In the garden, when Jesus was praying, what did He mean when He said, *"not my will, but yours?"* Was Jesus talking to Christ, or to God? We must assume that what was about to happen was always Christ's will, as part of the Trinity who planned this event. If so, why ask?

On another occasion when Jesus was twelve, He disobeyed His parents and wandered off, lost for several days. Clearly, the Apostle Paul wrote in Colossians 3:20, "*Children, obey your parents in all things, for this is well pleasing to the Lord.*" And the 4th Commandment of ten in Exodus 20:12 is *"Honor your father and your mother, that your days may be long upon the land which the Lord your God is giving you."* Which will—Christ's or Jesus'—controlled His action and His response to His parents in this situation?

If there were two independent wills, did they ever argue?

Here is Dr. McKinley's conclusion:

Instead of the imperial theological formulation that became Monothelitism, the Dyothelite model of the will as a natural property works wherever we look theologically. The process of willing is mysterious for us, but we may see clearly that neither do we choose independently of our natural desires, nor do we lack the capacity to transcend them... we make choices under the influence of our beliefs, experiences, perceptions, and the Spirit of God... Whatever conflicts we struggle against in denying particular desires so as to choose the Father's calling to us, Jesus entered the same struggle for us and for our salvation.[130]

After comparing the thoughts of Catholics and Southern Baptists, it was interesting to me to find a slightly altered interpretation voiced in an essay by Allison Low in the stphilipinstitute.org, on February 7, 2020. In this publication of the St. Philip Institute Catechesis and Evangelization, her position supports the statement that Jesus is uncreated and is of the same divine substance as God the Father. In other words, Jesus is not a human person with a human soul. I understand this to mean that Jesus did not possess two intellects and two wills. This holds that at conception, Christ became incarnate—*was made flesh*—in other words, Jesus was a Divine Person.[131]

Interestingly, in this essay, the suppression of Jesus the man's will, is never addressed. Although this writer disagrees that the Doctrine of Adoptionism would be possible, she seems to take the position that the human body of Jesus was a vessel (my words), used and discarded when Christ ascended into Heaven at the ascension. If I were smart enough to be a part of the scientific community, I would probably be saying to people of faith, "Please get your thoughts together."

The United Church of Christ (Disciples of Christ)

Before leaving this subject, let's take a look at an interesting viewpoint held by the United Church of Christ, an inter-faith religion that partnered with the Christian Church (Disciples of Christ) in 1989. On their website, ucc.org, under a section titled "Testimonies, not tests of faith," you will find the following as a quote from their statement of faith.

... one and the same Christ, Son, Lord, Only-begotten, recognized

> *in two natures, without confusion, without change, without division, without separation; the distinction of natures being in no way annulled by the union, but rather the characteristics of each nature being preserved and coming together to form one person and subsistence, not as parted or separated into two persons, but one and the same Son and Only-begotten God, the Word, the Lord Jesus Christ...*

They state that this complete statement was the end result of the 451 AD *"Definition of the Council of Chalcedon,"* to fully understand the struggle and the relationship of the three persons of the Holy Trinity. They state that, *"It is accepted as a symbol of Christian doctrine by the Roman Catholic, Eastern Orthodox, Anglican, Reformed and Lutheran churches."*[132]

Here is what I find interesting is their assertion, *"...the distinction of nature's being in no way annulled by the union, but rather the characteristics of each nature being preserved..."* The Catholic Church, whom they say adopted this position, seems to believe that the human nature, or will, was overtaken by the Divine Nature. Which one is it? Who speaks for whom? Which one is right?

It fascinates me that none of these doctrines say a word about dinosaurs. How can these faith communities be so forceful about things they cannot prove while ignoring actual bones of discovered creatures that do not fit in their scriptural timeline? My answer? Honestly, I don't have one. Instead of an opinion, I have a testimony about how God has intervened in my human sinful nature, making it possible for me to have Holy moments, or completeness, at times in my life, as outlined in a wonderful book, *The Biggest Lie in the History of Christianity* by Matthew Kelly.[133] My testimony centers firmly on being introduced to Christ's sacrifice by allowing Jesus, a human with a body like mine, to be killed by other humans, like myself, on a Roman cross. I am certain, had I not had that introduction and acceptance of His sacrifice for me when I was in the third grade, attending a Vacation Bible School, I would not be the person I am today.

CHAPTER 14

THE ENTREPRENEUR

... who needs a business plan anyway?

... Client: "I want to start a business that will upset all established business practices, operate with a small staff, very little overhead expense, no assets, no budget, and have someone publish what I say and do that will become the best-selling book in the history of the world."
... Consultant: "Do you have a business plan?"
... Client: "Yes, but it isn't finished"
... Consultant: "Any venture capital?"
... Client: "No, but I have a fairly influential father and a friend with a very spirited outlook."
Consultant: "I'm sorry, but in my opinion, I don't think you have a very good shot at success."
Client: "Did I mention I am committed to die for the sake of the enterprise?"

1 In the beginning was the Word, and the Word was with God, and the Word was God.

John 1:1 NKJV

14 And the Word became flesh and dwelt among us, and we beheld His glory, the glory as of the only begotten of the Father, full of grace and truth. J

OHN 1:14 NKJV

18 And Jesus, walking by the Sea of Galilee, saw two brothers, Simon called Peter, and Andrew his brother, casting a net into the sea; for they were fishermen. 19 Then He said to them, "Follow Me, and I will make you fishers of men." 20 They immediately left their nets and followed Him. 21 Going on from there, He saw two other brothers, James the son of Zebedee, and John his brother, in the boat with Zebedee their father, mending their nets. He called them, 22 and immediately they left the boat and their father, and followed Him.

MATTHEW 4:18-22 NKJV

28 After this, Jesus, knowing that all things were now accomplished, that the Scripture might be fulfilled, said, "I thirst!" 29 Now a vessel full of sour wine was sitting there; and they filled a sponge with sour wine, put it on hyssop, and put it to His mouth. 30 So when Jesus had received the sour wine, He said, "It is finished!" And bowing His head, He gave up His spirit.

JOHN 19:28-30

TO THE WORLD He looked like any other local. His appearance would broadcast common, working-class, dark complexion from the sun, unkept hair, dark eyes, and rough hands from hard work. But to some, there was something else, something different. To some, there were stories whispered in small groups about this Nazarene laborer as He passed by, including unusual things He had said and done over the past few months. It may not have been as unusual as some may have speculated concerning His sudden approach and invitation to select fisherman tending their nets and boats along the seashore to further His cause.

Were these encounters random, or had they been planned for thousands of years, before even the Earth was formed from a dark, shapeless void? Was *anything* that would happen over the next approximately three years random or by chance? Maybe, maybe not. There may have been those who refused His invitation to follow Him, but it wasn't important enough to record or be included in the yet-to-be-written New Testament or secular history. It is now clear, from a distance of time, that what occurred during the last years of this young man's life were not remarkable, but rather a disruptive shift in the timeline of all history; before and after the dash on His useless headstone that was never ordered, completed, or delivered to a borrowed grave.

Who Jesus was had a lot to do with *what* Jesus was before, during, and after his ministry. Here are a few key events. All of these events include elements contrary to scientific understanding and research.

- The Incarnation
- Jesus' Baptism
- His time spent in the wilderness
- The Crucifixion
- In the tomb—the Resurrection
- After the tomb
- The Ascension

All of these events were accompanied by unearthly circumstances and by key changes regarding both His body and His being. From the incarnation to the ascension, Christ, the Son of God, and divine partner with the Holy Spirit, lived in the same body as a human named Jesus, Yeshua, meaning, "the Lord is Salvation" in Hebrew. When reading that part of Scripture regarding His calling the twelve disciples, you get a sense that each disciple He called inherently knew there was something different about the man and therefore… His invitation… and it is possible Jesus didn't call all of His disciples, but some just followed Him. There *were* more than twelve.

The following events represent acts of the Trinity that were challenging, and equal to the power of original Creation. Science may one day create life in some form, but man will never create these once-in-an-eternity changes to our existence. All of these forces of the Trinity were powerfully and awesomely concentrated within one man who became known as Jesus Christ. Or maybe we should say, Christ Jesus. The Apostle Paul used both titles in the Book of Philippians, chapters 3 and 4. [134] Let's take a look at these seven events and why they are so special.

The Seven Miraculous Acts of the Holy Trinity, Within the Carnal Soul of a Man

The Incarnation

> 18 *Now the birth of Jesus Christ was as follows: After His mother Mary was betrothed to Joseph, before they came together, she was found with child of the Holy Spirit.* 19 *Then Joseph her husband, being a just man, and not wanting to make her a public example, was minded to put her away secretly.* 20 *But while he thought about these things, behold, an angel of the Lord appeared to him in a dream, saying, "Joseph, son of David, do not be afraid to take to you Mary your wife, for that which is conceived in her is of the Holy Spirit.* 21 *And she will bring forth a Son, and you shall call His name Jesus, for He will save His people from their sins."* 22 *So all this was done that it might be fulfilled which was spoken by the Lord through the prophet, saying:* 23 *"Behold, the virgin shall be with child, and bear a Son, and they shall call His name Immanuel," which is translated, "God with us." Matthew 1:18-23 NKJV*
>
> 26 *Now in the sixth month the angel Gabriel was sent by God to a city of Galilee named Nazareth,* 27 *to a virgin betrothed to a man*

whose name was Joseph, of the house of David. The virgin's name was
Mary. 28 *And having come in, the angel said to her, "Rejoice, highly*
favored one, the Lord is with you; blessed are you among women!" 29
But when she saw him, she was troubled at his saying, and consid-
ered what manner of greeting this was. 30 *Then the angel said to her,*
"Do not be afraid, Mary, for you have found favor with God. 31 *And*
behold, you will conceive in your womb and bring forth a Son, and
shall call His name Jesus. 32 *He will be great, and will be called the*
Son of the Highest; and the Lord God will give Him the throne of
His father David. 33 *And He will reign over the house of Jacob for-*
ever, and of His kingdom there will be no end." 34 *Then Mary said*
to the angel, "How can this be, since I do not know a man?" 35 *And*
the angel answered and said to her, "The Holy Spirit will come upon
you, and the power of the Highest will overshadow you; therefore,
also, that Holy One who is to be born will be called the Son of God.
Luke 1:26-34 NKJV

Let's cover some basics, and one particular myth about this event. First, what is the definition of this word, i*ncarnation,* in the Christian faith? In Latin, this word is interpreted as a noun of action from past participle stem of Late Latin, *incarnari,* "be made flesh."[135] Second, what does that mean for this discussion? If we are to believe that Christ and Jesus were one person in soul and body, we must understand how that occurred. It is believed that at the moment of conception, when and however that occurred, the angel touched Mary's womb with the being of Christ and that being became flesh, allowing Christ to be born of woman as any other human, entering the world gasping for that first breath of the air that sustains all humanity. Reading about the birth of Christ today, many suppose that Christians have always believed in a virgin birth, or a birth without the intervention of another human, a man. They would be wrong.

It wasn't until the 19th century, after a lot of Middle-Ages disagreements, that Christianity politically and dogmatically accepted the virgin birth, or as it has come to be called, "The Immaculate Conception" by the Catholic Church. Ah, another word to define, "immaculate." The Apostolic Constitution of Pope Pius IX solemnly defined the dogma of the Immaculate Conception on December 8, 1854. The Immaculate Conception is a dogma of the Roman Catholic Church which states that the Virgin Mary was free of original sin from the moment of her conception.

> *We declare, pronounce, and define that the doctrine which holds that the most Blessed Virgin Mary, in the first instance of her conception, by a singular grace and privilege granted by Almighty God, in view of the merits of Jesus Christ, the Savior of the human race, was preserved free from all stain of original sin, is a doctrine revealed by God and therefore to be believed firmly and constantly by all the faithful.* [136]

Most Protestant Churches simply refer to this event as the "virgin birth."

Why is this event so special? For that matter, why are any of these events outstandingly special? If any one of these events did not occur, the entire Christian Faith is useless. Science only has to disprove one of these events. Christianity must defend them all. Having a human child without the assistance of at least an artificial insemination has never been proven possible by science. Science might go a step further and assert that not yet proving something does not necessarily disprove it forever. While there are some species on Earth who can reproduce without the assistance of one of their counterparts, humans have yet to achieve that status.[137]

Let's back up just a bit to the part where Mary was free from original sin. We have discussed whether Jesus Christ was one person or two. This event approaches the answer to that question, but only partially, in my understanding. It is important to determine that allowing a God, the very divine being Christ, to occupy the same space as a sinful human is impossible. One of the arguments for Christ coming to the Earth in human form was that God cannot look on sin. If that is true, then neither can Christ, as a part of the Trinity, become united with a sinful creature in the flesh. We have a solution for Christ through the virgin birth—but what about the third divine member of the Trinity, the Holy Spirit? Due to the sheer volume of information, we will not cover this question in this book. So as not to leave you hanging on this subject, I suggest you begin your research using the link in this footnote below referencing Habakkuk 1:13.[138] While I am not endorsing this reference, every rock climber needs their first good hole in the rock-face to begin a very long, arduous climb. Like rock climbing, if you decide to make this attempt, make sure you are securely tethered at all times, you have the proper equipment and a trusted, faithful partner. Enjoy the adventure!

Sorry about the detour, let's get back on the road.

If Jesus and Christ are two persons, then Jesus being born of a sinless human would make Christ's occupation of the same body possible. But was Mary sinless prior to conception? Likely, not. The above-cited proclamation of the church states her sinlessness began at the moment of conception. Did she come from not one, but two sinless parents? Or was it a separate act of God that made her sinless? It would seem that her sinful life prior to conception would not require a sinless body, because Christ was not yet involved. This event, more than any of the others, is the most difficult to create or fake. It either happened or it didn't. It may be easier to look back at this event through the eyes of the other events listed. The only way they can be true is if this one actually happened.

In summary, I appreciate the words of Stephen J. Wellum posted in an article on crossway.org, December 21, 2016 titled, "10 Things You Should Know About the Incarnation."

> *The incarnation is an act of addition, not subtraction. In the incarnation, the eternal Son who has always possessed the divine nature has not changed or set aside his deity. Instead, he has added to himself a second nature, namely a human nature consisting of a human body and soul (Phil. 2:6-8). As a result, the individual Jesus is one person—the Son—who now subsists in two natures, and thus is fully God and fully man.*[139]

Jesus' Baptism

Before getting to the actual event of Jesus' baptism, let's first walk around inside the definition of this term… which baptism are we discussing, and how do these variations affect the reason Jesus asked John the Baptist to perform this act on his Divine, human body?

You have probably heard, especially in the faith community, the term Baptism of the Holy Spirit, or in some denominations, Baptism of the Holy Ghost. If you are even a minimal student of how and why various denominations and faith groups were formed, you will know that one of the major reasons for forming a new believer group involves disagreements over baptism's mode and importance in salvation, or belief in Christ. Also looming within this discussion are those who wonder why this is important, since there is little, if any, evidence—historical or scriptural—that Jesus Christ ever baptized anyone. So, what's the big deal? I know that last question will additionally brand me as a heretic by some in the faith community. Please don't judge too quickly. Thanks.

Baptism of the Holy Spirit (Ghost)

> 13 *For by one Spirit we were all baptized into one body—whether Jews or Greeks, whether slaves or free—and have all been made to drink into one Spirit.* 1 Corinthians 12:13 NKJV

> 8 *But you shall receive power when the Holy Spirit has come upon you; and you shall be witnesses to Me in Jerusalem, and in all Judea and Samaria, and to the end of the earth.* Acts 1:8 NKJV

> 17 *The Spirit of truth, whom the world cannot receive, because it neither sees Him nor knows Him; but you know Him, for He dwells with you and will be in you.* John 14:17 NKJV

> 22 *And when He had said this, He breathed on them, and said to them, "Receive the Holy Spirit."* John 20:22 NKJV

> 2 *He said to them, "Did you receive the Holy Spirit when you believed?" So, they said to him, "We have not so much as heard whether there is a Holy Spirit."* 3 *And he said to them, "Into what then were you baptized?" So, they said, "Into John's baptism."* 4 *Then Paul said, "John*

indeed baptized with a baptism of repentance, saying to the people that they should believe on Him who would come after him, that is, on Christ Jesus." [5] *When they heard this, they were baptized in the name of the Lord Jesus.* [6] *And when Paul had laid hands on them, the Holy Spirit came upon them, and they spoke with tongues and prophesied.* Acts 19:2-6 NKJV

Okay, confused yet? Bad news is, we haven't even gotten to the *mode* discussion. I don't have a position on this subject other than if Jesus felt compelled to be baptized, I should follow His example if I believe in Him. The Scripture is there for all to read and interpret. What I don't agree with is taking the stance that I am right and everyone else is wrong, and is going to Hell for not taking my position concerning baptism as infallible. Just so you don't feel inadequate, not totally understanding this topic, consider that the word for this act sometimes uses an "s" and other times a "z." And try to get the word "baptismals" past spellcheck in Microsoft Word.

With research, you will find beliefs regarding Baptism of the Holy Spirit (Ghost) to include:

- It gives the power to heal and handle snakes (vipers) without being harmed
- A connection to *sanctification* (another controversial word)
- Gives the ability to speak in unknown tongues, either other languages or unrecognizable words, and an interpreter should be present
- Salvation is incomplete until you are baptized
- The Holy Spirit will not come into your life without this act of baptism

Modes of Baptism—Who Is Right?

Another controversial aspect of baptism, within the faith community, concerns the way baptism is administered, and by whom. There are basically three methods: pouring, sprinkling, and immersion. All require water of some sort. The amount of water used runs the gamut from dirty, snake-infested drainage canals in rural southern US locations; tropical sites around the world; swimming pools; horse watering troughs; formal, built-in immersion baptismals; baptismal fonts and fountains; cups of water; and water that has been miraculously converted to Holy Water.

The three basic modes of baptism have reason and history. Pouring and sprinkling are mostly connected to Old Testament purification and animal sacrifice. Immersion is emphasized because it is assumed John the Baptist used this method. Let's look at Scripture related to Jesus' baptism by John.

> 13 *Then Jesus came from Galilee to John at the Jordan to be baptized*
> *by him.* 14 *And John tried to prevent Him, saying, "I need to be bap-*
> *tized by You, and are You coming to me?"* 15 *But Jesus answered and*
> *said to him, "Permit it to be so now, for thus it is fitting for us to ful-*
> *fill all righteousness." Then he allowed Him.* 16 *When He had been*
> *baptized, Jesus came up immediately from the water; and behold, the*
> *heavens were opened to Him, and He saw the Spirit of God descend-*
> *ing like a dove and alighting upon Him.* 17 *And suddenly a voice*
> *came from heaven, saying, "This is My beloved Son, in whom I am*
> *well pleased."* Matthew 3:13-17 NKJV

Those who proclaim immersion as the only mode of recognized baptism cite verse 16 above, "*When He had been baptized, Jesus came up immediately from the water.*" They explain "coming up from the water" to mean that John guided him under the water, symbolizing His entire body (sins) were washed away when He went under. Coming up out of the water from being completely washed all over at one time leaves nothing untouched by the act. Thus, immersion.

John 3:22-23 contains another set of verses used to support immersion. These verses seem to indicate Jesus was also baptizing people.

> 22 *After these things Jesus and His disciples came into the land of*
> *Judea, and there He remained with them and baptized.* 23 *Now John*
> *also was baptizing in Aenon near Salim, because there was much*
> *water there. And they came and were baptized.* John 3:22-23

In verse 23 above, the use of "much water there" would seem to indicate John needed enough water to get an entire body submerged. Due to the dryness of the area, there were times when finding a lot of water deep enough to cover a human would have been difficult. Verse 22 has also created a lot of debate over whether Jesus actually baptized as John did, with water. While you can assume literally that what the verse says means Jesus did baptize with water, John himself, in his preaching, clearly says he, John, will baptize with water, but He who comes after me will baptize with fire and the Holy Spirit. Hopefully, with this background, the following will make more sense when examining why Jesus felt it necessary to be baptized.

Why Was Jesus Baptized?

Before entering this topic, concerning Jesus' motive to be baptized, let's again read about the encounter between Jesus and His cousin, and also some background Scriptures.

> 3 *The voice of one crying in the wilderness: "Prepare the way of the Lord; Make straight in the desert a highway for our God."* Isaiah 40:3
>
> 23 *He said: "I am 'The voice of one crying in the wilderness: Make straight the way of the Lord,' as the prophet Isaiah said."* John 1:23
>
> 13 *Then Jesus came from Galilee to John at the Jordan to be baptized by him.* 14 *And John tried to prevent Him, saying, "I need to be baptized by You, and are You coming to me?"* 15 *But Jesus answered and said to him, "Permit it to be so now, for thus it is fitting for us to fulfill all righteousness." Then he allowed Him.* 16 *When He had been baptized, Jesus came up immediately from the water; and behold, the heavens were opened to Him, and He saw the Spirit of God descending like a dove and alighting upon Him.* 17 *And suddenly a voice came from heaven, saying, "This is My beloved Son, in whom I am well pleased."* Matthew 3:13-17 NKJV

Looking at this situation from thirty-thousand feet, this question—"Why was Jesus Baptized?"—is perfectly logical. Why? Jesus, (or more correctly at this time, Jesus Christ), was sinless! He needed no purification, no redemption, and no washing away of His sins. There must have been another reason. There was. Earlier, we looked at Isaiah 40:3. Let's look at Isaiah 40 verses 4 and 5.

> [4] *Every valley shall be exalted and every mountain and hill brought low; the crooked places shall be made straight and the rough places smooth;* [5] *The glory of the Lord shall be revealed, and all flesh shall see it together; For the mouth of the Lord has spoken.* Isaiah 40:4-5

Verse 4 further addresses how straight and level the desert highway will be. Before Christ, the connection to God was remote and distant to man. In fact, there was no direct connection to God. Verse 5, when taken in context with verse 3—*"The voice of one crying in the wilderness"*—places John the Baptist at the scene when "The glory of the Lord shall be revealed." What better place and event to announce the revelation of the glory of the Lord than when the prophesied one meets the coming Messiah? Jesus' baptism was not to wash away His sins. It was to announce Christ's arrival and the beginning of His ministry on the earth. Coming out of the water, Jesus, John the Baptist, and many others heard the voice of God fulfilling Isaiah's prophecy. It was truly the revelation of the glory of the Lord.

> [16] *When He had been baptized, Jesus came up immediately from the water; and behold, the heavens were opened to Him, and He saw the Spirit of God descending like a dove and alighting upon Him.* [17] *And suddenly a voice came from heaven, saying, "This is My beloved Son, in whom I am well pleased."* Matthew 3:16-17 NKJV

If you haven't yet wondered, let me ask another important question. Was John the Baptist ever baptized? Remember when he first met Jesus by the Jordan River, it was John's desire that Jesus baptize him. He was very surprised when Jesus instead insisted that John should baptize Him. So, did John ever get around to being immersed, poured, or sprinkled? All the sources I found concluded that there is no evidence of John the Baptist ever having been baptized. It seems strange that John the Baptist insists you must be baptized, while he appears to be exempt. However, I found one explanation on thywordistruth.com that seems to address the issue. It is listed as, Question #86, with the title, "Who baptized John the Baptist?"

The answer follows below:

> *The short answer is "No." The Bible does not say who baptized John and any other answer would be pure speculation. As a matter of fact, there is no record of John's being baptized at all. John was the divinely appointed forerunner of Christ. Malachi 3:1... His baptism was from God. Matthew 21:24-26. There is no reason to believe that the divinely prophesied and appointed forerunner needed to receive the baptism that he was sent to administer.*[140]

The summary of other sources also agree John the Baptist was not under New Testament law, since it had yet to be written. Prior to John's introduction of this new process, instructions concerning one's relationship to God were tied to Old Testament process, or at least the first five books, called the Torah and the Pentateuch—the Hebrew and Greek names, respectively. Although John the Baptist could not claim to be sinless, as could his cousin, Jesus, he was chosen and appointed by God before he was born or even thought of by his parents. He was also born of a miracle from significantly elderly parents.[141]

There may be other reasons Jesus was baptized. One scholar wrote that Jesus' baptism was part of his preparation for Satan's temptation in the wilderness. Baptism was a foreshadowing of Christ's death, burial, and resurrection. He also pointed out that this event announced the beginning of His ministry on Earth.[142]

Another writer, Steve Hartlaub, posting on givemechrist.com, October 10, 2016, suggests the following when Scripture says, "*The heavens were opened unto him, and he saw the Spirit of God descending like a dove*," in Matthew 3:16-17:

> *The heavens being opened unto Jesus means that he saw spiritually for the first time. Until that time, though he was born of the spirit of God, his only approach to his Father was via the (Old Testament) Scriptures. In this he was just like any other man who lived at that*

> *time. His baptism by John fully represented what happens to us when we are baptized by the Spirit. He went in a physical being, limited by his mind and understanding, but he came out "reborn" as a spiritual being, with the heavens (the spiritual realm) opened to him, with direct access to his Father, God.*

Hartlaub also cites 1 John 1:32-34 as evidence that Jesus (Christ) was not the only one who saw the Spirit descending like a dove on to His person. John the Baptist must have also been witness to the event. [143]

> 32 *And John bore witness, saying, "I saw the Spirit descending from heaven like a dove, and He remained upon Him.* 33 *I did not know Him, but He who sent me to baptize with water said to me, 'Upon whom you see the Spirit descending, and remaining on Him, this is He who baptizes with the Holy Spirit.'* 34 *And I have seen and testified that this is the Son of God."* 1 John 1:32-34 NKJV

If they were the only two who visibly saw this happen, those standing nearby must have thought the two had momentarily gone mad. Imagine one of the disciples asking Jesus where he was going when he left the crowd and receiving a blank stare. Apparently, He didn't have a clue where He was headed. Jesus Christ walked into the desert with nothing except the wet clothes on His back, which were likely dry before He reached His destination. It would have been a hike.

This revised act of baptism, replacing circumcision, will eventually become a powder keg of disagreement during the birth and growth of the New Testament Church.

In closing this section, we cannot overlook the awesome physical enormity of what occurred. This was another event science cannot recreate. This was not a quiet outing with the breeze blowing through the fig leaves and birds singing. For this example, I am assuming Jesus' baptism was by immersion where the heavens opened to him, and he saw the Spirit of God descending like a peaceful dove, although for some, pouring and sprinkling could have the same effect on a human. Just before John placed Jesus Christ under the water, the human in the body Divine would have taken a breath… and held it. Upon coming out of the water, there would have been a gasp of new, fresh air.

This second breath would have filled the lungs of a new being. This breath would be the first of a series of breaths Jesus Christ would take over about three years. His lungs would breathe this air until one day, on a cross, he would breathe His last human breath. That last breath was likely in a hushed whisper no more audible than when He took His first breath as a newborn. That breath was used by His vocal cords to proclaim: "It Is Finished." It was a whisper heard all the way to Heaven. Those words would bring the physical world as close as it had ever been to the power generated at Creation's initiation. The Trinity's combined power would then explode when Christ on the cross, the sacrificial lamb, channeled all sin ever committed in the history of time—past, present, and future—through His very being at His last breath. At that moment something supernatural occurred. That something began a few short years earlier when Jesus came out of the water with the supporting hands of His cousin, John the Baptist… the Proclaimer!

Jesus' Time in the Wilderness

Some may refer to Jesus Christ's time in the wilderness as His "boot camp" in preparation for His coming ministry, which would include fighting Jewish Temple monarchs, the Roman Government, and Satan with all His army of demons. Oddly, this part of Jesus Christ's life is often omitted from possible sermons, Bible lessons, Bible studies, and small group sessions. What is odd to me is that this is one of the most dramatic and impactful periods in the history of man on this Earth… beginning to end, whenever the end occurs. It is possibly left out of lessons, discussions, and sermons because it is so often misunderstood. We occasionally hear about the temptations, but did they take forty days and forty nights? Probably not. The narrative seems fairly seamless. Like other parts of Scripture, there are days of silence. While this writing style frustrates some who like everything spelled out, many enjoy the freedom to fill in details for themselves, details that better fit their life situation. This writing style continues today as in centuries past.

So, you thought the Immaculate Conception and the baptism of Jesus Christ was a big deal… well, you ain't seen nothing yet! Let's begin with the story from Scripture in Matthew, Mark, and Luke.

> [1] *Then Jesus was led up by the Spirit into the wilderness to be tempted by the devil.* [2] *And when He had fasted forty days and forty nights, afterward He was hungry.* [3] *Now when the tempter came to Him, he said, "If You are the Son of God, command that these stones become bread."* [4] *But He answered and said, "It is written, 'Man shall not live by bread alone, but by every word that proceeds from the mouth of God.' "*
>
> [5] *Then the devil took Him up into the holy city, set Him on the pinnacle of the temple,* [6] *and said to Him, "If You are the Son of God, throw Yourself down. For it is written: 'He shall give His angels charge over you,' and, 'In their hands they shall bear you up, Lest you dash your foot against a stone.' "* [7] *Jesus said to him, "It is written again, 'You shall not tempt the Lord your God.' "*
>
> [8] *Again, the devil took Him up on an exceedingly high mountain, and showed Him all the kingdoms of the world and their glory.* [9] *And he said to Him, "All these things I will give You if You will fall down and worship me."* [10] *Then Jesus said to him, "Away with you, Satan! For it is written, 'You shall worship the Lord your God, and Him only you shall serve.' "* [11] *Then the devil left Him, and behold, angels came and ministered to Him.* Matthew 4:1-11 NKJV
>
> [12] *Immediately the Spirit drove Him into the wilderness.* [13] *And He was there in the wilderness forty days, tempted by Satan, and was with the wild beasts; and the angels ministered to Him.* Mark 1:12-13 NKJV
>
> [1] *Then Jesus, being filled with the Holy Spirit, returned from the Jordan and was led by the Spirit into the wilderness,* [2] *being tempted for forty days by the devil. And in those days He ate nothing, and afterward, when they had ended, He was hungry.* [3] *And the devil said to Him, "If You are the Son of God, command this stone to become*

> *bread." 4 But Jesus answered him, saying, "It is written, 'Man shall*
> *not live by bread alone, but by every word of God.' "*
>
> *5 Then the devil, taking Him up on a high mountain, showed Him*
> *all the kingdoms of the world in a moment of time. 6 And the devil*
> *said to Him, "All this authority I will give You, and their glory; for*
> *this has been delivered to me, and I give it to whomever I wish. 7*
> *Therefore, if You will worship before me, all will be Yours." 8 And*
> *Jesus answered and said to him, "Get behind Me, Satan! For it is*
> *written, 'You shall worship the Lord your God, and Him only you*
> *shall serve.' "*
>
> *9 Then he brought Him to Jerusalem, set Him on the pinnacle of the*
> *temple, and said to Him, "If You are the Son of God, throw Yourself*
> *down from here. 10 For it is written: 'He shall give His angels charge*
> *over you, To keep you,' 11 and, 'In their hands they shall bear you*
> *up, Lest you dash your foot against a stone.' " 12 And Jesus answered*
> *and said to him, "It has been said, 'You shall not tempt the Lord*
> *your God.' " 13 Now when the devil had ended every temptation, he*
> *departed from Him until an opportune time.* Luke 4:1-13 NKJV

After Jesus Christ came up from watery baptism, the Holy Spirit descended on Him like a dove. Notice it does not say it was a dove, as portrayed in numerous famous depictions of the event. It says the Spirit of God descended like a dove. In other words, gentle, graceful, and beautifully. Quite unlike the Holy Spirit, actually. The Spirit often appears in scenes of dramatic action or God's anger. It is beautiful to see the Holy Spirit in this personality. As I understand, this same Holy Spirit was sent to be a comforter after Christ left the Earth. It is reassuring to know that this same Holy Spirit will be by my side when I need a calming touch. My compliments to the Trinity for the contrast experienced in this passage.

After they hear God tell His Son of His pleasure for acceptance of this task, this same gentle, graceful, beautiful Holy Spirit leads Jesus Christ into the wilderness to be starved, baked, and tempted, or tested, depending on your favorite version of the Scripture. The length of time spent in the wilderness—forty days and forty nights—is consistent with several other time periods in the Old Testament. During the flood of Noah, it rained for forty days and forty nights. Moses was raised and lived in the house of the Pharaoh for forty years, ran away to live as a shepherd for forty years, and then spent his last forty years leading the Children of Israel to the Promised Land. And, as in the life of Jesus, parents were expected to go to the Temple and make a sacrifice when the child was forty days old.[144]

The events at the end of the baptism are not what *everyone* thinks happened. Some feel the only one who heard God's voice was Jesus Christ, His Son. To their credit, there is no indication in these three gospel descriptions that it was a public address for all who could hear. It does say the Heavens were opened to Him, and He heard… and there is no reaction from John or the crowd recorded in these accounts. However, there is a reference in 1 John 1:32-34 indicating that John saw the Holy Spirit descend as proof to him that this was the Messiah he had been announcing to those who would listen. After God spoke, He was *immediately* led to the wilderness by the Holy Spirit.

The first four gospels of the New Testament are similar, but not exactly the same. The previously highlighted Scripture from three of these books is a prime example. Obviously, no reader of the New Testament is surprised that Mark's account is the "Cliff Notes" of the event. It is just the way he wrote. However, the difference between Matthew and Luke has a striking difference, and there are a few things of interest.

The order of the temptations is different between these two accounts. I am biased, but I would take the order from Luke, based on his other attention to detail. Also, Matthew was a tax collector, after all! In Matthew, Jesus Christ is led by the Holy Spirit into the desert. Mark uses the term, *"Immediately the Spirit drove Him into the wilderness"* and Luke says, "*Then Jesus, being filled with the Holy Spirit, returned from the Jordan and was led by the Spirit into the wilderness.*" These are three very different descriptions of the same event that leave the reader with three versions and possible motives. Here again, we must look for commonality. The central figures in all three descriptions are the same: Jesus Christ and the Holy Spirit.

What was the motive? Had Christ forgotten the plan developed by the Trinity, eons before? Did Christ miss that meeting? Was He in another "Not My will but Thine…" mode? The question remains, "Why did Jesus

Christ need any guidance or persuasion from the Trinity to take this detour?"

In a response in catholic.com,[145] the writer suggests the difference between Matthew and Luke's responses, compared to Mark's, had to do with writing style and to whom Mark was addressing his remarks. The article states it had to do with the immediate audience of these three writers. While Matthew and Luke were addressing a Jewish audience, Mark was primarily addressing Romans. It is suggested that Mark's audience, Romans, were not so much interested in philosophical Greek thoughts, but rather in descriptions of action, which would have appealed to them. Either way, the focus must be on the characters of the story.

I find Jesus Christ's recorded use of the word "you" in Satan's offer to give power to rule all nations interesting. This one word covers a lot of ground. Today, we take Jesus Christ's statement here, and in other verses, as a command to us not to sin and honor God only. I read this as a more personal instruction—and reminder—to Satan only. I hear Jesus Christ's direct assault on Satan for this fallen angel's past misdeeds. Let's look back to a very significant and important event in our timeline.

Many point to Ezekiel 28 as a metaphor explaining Satan's misdeeds and the consequences he brought on himself. Isaiah 14:12-15 is more to the point.

> [12] *How you have fallen from heaven, morning star, son of the dawn! You have been cast down to the earth, you who once laid low the nations!* [13] *You said in your heart, "I will ascend to the heavens; I will raise my throne above the stars of God; I will sit enthroned on the mount of assembly, on the utmost heights of Mount Zaphon.* [14] *I will ascend above the tops of the clouds; I will make myself like the Most High."* [15] *But you are brought down to the realm of the dead, to the depths of the pit.* Isaiah 14:12-15 NKJV

In Luke 10:18 NKJV, Jesus Christ makes the following statement: "*18 And He said to them, 'I saw Satan fall like lightning from heaven'.*" In context, the previous chapter marks His sending out the Twelve into ministry and the instructions that went with His release of these folks into the known world to tell people what has and is taking place. In chapter 10, we discover that Jesus Christ also sends out another group of seventy with similar instructions. A major difference in the launch is that these seventy individuals are paired in sets of two. It is in His discussion with this group that He mentions what He witnessed in Heaven. While we are not told when Satan fell from Heaven, we must assume it was before Jesus Christ was born on Earth.

The most revealing description is in Revelation 12:7-12. If you have not spent much time in the book of Revelation, I offer the advice of N.T. Wright and Michael F. Bird in their excellent study series, *The New Testament You Never Knew.* [146] They suggest not merely reading sections of this New Testament book, but rather reading it in its entirety each time. Like this Scripture reference in chapter 12, it can get lost if not supported by everything meant to be taken away from this writing. This example, like many others in Revelation, although references, have substance nonetheless.

Assuming, as I do, that Satan has a history before Creation, this fallen angel's appearance in the Garden of Eden early in Genesis answers a lot of questions. Going back to my reason for taking this route of explanation in the first place, I believe Jesus Christ's remarks to Satan in the wilderness were personal. This scene between these two rivals sets the stage for the continuing conflict between good and evil. Until this moment in the wilderness, I believe Satan still mistakenly believed that, because God did not kill him when he tried to ascend up to the level of God himself, he still had a chance to defeat God—and become God. Satan could have thought getting Christ to sin just one time would make God's Son ineligible to be the sacrifice for all humanity, thus ruining the Trinity's plans.[147] It could be that Satan, like us, still had a hard time understanding *Who* and *What* the Trinity was and the ultimate power They possessed. In effect, in the three referenced Scripture passages regarding the wilderness journey, Jesus Christ simply told Satan to stop cutting in line—and like Satan's invasion of Peter's personal space later on, to get behind Him.

Evil has a way of letting selfish behavior, and the ego that engulfs it, create a false sense of self-worth and power. What I get from endless studies of Scripture, and interpretations of others, is that even after

Jesus overcame death and was transfigured and lifted back into Heaven, where Satan is no longer welcome, either the evil one still didn't get it, or he did and was holding onto the power, given to him by God, until the end he must have known was coming. Perhaps at that time he didn't realize that the coming end will consume him. It is important to note that Jesus Christ's death, burial, and resurrection did not do away with death or remove death's controlling power from Satan. Instead, Jesus Christ's actions *overcame* death. There is a huge difference! Satan may think He will have this power for eternity and, depending on your reading of Scripture, He may very well still control death at that juncture. Problem is, THERE WILL BE NO MORE DEATH TO CONTROL. Even those who will live with the Evil One for eternity can't die. Satan will become… the King of Misery.

Speaking of the Evil One, for the coming ministry of Jesus Christ, it was critical that His human side get a personal introduction to Lucifer, Satan, the Devil, the Evil One. Jesus Christ would see His adversary at least one more time during the next three years. Satan would take possession of one of Jesus Christ's closest friends, Peter the Rock. Here again, Jesus Christ addresses Satan directly: *"But He turned and said to Peter, "Get behind Me, Satan! You are an offense to Me, for you are not mindful of the things of God, but the things of men."* Matthew 16:23 NKJV

In a genuine effort to keep the scientific community engaged, I did find a well-researched article, "Where Did Satan Come From?" on livescience.com. It was written by Laura Geggel, assistant editor, on October 2, 2016. The thrust of her article was an effort to find out how the story of Satan originated. Her research indicated many ancient religions detailed the struggle between good and evil. She found that this same ancient world also struggled with the *coexistence* of good and evil. "They hypothesized a kind of demonic, divine force that was responsible for evil, arising out of the notion that a good god could not be responsible for bad things."

She points out that there were few demon-like figures in Hebrew Scripture, suggesting that Satan was not a prominent figure in Judaism. She mentions brief examples in the books of Job and Daniel. Even then, the reference to these figures uses other names than Satan, and they are weak compared to their adversary, God. Her understanding was that for

Judaism to promote a powerful evil being, created by a god, would play down the religion's emphasis on their god's omnipotence and even his omniscience.

Ms. Geggel discovered that Judaism started to give more credence to the Evil one about the time Jesus was born. Still, Judaism's mystical teachings, called the Kabbalah, mention a light side and a dark side, but the dark side is never given equal power to the light. At this point in the article, the research gets a little out of order because she admits most Christians today accept that Satan was a fallen angel, while claiming proof of Satan's fall is not in the New Testament. Overlooking Revelation 12:7-12, she jumps to the sudden appearance of Satan tempting Jesus Christ in the wilderness. The faith community has a lot of homework to do in preparing to answer tough questions. Perhaps this community can take away from this article an example of how important it is to be prepared to defend your faith with facts.[148]

To strengthen her position, Ms. Geggel quotes Jerry Walls, a professor of philosophy at Houston Baptist University:

> *"The only thing that can go bad by itself is a free being," Walls said. "Since there was evil before human beings came on the scene, the inference is [Satan] must have been a fallen angel."*[149]

"I thought of Satan as a kind of a joke, kind of a throwaway character," said Elaine Pagels, a professor of religion at Princeton University. She went on to say, *"In the Book of Job, he's practically a device to explain what happened to Job."*[150]

Ms. Geggel ends her article quoting Ms. Pagels:

> *"Turning an enemy into Satan is useful," she added. It suggests that "our opponents are not just people we disagree with—they're bad. You can't negotiate with them. You can't do anything with them, because they're essentially evil."*

I must admit, these are some folks I would have trouble engaging in a serious discussion.

Just for clarity—no, the desert wasn't across the street... and what happened to all the disciples who were with Jesus when He and John encountered each other? Did they go with Him for forty days and nights into the wilderness? Were they also baptized? Remember me talking about writing styles and leaving certain things to the imagination of the reader. Guess what? You could fill in with your imagination. Here is an example.

You may have wondered, as you read about this forty-day ordeal in the wilderness, how the Gospel writers learned of the details included in their New Testament books. As suggested, go ahead and fill in the details from your imagination. Become a part of the story. What makes sense to you? Here is how I imagine the answer to this question developed.

I speculated that a while after Jesus looked at Peter and told Satan to get behind Him, He again addressed Peter as the two joined the other disciple sitting around a camp fire. They had just finished a meal of freshly caught fish. Jesus addressed the group, for Peter's benefit in particular, and to satisfy the curiosity of the others concerning His forty-day trip into the wilderness.

Is that what happened? Probably not, but this exercise will allow you to be there sitting around that fire. You could also add some humor to the story. Don't be afraid to try it. It is obviously important that you footnote where this came from if written and, if spoken, declare you are making it up. I believe God encourages us to work with Scripture in this way. Why else would these details be missing?

But I encourage you to do the better thing: *"Be diligent to present yourself approved to God, a worker who does not need to be ashamed, rightly dividing the word of truth."* 2 Timothy 2:15 NKJV

The Crucifixion

The act of the crucifixion can be broken down into seven individual incidents. Each of these are critical to the understanding of Jesus Christ's death on the cross. Below is a Scripture reference and a brief explanation of each. I have reprinted only the Book of Matthew references to describe the first six incidents, but have included references from all four Gospels when discussing the crucifixion.

The Betrayal

> 14 *Then one of the twelve, called Judas Iscariot, went to the chief priests* 15 *and said, "What are you willing to give me if I deliver Him to you?" And they counted out to him thirty pieces of silver.* 16 *So from that time he sought opportunity to betray Him.* Matthew 26:14-16 NKJV

It is important to note that, unlike most of the other events leading up to the crucifixion, this event is an integral part of several of the others. Let's begin here. Without Judas, or someone like him, the crucifixion would probably not have happened. Judas Iscariot, like Benedict Arnold in the US Revolutionary War, forever became synonymous with the term "traitor." This piece of the story has raised countless questions over centuries of recorded history. Let's explore one of these assertions. You may have heard this statement and follow-up questions.

"Surely Jesus Christ knew there was something wrong with this man. Why didn't He get rid of him, or why did He choose him in the first place?"

If, like everything else we have said, was planned long before Christ was born, then surely this man's presence in the Messiah's life was a part of that ancient plan.

> *"But there are some of you who do not believe." For Jesus knew from the beginning who they were who did not believe, and who would betray Him.* John 6:64 NKJV

Unlike the early example of Abraham and Isaac on the mountain, where it would take only one man to plunge a knife into the chest of His son for the sacrifice to work, Christ's execution was complex, involving multiple layers of Jewish Law, Roman Law, and Jewish and Roman tradition.

The Jewish leaders, whose goal was to rid themselves of this Nazarene nuisance, must begin by having a traitor's voice in their testimony from inside the inner circle of Christ's small, tight-knit group. Yes, Jesus Christ surely knew Judas' plans before he made them, and recognized Judas' probable withdrawal and small questioning moments from within the group. In a way, Judas' betrayal made the process faster. Without him, Jewish leaders would have had to be the one's originating the accusations that would lead to the accused's death. Too messy.

Jesus Christ knew Judas' character, but He never attempted to get rid of him.

> *But one of His disciples, Judas Iscariot, Simon's son, who would betray Him, said, "Why was this fragrant oil not sold for three hundred denarii and given to the poor?" This he said, not that he cared for the poor, but because he was a thief, and had the money box; and he used to take what was put in it. But Jesus said, "Let her alone; she has kept this for the day of My burial. For the poor you have with you always, but Me you do not have always."* John 12:4-8 NKJV

There were probably other unrecorded incidents between Jesus Christ and Judas over the three-year period of the ministry. One, however, is to me dark and unlike Jesus Christ. It happened at the Last Supper, their final meal together. The book of John chapter 13 paints the scene this way. In a trailing conversation with Peter, while washing the disciples' feet to demonstrate humility in verses 9 through 11, Christ's words are spoken to Peter, but not at him.

> *Simon Peter said to Him, "Lord, not my feet only, but also my hands and my head!" Jesus said to him, "He who is bathed needs only to wash his feet, but is completely clean; and you are clean, but not all of you." For He knew who would betray Him; therefore He said, "You are not all clean."'*

In my imagination, before he says the last five words, I see Jesus' head turn away from Peter without moving His shoulders. Judas was now looking squarely into the eyes of the Savior, the Son of God, the creator of the universe.

Later, in the 13th chapter of John, verses 18 through 27, Jesus Christ culminates His relationship with Judas Iscariot during a very quiet, meditative moment while everyone is gathered around a table sharing bread and wine. It was their last supper, but only two people in the room knew it. Jesus is speaking, sharing words of prophecy from the Old Testament. The words suddenly seem to come from a different place than His previous thoughts. If the group had not been listening closely, they would have probably thought they dosed off briefly and came back awake during another part of a different conversation. This prophecy was talking about someone lifting their heel against Him. Surely, they thought, this was referring to one of the Temple leaders or some Roman soldier.

His voice became troubled and pointed. In verse 21 He says, *"Most assuredly, I say to you, one of you will betray Me."* The room is stunningly silent. Eyes shoot stares back and forth between the listeners looking for someone to say something. Without surprise, Peter motions to John, whose is reclining near Jesus Christ, to ask who it is. John accommodates. The answer is not immediate, but rather Jesus Christ decides, for whatever reason, to indicate the betrayer through symbolism. With the same bread and wine used to symbolize His broken body and shed blood, Jesus Christ tells the group that whoever He gives a piece of dipped bread is the one who will betray Him. Then He exposes Judas.

After hearing Jesus Christ say that it would be better for the man who betrayed Him not be born than to suffer what was to come, Judas left the group. He would later show up in the garden where he would be asked to identify the blasphemer with a kiss on His cheek. Then, as written in Matthew 27:3-4, Judas would return to the Temple and beg the priests to take back the silver he was paid for the betrayal. They refused and ignored his pleas. Judas left the money with them and his last breath was taken below an overhanging branch as recorded in Matthew 27:5: *"Then he threw down the pieces of silver in the temple and departed, and went and hanged himself."* NKJV

Endless debates have ensued since that day speculating as to the fate of Judas Iscariot. Would he suffer the destiny of all unbelievers in Hell, absent of all good and love and the presence of Almighty God? Or did God have mercy on him, since someone had to commit this act to get Christ to the cross? Did Judas, before taking his last breath, ask God for forgiveness and received His grace as any murderer would on Death Row?

In the upper room during dinner, Jesus Christ used the words, *"One of you does not believe."* If you also read the description of this scene in

John 6:64, you will remember it says, "'*But there are some of you who do not believe.' For Jesus knew from the beginning who they were who did not believe, and who would betray Him.*" Was that Judas? Was there more than one? And, as with Adam and his wife, and every other person born after, including Jesus, did Judas have the free will to decide not to go through with it?

While there is no conclusive evidence as to the order the disciples were chosen, in the Book of Matthew chapter 10 verse 2, Matthew uses the word "first" in front of Peter's name. In that same listing, in verse 4, Judas Iscariot is listed last. Mark 3:19 and Luke 6:16 both list Judas last. If he was chosen last, could it be that Jesus Christ delayed his selection, knowing who Judas was and how His selection of this man would end?

The Last Supper

> 17 *Now on the first day of the Feast of the Unleavened Bread the disciples came to Jesus, saying to Him, "Where do You want us to prepare for You to eat the Passover?"* 18 *And He said, "Go into the city to a certain man, and say to him, 'The Teacher says, "My time is at hand; I will keep the Passover at your house with My disciples." ' "*
> 19 *So the disciples did as Jesus had directed them; and they prepared the Passover.*
>
> 20 *When evening had come, He sat down with the twelve.* 21 *Now as they were eating, He said, "Assuredly, I say to you, one of you will betray Me."* 22 *And they were exceedingly sorrowful, and each of them began to say to Him, "Lord, is it I?"* 23 *He answered and said, "He who dipped his hand with Me in the dish will betray Me.* 24 *The Son of Man indeed goes just as it is written of Him, but woe to that man by whom the Son of Man is betrayed! It would have been good for that man if he had not been born."* 25 *Then Judas, who was betraying Him, answered and said, "Rabbi, is it I?" He said to him, "You have said it."*
>
> 26 *And as they were eating, Jesus took bread, blessed and broke it, and gave it to the disciples and said, "Take, eat; this is My body."* 27
> *Then He took the cup, and gave thanks, and gave it to them, saying,*

> *"Drink from it, all of you.* [28] *For this is My blood of the new covenant, which is shed for many for the remission of sins.* [29] *But I say to you, I will not drink of this fruit of the vine from now on until that day when I drink it new with you in My Father's kingdom."* [30] *And when they had sung a hymn, they went out to the Mount of Olives.*
> Matthew 26:17-30 NKJV

Scientists will have to admit the Last Supper really did happen. There's even a famous portrait of the occasion. There's also the famous line spoken by—no one seems to know who said it—"All right, time for a picture, everyone on one side of the table!" There are actually a lot of things wrong in that famous picture. If you are curious, just Google it.

There are two divine sacraments brought into worship within these seven elements leading up to and including the crucifixion. These two sacraments are still held in high esteem, as they should be in today's worship, regardless of Christian Denomination. Those two sacraments are baptism and the Lord's Supper. Actually, there were three. Some faith groups still honor and perform the third sacrament on a regular basis. That sacrament is the washing of feet. I grew up in a small, Free-Will Baptist Church in Esserville, Virginia. It is a very small community in the southwest tip of the state. I continued to attend a church in that denomination after we moved to Florida in 1959. I moved to Tennessee to attend college in 1968, and eventually began attending a Cumberland Presbyterian Church. As a child and teenager, foot-washing services were just normal to me. As in the time of Christ, and as recorded in Scripture, the purpose is to demonstrate humility and respect for the welfare of your brother and sister.

In Jesus Christ's time, it was also practical. Most wore sandals or went barefoot. As you can imagine, walking on dusty or muddy streets and paths, a person's feet were always dirty when they entered a house. Water would have been kept at the door to clean your feet before entering the living quarters. This was usually performed on one's own feet. However, in their culture, for their leader, teacher, or Master to perform this menial task on someone else was unheard of, almost embarrassing to the receiver.

We usually remember something someone did or said just before they die. I have always felt Jesus Christ performed this ceremony at the particular time He did because it was the image He wanted burned into His followers' memories. After His death, I believe there were many sleepless nights among all of these remaining eleven disciples. I truly believe, of all the images they would have run through their minds, this image continued to motivate them to servant leadership. What if we all could imitate Jesus Christ's acts of humility?

Nothing in science can duplicate that image.

A lot is made of Jesus Christ being buried in a borrowed tomb. And why not? He only needed it for less than three days and nights. But He also held the last supper in a borrowed upper room. How did He acquire the space? All we are told is that He knew a man. He sent a couple of His disciples on ahead to see the man and prepare the room for this quickly called meeting of His chosen. They had met several times this way and it was probably not thought to be unusual to any of the twelve. In fact, things were very normal until the meeting was called to order.

A lot happened in a very short period of time; the establishment of two acts that would become sacraments for thousands of years, an announcement that He was going to die, betrayal, promise of denial and of course... they sang a hymn. It was going to be a very busy night. Not just busy, but rather twenty-four hours of humanity-changing events. When all three of the Trinity show up, something very big is about to happen.

Those remaining in the borrowed upper room, after Judas departed, had become disturbed and nervous. After the singing was over, Jesus headed to what was probably a favorite place for Him to pray. The first Adam came alive in a Garden. The second Adam would be arrested, crucified, and buried in one.

The Prayer in the Garden

> 36 *Then Jesus came with them to a place called Gethsemane, and said to the disciples, "Sit here while I go and pray over there."* 37 *And He took with Him Peter and the two sons of Zebedee, and He began to be sorrowful and deeply distressed.* 38 *Then He said to them, "My soul is exceedingly sorrowful, even to death. Stay here and watch with Me."*

[39] *He went a little farther and fell on His face, and prayed, saying, "O My Father, if it is possible, let this cup pass from Me; nevertheless, not as I will, but as You will."*

[40] *Then He came to the disciples and found them sleeping, and said to Peter, "What! Could you not watch with Me one hour?* [41] *Watch and pray, lest you enter into temptation. The spirit indeed is willing, but the flesh is weak."* [42] *Again, a second time, He went away and prayed, saying, "O My Father, if this cup cannot pass away from Me unless I drink it, Your will be done."* [43] *And He came and found them asleep again, for their eyes were heavy.*

[44] *So He left them, went away again, and prayed the third time, saying the same words.* [45] *Then He came to His disciples and said to them, "Are you still sleeping and resting? Behold, the hour is at hand, and the Son of Man is being betrayed into the hands of sinners.* [46] *Rise, let us be going. See, My betrayer is at hand."* Matthew 26:36-46 NKJV

Jesus Christ began His suffering in a Garden. Conversely, when He first began His ministry, after His baptism, He was in a wilderness. Satan was present in both. Adam, the Genesis human, committed the first sin in a Garden by eating forbidden fruit. Jesus Christ first suffered physically in a garden adjacent to a mount on which grew olive trees. Olive trees are known to live thousands of years. The oldest tree in the world is thought to be on the Isle of Crete. Scientists have discovered this olive tree is in excess of 4,000 years old.[151]

The Garden of Gethsemane was located at the foot of the Mount of Olives and likely a place often visited by Jesus Christ to think, pray, and teach small groups of disciples. It seems God has an affinity for gardens. In fact, John 15:1 says that He is the keeper of the garden! Many descriptions about Heaven talk of gold, silver, and precious jewels but, I want to see God's victory garden. Some may prefer all that metal and glass, but me, I'll take a calm, late afternoon, blooming flowers, a sunset, a water fountain, and lightning bugs, thank you. I believe precious metal and glass were simply byproducts of His creation. Things that live were… intentional.

I believe the use of gardens on these key spots in the timeline are also intentional. Among other things, they remind us there is no safe place to hide from a Satan who roams the Earth looking for those He may devour. Let's look at another garden.

Jesus Christ's borrowed tomb was located... in a garden. That said, He was buried in a tomb, in a garden, and He rose from the dead in a garden. It would be safe to say that gardens played a significant role in our timeline from alpha to omega. Almost as an afterthought, it is precious that Mary mistook a resurrected Jesus as a... gardener. John 20:15 She wasn't too far off. He was the Son of *the* Gardener!

More than a casual reading of this scene reveals the human and divine make-up of Jesus Christ. In Matthew 26:36, upon arrival in the garden, Jesus Christ picks a spot and asks His disciples, as a whole, to take up a position. At this time, He doesn't ask them to do anything but sit. At first it seems he is asking them to stay put while He goes on a little further to be alone. Instead, He asks three of the disciples to go on ahead with Him.

As the foursome move on a little distance from the rest of the disciples, Jesus Christ begins to break down emotionally and express those feelings to just these three. It was as if He didn't want all to see His raw emotion. Through that emotion, He asks them also to stay in that spot and He went on a little ahead of them. Interestingly, He did not ask them to pray; He asked them to *watch*. Watch for what?

After he moved further away from the two groups, Scripture recounts His falling to the ground, face first, as He began to pray through His open sorrow. It was a short prayer, and He was the only one praying. The larger group was sitting and the smaller group watching. Jesus Christ had already given the disciples hints about what was coming, but none really got it. They were about to either understand or just become even more confused.

The last half of verse 39 has caused extended conversations and debates. Just as asking why Jesus Christ needed to be baptized, why would Christ, if He was part of the long-range planning team, make the following statement: *"O My Father, if it is possible, let this cup pass from Me; nevertheless, not as I will, but as You will"*?

Returning to the discussion related to two wills, is it possible that Christ, having now been a part of a living human being since immaculate conception, realized something He didn't completely understand in the planning phase? Is this His realization of the suffering the human side of Him was being asked to sacrifice? It was one thing for a divine being to go through this, but He now has to feel what Jesus, the man, is experiencing. In Christ's compassion, is He doubting His and the rest of the Trinity's decision concerning sacrifice? It was the human body of Jesus that was about to be torn into shreds. If this is accurate, it only endears us as carnal human sinners to the unselfish action of Jesus.

In the book of Luke, chapter 22 verse 44, we find a detail only a physician could fully understand.

> *And being in agony He was praying very fervently; and His sweat became like drops of blood, falling down upon the ground.*

In the garden, when He was praying, He must have suffered from a condition called *hematohidrosis*. This condition manifests itself when the capillary blood vessels that feed the sweat glands break down. Blood released from the vessels mixes with the sweat; therefore, the body sweats drops of blood. Mental anguish or high anxiety was likely the cause of this reaction. Matthew 26:28 says, *"My soul is deeply grieved to the point of death."* [152]

After an unknown period of time, Jesus Christ returns to the smaller group and finds that instead of watching, they have fallen asleep. We may find it unacceptable that they were bored or disinterested enough to fall asleep and not watch as they were asked. Again, I ask—watch for what? Sure, there was something that happened during the meal and one of their brothers ran out of the room, but watch for what? Still addressing the smaller group, He indicates He has been alone for an hour and is upset that they couldn't do what He had asked, to watch.

Instead of clearing the air, Jesus Christ may have added to their confusion when He inserts yet another piece of the metaphor. Verse 41 has a vague reference that could be related to Adam in the Garden of Eden or Himself in the wilderness when confronted by Satan: "*Watch and pray, lest you enter into temptation. The spirit indeed is willing, but the flesh is weak.*"

In addition to the metaphor, now they are being asked to not just watch, but to pray.

Watch and pray for what? Not to be tempted... by whom? As an aside to Jesus Christ's above statement concerning the spirit and the flesh, is it possible He could have been referring to a struggle between Jesus the man and Christ the Son of God over the torture He was about to endure? Then again, He leaves and returns to his isolated prayers.

Before continuing let's talk about the rock. What rock? Good question...

Like the painting of everyone on one side of the table in the upper room, there is another painting I remember from my Sunday School days as a small child. They handed out little cards with a picture on one side and something written on the other side. None of us at that age could read, so we mostly stared at the picture while the teacher read and commented from the text. One of those pictures was about the scene we have been discussing; Jesus praying in the Garden with His hands clasped, looking up into a bright ray of light coming from the clouds... and His elbows supported by a very large rock. What rock?

I was privileged to be a part of a team that co-wrote a musical drama about the birth of several denominations during the Second Great Awakening occurring in the Cumberland Valley between Kentucky and Tennessee in the late 1700s early 1800s. In particular, the Cumberland Presbyterian Denomination. In my research, I discovered that it was common practice for groups of people to go out into the woods and lay prostrate, face down, on the ground to pray. This was to bow as low as possible before addressing God in prayer. It is assumed this tradition stemmed from verse 39 where He falls on His face before addressing His Father. No room for a rock in this verse.

The second time Jesus Christ retires to pray by himself, He almost repeats the same request. We don't know how long this prayer lasted. Upon completion, He returns and finds the small group again asleep; this time with a scriptural note, "because their eyes were heavy." He says nothing and leaves for a third time. Once again, repeating His same request closing with the fact that He will do whatever is the will of the Father.

Christ knew what that will would be, Jesus maybe not. He apparently needed to ask.

Something I don't often hear said when someone is preaching on these verses is any speculation about why Jesus Christ had to ask three times. And what did God answer? Sound familiar? How many times have you prayed the same request with no seeming answer? Sometimes the answer, as was probably true in this situation, is, "You know the answer, why are you asking me?"

There is possibly one other aspect of these unanswered prayers that may also be true of yours. Could it be that God is crying for you, and is so upset that He cannot speak? Yes, He cares that much!

Toward the end of this passage Jesus Christ tells His disciples, all of them I believe, "*...Behold, the hour is at hand, and the Son of Man is being betrayed into the hands of sinners. 46 Rise, let us be going. See, My betrayer is at hand.*" Matthew 26:45b-46 NKJV

The Arrest

> 47 *And while He was still speaking, behold, Judas, one of the twelve, with a great multitude with swords and clubs, came from the chief priests and elders of the people.* 48 *Now His betrayer had given them a sign, saying, "Whomever I kiss, He is the One; seize Him."* 49 *Immediately he went up to Jesus and said, "Greetings, Rabbi!" and kissed Him.* 50 *But Jesus said to him, "Friend, why have you come?" Then they came and laid hands on Jesus and took Him.*
>
> 51 *And suddenly, one of those who were with Jesus stretched out his hand and drew his sword, struck the servant of the high priest, and cut off his ear.* 52 *But Jesus said to him, "Put your sword in its place, for all who take the sword will perish by the sword.* 53 *Or do you think that I cannot now pray to My Father, and He will provide Me with more than twelve legions of angels?* 54 *How then could the Scriptures be fulfilled, that it must happen thus?"*
>
> 55 *In that hour Jesus said to the multitudes, "Have you come out, as against a robber, with swords and clubs to take Me? I sat daily with*

you, teaching in the temple, and you did not seize Me. [56] *But all this was done that the Scriptures of the prophets might be fulfilled." Then all the disciples forsook Him and fled.* Matthew 26:47-56 NKJV

In a measure of time, this scene will be very short. In a measure of dynamics, this scene must have seemed to be in slow motion to Jesus Christ. This was the culmination of thousands of years of prophecy, and over thirty-three years of human living, for this unique, but very divine being called Jesus Christ. Two lives were about to embrace… with a kiss, a method chosen by the betrayer to identify the one to be taken. There is a myriad of questions raised in this very short interaction, but first, let's go back and set the scene.

You might expect that there were two groups: Jesus Christ and His disciples, and the soldiers. This casting would depend on how you define the words, "…*a great multitude*…" In the crowd there would obviously have been twelve disciples and Jesus Christ. All those folks are mentioned. How many, and who are in the "*great multitude*" is left to the reader in the Book of Matthew. The Gospel of John gives better definition and we will address that a little later. Matthew does mention that the multitude came with swords and clubs. It was a simple assignment, one that any band of soldiers would have been able to easily control. So why all the weapons? Why clubs? The most important question would be, "Who was missing?"

It seems this multitude, whoever they were, was sent by the "Chief Priest and the Elders of the people." These priestly accusers could not be seen during this arrest. Why? They had no authority; that belonged to the Roman Government. In Matthew it doesn't say the group arrested Him, or even bound Him, but just that they laid hands on Him. The Gospel of John is a little more informative. You don't have to draw a weapon to use it, especially if the one you are dealing with has no defensive weapons of their own.

But, why the kiss? A kiss—not referenced in the Gospel of John—usually in the area of both cheeks, was a common greeting between two people, even of a casual acquaintance. This was a curious means of identifying the target. Why not just walk up to Jesus Christ and point straight at His face? Didn't this great multitude know who this guy was? Surely, he was a rock star and everyone recognized Him! That is where most folks have a break in reality regarding the life and ministry of Jesus Christ. We tend to transport His popularity in the world today to His presence in Jerusalem in the 1st century. Nothing could be further from the truth.

Remember that earlier in Matthew, and other Gospels, Jesus Christ was sending out His select twelve disciples and later seventy-two other disciples, in pairs, to begin spreading seeds throughout the region about His ministry and who He was. Just for clarity the number of disciples sent out in pairs, as noted in Luke 10:1, varies between seventy and seventy-two. As an example, all King James versions use seventy. Other translations, including all New International versions and the New American Standard Bible, use seventy-two disciples. Aside from those who were the twelve disciples chosen by Jesus, plus another seventy to eighty mentioned elsewhere in Scripture, few knew anything about Him. Of course, there was His family and close relatives, including His cousin, John the Baptist and his family, the elite of the Temple, and a few thousand who were healed, fed by Him, or attended one of his massive sermons. All of these were aware of who He was. But there were several hundred thousand in this large metropolis who had no reason to know who He was, or even care—especially soldiers, even the Temple soldiers. So, whose soldiers were they?

A more complete description of these events is located in the Gospel of John 18:1-12. Let's walk though these verses and fill in some blanks, including better identifying who made up the great multitude.

Verse 2 of chapter 18 of John tells us that Judas was very familiar with this Garden of Gethsemane, which reinforces that Jesus Christ and His select disciples often visited there. The location was either an educated guess on the part of Judas or he was told about the after-dinner plans and it is not mentioned in Scripture. Verse 3 of chapter 18 inserts an interesting twist. It says that Judas "*...procured a band of soldiers and some officers from the chief priests and the Pharisees...*" We can speculate on what is meant by *procured*, but the key is that there were two very different groups within this definition of soldier.

In a verse-by-verse analysis of John 18, the writers for trustinginjesus.com made the following observation: "*The officers of the Chief Priest and Pharisees arresting Jesus were most likely temple guards who had authority under the Jewish leaders to make arrests. The Roman soldiers did not arrest Jesus; they merely accompanied the temple guard for added security."* [153]

It could be that Judas used some of his pay-off to procure their services. The Temple guards may have seen the man they were now facing in the light of torches around the Temple, but they would have no reason to know who He was and He certainly didn't appear to present any danger. The Roman guards were clueless, probably invited as muscle should the scene get out of hand. At this point, it would have been the Temple guards who would have secured, or as John states, arrested Him. Matthew does not use this term. The most amazing part of this Scripture may pass you by if you don't pause and consider what just happened.

The kiss mentioned in Matthew must have occurred somewhere around John's verses 4 and 5 of chapter 18. During these verses, instead of it being important that Judas identify Him with a kiss, John points out, based on Jesus' knowledge of prophecy and His omniscience that He speaks out, "*Whom do you seek?*" They answered him, *"Jesus of Nazareth."* Jesus said to them, *"I am he."*

A very important thing occurs in verse 6 after He makes this last statement. It says they drew back and fell to the ground! Why would they do that? The answer might be found in another Scripture dealing with Peter trying to walk on water.

In Matthew 14:27, after the frightened disciples, riding the waves of a terrible storm out on the sea, mistake Jesus Christ walking toward them on the water as a ghost, He makes the following statement: "*But Jesus immediately said to them: 'Take courage! It is I. Don't be afraid.'*" It was likely that these same words gave Peter the power to step out of the boat and suspend him momentarily on the surface of the waves. What words, what power? In Greek, these three words—"It is I"—are only two words. It is the same two words translated from Hebrew in Exodus 3:13-14 in the conversation between Moses and God. Moses asks God who he should tell the Israelites said these things. God replied that he should just tell them His name is "I Am." Those two Greek words are *ego eimi.*

Scripture does not say Judas fell down, only that he was standing with the crowd Jesus Christ addressed. We might assume the disciples did not fall from the blast of power by Him just mentioning His divine nature.

As in the example of Peter and Moses, the mention of this phrase seemed to strengthen believers. This incident does create for me a moment, or maybe minutes, or hours of silence. Why? Consider what just happened.

This group was made up of some of the most notorious men in this urban setting. There were probably curious onlookers picked up as the odd mixture of Jewish and Roman soldiers marched toward their destination. They could have been described in several ways, but none of those descriptions would have included the word meek! They didn't know enough about the man to simply step back and fall to the ground when He identified Himself as who they were looking for. So, think about the reaction you would have expected from this group after He admitted who He was. Would one of those reactions have been to step back, or be blown back and then fall to the ground? Unlikely.

Now consider what you think their reaction would have been after being blown back and forced to the ground. Here is the very next thing the Scripture recounts, whether they were still on the ground, beginning to rise, or they had recovered and were standing again, "*So he asked them again, Whom do you seek? And they said, Jesus of Nazareth. Jesus answered, I told you that I am he. So, if you seek me, let these men go.*" John 18:7-8

Somewhere in this field of action there must have been a pause, confusion even. What a perfect opportunity for Jesus Christ and His disciples to flee. A hint about the soldier's reaction in the garden can be found in the scene where other soldiers are guarding the Apostle Paul and his associates on one of their missionary trips. After an earthquake opened the cell door, Paul begged the soldier in charge not to kill himself, because they were still there. Like Paul and his group, Jesus Christ, nor any of the eleven disciples, left the scene… at least not yet. Instead of fleeing, Jesus Christ asks that the eleven be allowed to leave. Think about it. If you were in the *great multitude,* what would you have done?

This brief transition to Jesus Christs' trial has one last surprise—that is, if you don't know the character of Peter, the "Rock," very well. Peter's reactions to several situations throughout the Bible lead us to expect the unexpected. We are not disappointed. Peter apparently has a sword. Why, I don't have a clue. As a fisherman by trade, he may have normally carried a very sharp knife used to filet and clean fish. Nothing else would lead me to believe he was a card-carrying sword owner. Whatever the instrument, he draws it and slices off the right ear of Malchus, the servant of the high priest, Caiaphas. Yes, that adds another probable group to the great multitude. Although not present, the Temple needed some unbiased feedback after the deed had been accomplished. Who better than a devoted servant of the high priest? It was part of his job description. Boy, did they get an ear full. Sorry, couldn't resist the temptation.

It is worth noting that all four Gospels mention the incident of cutting off the ear of the servant. Details are a little different in each passage, however, only the Book of Luke records that Jesus, after scolding the sword-bearer, reaches up and heals the servant's ear. Who else but a physician would note such a detail? This is also considered the last chronicled healing of Jesus Christ prior to His crucifixion.

Most things in the Bible are explained through an analysis of Greek words, Biblical or historical evidence, or other references in Scripture itself. Many scholars will agree that the most baffling verses in the New Testament are found as the Book of Mark closes out the arrest of Jesus Christ. Verse 50 of the 14th chapter of Mark reads, *"Then everyone deserted him and fled."* Verse 53 reads, *"They took Jesus to the high priest, and all the chief priests, the elders and the teachers of the law came together."* Now, what would you expect to be the text between verses 50 and 53? Verse 50 is obviously the end of the scene in Gethsemane and verse 53 is the introduction to the trial before the Sanhedrin. What we find in these two short interim verses is the introduction and ending of a story about a young naked man who turns into an ancient version of Ray Stevens' hit song, "The Streak."

> 51 *A young man, wearing nothing but a linen garment, was following Jesus. When they seized him,* 52 *he fled naked, leaving his garment behind.* Mark 14:51-52 NKJV

That's it. That's the whole story. Many scholars have tried to make sense of this passage. They have speculated as to who this young man might be, including the possible owner of the garden; John, the youngest disciple; Lazarus; a total stranger; and… Mark himself. Others connect the linen garment to a cloth or sheet called a *sindon*. This cloth was also used to wrap the body of the dead when they were laid in a tomb. This is an obvious attempt to introduce what is coming for Jesus Christ. The truth is, as affirmed by multiple theologians and Bible scholars, no one can explain why, of the four Gospels, only Mark records this incident.

It does give you and I a chance to use our imaginations and form our own conclusions. I have a fairly imaginative brain, so here is my shot in the dark. I recall another young, naked man, in Scripture. Unlike the young man in the Book of Mark, this young fellow had no idea he was naked. An encounter with a serpent, his young wife, and a bite of fruit grabbed his thin garment of innocence and stripped him naked. That young man, Adam, and his wife also ran away. This event was the introduction of sin into creation.

This incident in Mark could have been a signal to the same serpent, that the first Adam's sin was about to be erased. I am sure this wasn't Adam showing up in some divine re-creation of the Garden of Eden. What I am quite confident of is that Satan was there during that arrest. Not in the worldliness of the circumstances, but rather as a confused spectator. This may have been the first time in His long, defiled life Satan felt powerless and defeated. The Son of God was actually going to go through with it. Satan was speechless. Whatever air He was breathing disappeared. It was about to be Friday… but Sunday was coming!

The Book of John concludes this scene with an odd combination of soldiers binding Jesus Christ and escorting Him to the high priest's father-in-law, the former high priest, Annas. Earlier, Jesus Christ had asked the soldiers if He was the only One they were seeking, and if so, to allow the other men to go free. The Book of Matthew 26:56b ends this scene with, "Then all the disciples forsook Him and fled"… except, of course, Judas.

The Trial

Before the Sanhedrin Temple Court

> 57 *And those who had laid hold of Jesus led Him away to Caiaphas*
> *the high priest, where the scribes and the elders were assembled.* 58 *But*
> *Peter followed Him at a distance to the high priest's courtyard. And*
> *he went in and sat with the servants to see the end.*
>
> 59 *Now the chief priests, the elders, and all the council sought false*
> *testimony against Jesus to put Him to death,* 60 *but found none. Even*
> *though many false witnesses came forward, they found none. But at*
> *last two false witnesses came forward* 61 *and said, "This fellow said, 'I*
> *am able to destroy the temple of God and to build it in three days.' "*
>
> 62 *And the high priest arose and said to Him, "Do You answer noth-*
> *ing? What is it these men testify against You?"* 63 *But Jesus kept silent.*
> *And the high priest answered and said to Him, "I put You under oath*
> *by the living God: Tell us if You are the Christ, the Son of God!"* 64
> *Jesus said to him, "It is as you said. Nevertheless, I say to you, hereaf-*
> *ter you will see the Son of Man sitting at the right hand of the Power,*
> *and coming on the clouds of heaven."*
>
> 65 *Then the high priest tore his clothes, saying, "He has spoken blas-*
> *phemy! What further need do we have of witnesses? Look, now you*
> *have heard His blasphemy!* 66 *What do you think?" They answered*
> *and said, "He is deserving of death."* 67 *Then they spat in His face*
> *and beat Him; and others struck Him with the palms of their hands,*
> 68 *saying, "Prophesy to us, Christ! Who is the one who struck You?"*
> Matthew 26:57-68 NKJV

We often hear the term, *the Trial of Jesus*. When you hear that, what images does it evoke in your brain? Your brain has been saturated with every possible court scenario imaginable. To me, one of the funniest TV shows years ago was called *Night Court* with Judge Harry T. Stone played by Harry Anderson, John Larroquette playing Dan Fielding, and the unflappable court officer, Bull Shannon, portrayed by Richard Moll. To those my age, we remember *Perry Mason* played by Richard Burr always winning his case when his private investigator, Paul Drake played by William Hopper came running through the back-courtroom door with a last-minute piece of evidence. And, don't leave out Andy Griffith as *Matlock!* Today you can't watch TV for over thirty minutes without an ad by some attorney who can win you gazillions of dollars. Some of us have been fortunate enough to contribute to our community by serving jury duty as "real" jurors. All of these images are stuck in our memories… and then we hear *the Trial of Jesus*.

If I were to compare Jesus' *trial* to one of the melodramas above, the winner would have to be *Night Court*. That TV sitcom would mirror the characters Jesus Christ was left to deal with during his last few days as a living, breathing human being on Earth. Let's review some of the ancient characters, without disclosing their names, to protect their reputations.

There was a Roman City Manager whose wife had nightmares about her husband's involvement in Jewish affairs, and who became the ultimate judge by making it appear he let the people make the decision (but he had very clean hands). There were an endless number of vagrants looking for a quick buck by lying to the high court, two of which are chosen to bring false testimony. There was a half-Jewish king, enamored and jealous of his title to the point of insanity. There were a band of underpaid Jewish Temple guards who had no clue what civil liberties could possibly mean. There was a star witness who hung himself rather than testify. Finally, there was a lesser court of high-ranking Jewish aristocrats, whose chief job description was to keep their names out of the muddy streets of those they ruled but who refused to actually try the accused—this on the advice of the chief judge of that court, who loved to slap people around and manipulate the political system. Kind of makes those TV ads seem tame, doesn't it?

This was Jesus Christs' world. This is what he came to face. His life landed in the rag-tag army of the mighty Satan. It must have been effective, because those dark elements and their leader are still very much alive and working in the lives of humans to this very day. In retrospect, there really wasn't *a* trial of Jesus Christ. There were five or six trials, if they can be called that with a straight face. Let's see if we can fathom what happened?

Trial Number One

Where Jesus' first trial was held depends on which Gospel you are reading. Matthew moves from the arrest in the garden to the Sanhedrin Temple Court. John 18:13 mentions Jesus was first taken to the previous high priest, Annas, who was the father-in-law of the then-current high priest Caiaphas. Jewish history reveals Annas was a shrewd and very powerful priest when in office and for many years after. He was born the son of Seth, another Jewish high priest. Annas fathered five sons, all of whom would become high priest, and one daughter whose husband became the well-known scriptural nemesis, High Priest Caiaphas. Annas was appointed to the office by Quirinus, governor of Syria, to succeed Joazar. When he was thirty-seven, and after having held his position from the years 6-15 AD, he was deposed by Valerius Gratus, procurator of Judea. Or at least, that is what Rome was told.[154]

In the Book of Acts, chapter 4, verse 6 we find Peter and John during the beginning of their efforts to build the New Testament church. They have been arrested, and for whatever reason this passage refers to Annas as, *the* high priest followed by a list of other names of those in attendance including simply… plain old Caiaphas. It is clear in John 18:13 who is really in charge. It says after arresting Jesus Christ they "first" brought Him to Annas, and then they took Him to Caiaphas. Not a lot is said about what happened at this first meeting; sorry… trial. It can be assumed the retired high priest, lurking in the shadow, approved. Had he not, the entire ordeal would have taken an entirely different course. Poor Annas just didn't have a clue that he had no choice in the matter. Someone way above him was pulling the puppet strings. That puppet master was Satan. No footnote, just Bob's opinion.

Trial Number Two

Rome may have been building roads, viaducts, and killing lions in the Colosseum, but what they were not doing was controlling the underworld of the Jewish people. That was being accomplished with a wink and a nod from local Roman authorities to a group of wealthy religious elitists who formed the Sanhedrin Temple Court. High Priest Caiaphas presided over this body. As we will see, not everyone bought into the leader's obsession with this upstart Jesus Christ. But, majority rules and political vice triumphs. Today, we call these sins of omission… or keeping your mouth shut. Thus, began a show trial behind closed doors. This group had to keep their actions out of the newspapers and instead, gossip on the street.

In an earlier encounter that evening, Jesus Christ had warned Peter he would betray Him. Of course, Peter denied any possibility of that happening. Matthew's account gives us a back-seat view of the trip culminating in the closed meeting of the Sanhedrin. It seems not all of the disciples ran to hide, as written by this same author in Matthew 26:56b. In Matthew 26:58, just two verses later, we find Peter following the arrest crew and the hangers-on. You might not think it about Peter, but apparently, he could blend into a crowd. Seems he snuck in and blended in with the other servants. This again takes me back to visions of the TV sitcom, *Night Court.*

The scene playing out in front of Peter wasn't an accident. This was late at night. This esteemed group of court members didn't just happen to all be present when the renegade was brought into the courtyard. The group sent out to the garden to arrest Him didn't just come together. This entire operation was likely orchestrated for days or even weeks prior to this fateful night. Most telling is that a group of ill reputes suddenly line up outside the massive structure, waiting for their turn to earn money to buy a bit of food to stave off hunger by concocting a story worthy of the Court's need to condemn their captive.

It appears from Scripture the deceivers went through scores of willing liars to finally get to two who were believable. The stories both chosen accusers used included the magic word, *blasphemer!* This word alone they knew could support a death penalty. Scourging and incarceration alone would not work. They needed to silence Him… and all who supported Him. They had decided the only spectacle that could silence His supporters would be crucifixion. Their problem was simple but halting… they had no power to condemn Him to death… and… kill him. That would require a business partner; someone nobody could question. They had the perfect candidate.

After the accused undergoes tortuous questioning, pushing, slaps in the face, and demeaning gestures and slurs, Caiaphas is brought to a steaming boil because the accused will not confess. He doesn't say a word. Caiaphas uses his last hope, placing Him under oath. I am sorry, but I would find this humorous if it wasn't so sad. Caiaphas is placing Jesus Christ under oath to His own Father! This is the epitome of a man so full of himself he cannot see his feet melting into the Earth where he stands.

Scripture doesn't indicate here, but I see a very dramatic pause. I feel the air being drawn out of the room through the open mouths of everyone else gasping at what just occurred. There is an exclamation mark at the end of Caiaphas's apparent shout. I imagine him slowly turning away from his victim with a blank stare waiting on a sound he is confident will follow his show of authority. Just when he thinks no sound will follow there is a hushed, low, even calm voice from behind him. The words, while cutting him in half, are exactly what he wants to hear, *"It is as you said. Nevertheless, I say to you, hereafter you will see the Son of Man sitting at the right hand of the Power, and coming on the clouds of heaven."*

It is not clear whose clothes are torn at this point, but following a reflex and only hearing the first part of Jesus Christs' reply, Caiaphas turns to the court and screams, *"He has spoken blasphemy! What further need do we have of witnesses? Look, now you have heard His blasphemy! What do you think?"* As he asks the court for a verdict, they know they cannot carry out, Peter must have been in the back of the room in shock. For a large boisterous man of the sea this was one of the few times Peter was speechless. Unlike pulling a knife and cutting off the ear of a servant in defense of his teacher, Peter stands motionless, trembling, not for himself, but because he is beginning to understand some of the unusual things he heard his teacher say through the years about His going away. What he had been denying for years he must now accept. There is nothing he can do but stay close. He will not deny his master!

The night doesn't end well for Jesus Christ or for the one who betrayed Him. We can assume Judas was also present during this second trial before the Sanhedrin. He was the hero of the Court. They probably brought him back with them from the garden. He may have even been introduced to the legend, Annas. If he was present during Caiaphas' antics, he, more than anyone else in the room, knew how perverted the scene became.

Judas probably doubted his actions from the beginning. Maybe he was driven by a force he didn't understand. He was on the side of finding a warrior savior who would free Israel from their oppression. More than a blasphemer, he probably thought Jesus Christ to be a failure. Maybe in his disappointment he felt a need to rid Israel of this imitator so the real savior could arise. In that courtyard, watching the scene play out in front of him, Judas, like Peter, finally got it.

Judas During Court Recess

He had probably never seen his teacher under this kind of duress and suffering. Watching Jesus Christ's reactions and response to endless prodding by wealthy, religious Jews allowed his eyes to see Jesus Christ for who He really was, God's Almighty Son. With that realization, Judas felt hopeless. All he could do was to separate himself from everything he had done to bring God's Son to this place. The pieces of silver. The money, it suddenly felt like a heavy oxen's yoke around his waist. He must get rid of it. It was tainted, he couldn't just throw it away or give it to a stranger. He knew what he had to do. This was a small group, and Peter likely was watching Judas' movements and demeaner. He would have seen Judas drop his head and make his way to a secluded spot waiting for the crowd to leave and for the Temple guard to remove the prisoner to a dark filthy room for safekeeping till morning.

From his lonely place in the shadows, Judas waited for an opportune time. Probably as Caiaphas and a few others were getting ready to retire to be fresh for morning planned activities: a very public parade to the governor's palace with the blasphemer in tow. As the troop of Court Officials came near where Judas was standing, he jumped in front of them and immediately began to plead with them to take back the money. This is where he must return it, as if doing so will somehow absolve him from his deed.

He confesses his sin and proclaims Jesus Christ as innocent; a certain mistrial. His teacher has not yet been convicted, sentenced, or killed. Maybe he has acted in time to turn this around. He has confessed. Judas doesn't realize he has been part of a very large plan carried out for weeks or months during its development. He doesn't realize spies probably followed him for days, listening to him talk in small groups, with other disciples, and nonbelievers about the problems with this self-proclaimed Messiah. The leaders of the Sanhedrin had long ago selected Judas, an insider, to be the key to their entire plot to rid themselves of this threat to their livelihood.

Judas' metal health at that moment must have been at its lowest in his entire lifetime. He could do only one thing. Ripping the bag of silver from his waist belt, he threw it to the floor at the feet of his users… and ran. He ran until he was breathless. At that moment, only one thing mattered to him. He must find a piece of rope, a sheet, a belt, anything long enough to tie to a tree limb and hold firm around his neck. Scripture says it succinctly in Matthew 27:5: *"So Judas threw the money into the temple and left. Then he went away and hanged himself."*

No one knows what went through Judas' mind at the end. It seems he was very alone. His deed was pivotal to allow Jesus Christ to die for our sins. Could someone else have made it happen if, in his free will, he chose not to go through with his betrayal? Yes. But he did go through with it. He confessed his deeds to the high priest, who at that time was still eligible to receive such a thing. With the act of suicide, which is murdering a person, could he have received forgiveness for his sins? Who am I to say? Who am I to judge? Yet, as I write this, I have to ask myself… what would Jesus have done? Deep inside me, I can't imagine Jesus Christ doing anything else other than forgiving a repentant sinner.

Trial Number Three

Before Pilate

> 1 *When morning came, all the chief priests and elders of the people plotted against Jesus to put Him to death.* 2 *And when they had bound Him, they led Him away and delivered Him to Pontius Pilate the governor.* Matthew 27:1-2 NKJV
>
> 11 *Now Jesus stood before the governor. And the governor asked Him, saying, "Are You the King of the Jews?" Jesus said to him, "It is as you say."* 12 *And while He was being accused by the chief priests and*

elders, He answered nothing. [13] *Then Pilate said to Him, "Do You not hear how many things they testify against You?"* [14] *But He answered him not one word, so that the governor marveled greatly.*

[15] *Now at the feast the governor was accustomed to releasing to the multitude one prisoner whom they wished.* [16] *And at that time they had a notorious prisoner called Barabbas.* [17] *Therefore, when they had gathered together, Pilate said to them, "Whom do you want me to release to you? Barabbas, or Jesus who is called Christ?"* [18] *For he knew that they had handed Him over because of envy.*

[19] *While he was sitting on the judgment seat, his wife sent to him, saying, "Have nothing to do with that just Man, for I have suffered many things today in a dream because of Him."*

[20] *But the chief priests and elders persuaded the multitudes that they should ask for Barabbas and destroy Jesus.* [21] *The governor answered and said to them, "Which of the two do you want me to release to you?" They said, "Barabbas!"* [22] *Pilate said to them, "What then shall I do with Jesus who is called Christ?" They all said to him, "Let Him be crucified!"* [23] *Then the governor said, "Why, what evil has He done?" But they cried out all the more, saying, "Let Him be crucified!"*

[24] *When Pilate saw that he could not prevail at all, but rather that a tumult was rising, he took water and washed his hands before the multitude, saying, "I am innocent of the blood of this just Person. You see to it."* [25] *And all the people answered and said, "His blood be on us and on our children."* [26] *Then he released Barabbas to them; and when he had scourged Jesus, he delivered Him to be crucified.*
Matthew 27: 11-26 NKJV

The next day, after the accused was mocked and slapped around by the elite of the Sanhedrin Court, they met again to carry out their complex plan to murder a fellow Jew from the town of Nazareth. Although murder by them, where they were headed, would just be another judgment by their Roman conquerors, in particular a governor by the name of Pontius Pilate. Until now Pilate had performed well in keeping peace in this city… until this day.

Pilate would have been in his private quarters, since it was early in the morning. We can assume he met the Jewish leaders in an outer court, likely on the second level above the street since later we learn his address was to a crowd which would have never been allowed inside the residence. Nor would they or the priests have wanted to because it would have defiled the Jews and prevented them from participating in the Passover Feast Celebration. Some have referred to Pilate's location as a balcony overlooking the city. In this case, the Jewish leaders and their captive would have been on street level looking up at the governor.

The scene, wherever located, would have consisted of Pilate and his servants, the crowd of Jewish leaders and a few curiosity-seekers, and in between them all, a single dirty individual, bruised and bleeding. When Pilate was first informed that some of the elite Jews of the community requested his audience, he was told that they were accompanied by a blasphemer claiming to be the King of the Jews. After quieting the taunting crowds, and with no response from the accused, Pilate addressed the disheveled young man standing by himself, *"Are You the King of the Jews?"*

To which Jesus Christ did answer… kind of, *"It is as you say."* Matthew 27:11 NKJV

Again, the governor asks Jesus Christ if He doesn't hear the accusations coming from the crowd, to which He says nothing. Looking on this spectacle and observing the bravery and fortitude of the accused, the governor is actually impressed. Matthew uses the term "*…governor marveled greatly.*"

There is a missing trial in Matthew that is only found in the book of Luke, chapter 23. Shortly after Jesus Christ is brought before Pilate, and after hearing all of the accusations, Pilate says he finds no fault in the man! This raises the intensity of the efforts to prove the accused should be dealt with, and eventually brought shouts of crucify! Through the distorted, overlapping shouts, Pilate heard a phrase that tickled his ear. He paused and listened more carefully… there it was again… Galilee. This man was from Galilee!

The governor, doing what he did best to keep his job, discovered a way to pass the buck. Phase one of the Pilate trial was about to come to an end. Unfortunately for Pilate, they would all be back. But for now, the task has been transferred to a Jewish King who happened to be in town that day—and who incidentally was ruler over… Galilee. With a whisk of his hand, the governor sent the entire motley crew away. His day had been saved. Well, the interlude wasn't without drama. Pilate still had to deal with his daunting wife and her silly dreams.

Trial Number Four

Off they went, Jesus Christ in tow, falling and stumbling after very little sleep, occasional cracks of a whip, shouts to move along, and probably nothing to eat. It would have been a very public parade right through crowds of Jews who had heard, followed, and maybe even been healed or raised from the dead by this man who now looked nothing like the peaceful man they had come to love. None, however, could have come to His rescue without being in danger of losing their own life. The now mass of people were headed straight for the temporary living quarters of King Herod the Tetrarch, who was visiting Jerusalem.

There was no doubt Herod would see the Jewish leaders without an appointment as soon as he heard who was with them. It turns out that Herod had two motives for wanting to see this man from Nazareth. First, he wanted to witness one of His miraculous miracles for himself. He had seen some of the best magic shows in town, but this man, WOW! Second, and supremely, Herod wanted an opportunity to tell this imitator that he, Herod the Great, was the KING OF THE JEWS!! The meeting was probably brief, but long enough for Herod to have his way with this so-called Jewish leader.

Herod treated Jesus Christ with contempt and mocked Him. His soldiers abused Him and found a gorgeous robe and placed it on His shoulders to signify he was, indeed, a king. With this, they sent them all away. Scripture tells us that Pilate and Herod previously had a questionable relationship, but after this event, they would become very good friends. Herod was insane and probably was the instigator of the bad relationship. Pilate, recognizing that Herod did have a certain amount of authority, allowed him to relish in the glow of a Roman governor's accolade. Still, like the Jewish Court elite, Herod had to admit he also didn't have the authority to finish the job. All he could do was give his approval for whatever the governor decided. Time for the trip back across town.

As the mob made its way back to Pilate's residence, the leaders picked out some hecklers from the crowd who were ordered to inject themselves into the crowd gathered near Pilate's place. They were to begin shouts of "Crucify him, crucify him!" They would be joined by the priests, who would shout the same message. Pilate's day was about to turn on him.

Trial Number Five

Scripture doesn't give a strict timeline concerning what Pilate's agenda was to be if the mob were to return. Since he was the governor of this town, he had to be very aware of Jewish holidays, celebrations, and customs. One such custom could save him face and remove the need for him to make the decision of death.

These trials took place during the Jewish Feast of the Passover. Pilate had established a tradition during this annual festival of placing two prisoners before the public, allowing them to decide which one would go free. Perfect. He chose Barabbas, a disrupter and murderer as one and Jesus Christ as the other. After his first attempt to release Jesus Christ, because he found no fault in him, Pilate innocently brought out the murderer and placed him alongside Jesus Christ. He stated the purpose of the tradition and then gave the decision to the masses. The Jewish leaders and their hired screamers began their loud chants to crucify Jesus, drowning out the softer cry of Barabbas' victims that he should receive the death sentence.

There are two incidents covered in two separate Gospels that are either not mentioned in any of the others or are not clear. The first one is in John 18:33-38. It seems Pilate called for Jesus Christ to be brought into his headquarters, without the accusers. There Pilate has a rather intimate conversation with the accused. It seems from these verses that Pilate is genuinely interested in this man's story. In fact, it is recorded, after a very short conversation, Pilate replies with a question: *"What is truth?"* John 18:38 NKJV

Following this conversation, Pilate returns to the mob and proclaims he finds no fault in the man.

Somewhere in this timeline, Pilate has a second distraction. We are told in Matthew 27:19 about a message Pilate received from his wife: *"Don't have anything to do with that innocent man, for I have suffered a great deal today in a dream because of him."* This, after King Herod's refusal to handle the situation, the failed offer to release Jesus Christ instead of a murderer, and three attempts to shout down the mob, Pilate decides to relent. But as a final attempt to clear his name with whoever may look back on this day and accuse him of any misdeed, he calls for a pan of clean water and a towel. As he dips his hands in the cool water he begins to speak loudly as he turns and splashes water from the bowl and lifts his hands heavenward, *"I am innocent of this man's blood," he said. "It is your responsibility!"* Matthew 27:24 NKJV.

The Verdict

With that, Jesus Christ is seized and taken to be prepared for death. There is a word used here in connection with Jesus Christ's crucifixion that rarely gets the emphasis it deserves. Jesus' death may have occurred on the cross, but it didn't begin there. The often-underemphasized word is *flogged.* He is hours away from breathing His last breath as a divine human. He will exhale his last breath on a Roman cross in the presence of His mother after breathing His first in her arms, in a manger in Bethlehem.

CHAPTER 15

THE EXCLAMATION MARK ON THE TIMELINE OF ETERNITY

... the cross and the grave are inseparable
... more than any other time in the history
of the world, God would have had the
motive to end the Creation story.
... all other religious leaders who are
used as the basis and beginning of their
movement are either dead or alive
... no other religious leader recognized
as the basis for their movement ever
claimed to be the Son of God

... There is salvation in no one else; for there is no other name under heaven that has been given among men by which we must be saved.

ACTS 4:12

AS WE CONTINUE to look at The Seven Miraculous Acts of the Holy Trinity Within the Carnal Soul of a Man, these last few are at the core of the faith community's belief. It should be noted that, although one of the other largest faiths in the world, Islam, recognizes Jesus Christ as a real figure Who performed miracles, it does not believe He was crucified, and therefore had no reason to come back to life. Instead, Islam teaches Jesus Christ was caught up into heaven by God and never died. These two acts, dying on the cross and coming back to life after being buried, distinguish Christianity from literally everything else on our timeline. Since much of this Christian belief is based on simply faith, why is this so dramatically important? Science and logic.

Other faiths, interestingly, require... less faith. Notice I didn't say commitment and I didn't say their followers were insincere. Believing the crucifixion, coupled with the resurrection, takes more than faith. It requires a dismissal of logic and the laws of natural science. The faith community accuses science of not being able to create life from nothing and the science community accuses the faith community of suggesting that someone can die, lie in a grave for over thirty hours, and then raise Himself up; back to life. In any logic-based discussion, both have a valid point. At the end of such a debate the fact remains that the faith community is willing to *believe* the death and resurrection story is real while the science community must have evidence. Not only evidence, but the act has to be replicable. I would ask, replicable by whom?

As each community lives their story, both continue to build evidence. Science continues to be unable to create or reproduce what the faith community claims while the faith community, living in faith, build evidence within their personal relationship with the One they proclaim to be the risen Lord. Oddly enough, the evidence obtained by the faith community is replicable, dependent on living the life modeled by the One who rose from the dead.

As we approach the sunset of the Creation Season of Spring, let's take a look at these final elements of *The Seven Miraculous Acts of the Holy Trinity, Within the Carnal Soul of a Man.*

The Crucifixion (Scripture from all four gospels)

27 *Then the soldiers of the governor took Jesus into the Praetorium and*
gathered the whole garrison around Him. 28 *And they stripped Him*
and put a scarlet robe on Him. 29 *When they had twisted a crown of*
thorns, they put it on His head, and a reed in His right hand. And
they bowed the knee before Him and mocked Him, saying, "Hail,
King of the Jews!" 30 *Then they spat on Him, and took the reed and*
struck Him on the head. 31 *And when they had mocked Him, they*
took the robe off Him, put His own clothes on Him, and led Him
away to be crucified.

32 *Now as they came out, they found a man of Cyrene, Simon by*
name. Him they compelled to bear His cross. 33 *And when they had*
come to a place called Golgotha, that is to say, Place of a Skull, 34
they gave Him sour wine mingled with gall to drink. But when He
had tasted it, He would not drink. 35 *Then they crucified Him, and*
divided His garments, casting lots, that it might be fulfilled which
was spoken by the prophet: "They divided My garments among them,
And for My clothing they cast lots." 36 *Sitting down, they kept watch*
over Him there. 37 *And they put up over His head the accusation*
written against Him: THIS IS JESUS THE KING OF THE JEWS.

38 *Then two robbers were crucified with Him, one on the right and*
another on the left. 39 *And those who passed by blasphemed Him,*
wagging their heads 40 *and saying, "You who destroy the temple and*
build it in three days, save Yourself! If You are the Son of God, come
down from the cross." 41 *Likewise the chief priests also, mocking with*
the scribes and elders, said, 42 *"He saved others; Himself He cannot*
save. If He is the King of Israel, let Him now come down from the
cross, and we will believe Him. 43 *He trusted in God; let Him deliver*
Him now if He will have Him; for He said, 'I am the Son of God.' "
44 *Even the robbers who were crucified with Him reviled Him with*
the same thing.

45 *Now from the sixth hour until the ninth hour there was darkness*

over all the land. [46] *And about the ninth hour Jesus cried out with a loud voice, saying, "Eli, Eli, lama sabachthani?" that is, "My God, My God, why have You forsaken Me?"* [47] *Some of those who stood there, when they heard that, said, "This Man is calling for Elijah!"* [48] *Immediately one of them ran and took a sponge, filled it with sour wine and put it on a reed, and offered it to Him to drink.* [49] *The rest said, "Let Him alone; let us see if Elijah will come to save Him."* [50] *And Jesus cried out again with a loud voice, and yielded up His spirit.*

[51] *Then, behold, the veil of the temple was torn in two from top to bottom; and the earth quaked, and the rocks were split,* [52] *and the graves were opened; and many bodies of the saints who had fallen asleep were raised;* [53] *and coming out of the graves after His resurrection, they went into the holy city and appeared to many.* [54] *So when the centurion and those with him, who were guarding Jesus, saw the earthquake and the things that had happened, they feared greatly, saying, "Truly this was the Son of God!"* [55] *And many women who followed Jesus from Galilee, ministering to Him, were there looking on from afar,* [56] *among whom were Mary Magdalene, Mary the mother of James and Joses, and the mother of Zebedee's sons.* Matthew 27:27-56 NKJV

[16] *Then the soldiers led Him away into the hall called Praetorium, and they called together the whole garrison.* [17] *And they clothed Him with purple; and they twisted a crown of thorns, put it on His head,* [18] *and began to salute Him, "Hail, King of the Jews!"* [19] *Then they struck Him on the head with a reed and spat on Him; and bowing the knee, they worshiped Him.* [20] *And when they had mocked Him, they took the purple off Him, put His own clothes on Him, and led Him out to crucify Him.*

[21] *Then they compelled a certain man, Simon a Cyrenian, the father of Alexander and Rufus, as he was coming out of the country and passing by, to bear His cross.* [22] *And they brought Him to the place Golgotha, which is translated, Place of a Skull.* [23] *Then they gave*

Him wine mingled with myrrh to drink, but He did not take it. 24
And when they crucified Him, they divided His garments, casting lots
for them to determine what every man should take.

25 Now it was the third hour, and they crucified Him. 26 And the
inscription of His accusation was written above: THE KING OF
THE JEWS.

27 With Him they also crucified two robbers, one on His right and the
other on His left. 28 So the Scripture was fulfilled which says, "And
He was numbered with the transgressors." 29 And those who passed
by blasphemed Him, wagging their heads and saying, "Aha! You who
destroy the temple and build it in three days, 30 save Yourself, and
come down from the cross!"

31 Likewise the chief priests also, mocking among themselves with the
scribes, said, "He saved others; Himself He cannot save. 32 Let the
Christ, the King of Israel, descend now from the cross, that we may see
and believe." Even those who were crucified with Him reviled Him.

33 Now when the sixth hour had come, there was darkness over the
whole land until the ninth hour. 34 And at the ninth hour Jesus cried
out with a loud voice, saying, "Eloi, Eloi, lama sabachthani?" which
is translated, "My God, My God, why have You forsaken Me?" 35
Some of those who stood by, when they heard that, said, "Look, He
is calling for Elijah!" 36 Then someone ran and filled a sponge full of
sour wine, put it on a reed, and offered it to Him to drink, saying,
"Let Him alone; let us see if Elijah will come to take Him down." 37
And Jesus cried out with a loud voice, and breathed His last.

38 Then the veil of the temple was torn in two from top to bottom. 39
So when the centurion, who stood opposite Him, saw that He cried
out like this and breathed His last, he said, "Truly this Man was the
Son of God!"

40 There were also women looking on from afar, among whom were
Mary Magdalene, Mary the mother of James the Less and of Joses,
and Salome, 41 who also followed Him and ministered to Him when

*He was in Galilee, and many other women who came up with Him
to Jerusalem.* Mark 15: 16-41 NKJV

23 *But they were insistent, demanding with loud voices that He be
crucified. And the voices of these men and of the chief priests pre-
vailed.* 24 *So Pilate gave sentence that it should be as they requested.*
25 *And he released to them the one they requested, who for rebellion
and murder had been thrown into prison; but he delivered Jesus to
their will.*

26 *Now as they led Him away, they laid hold of a certain man, Simon
a Cyrenian, who was coming from the country, and on him they laid
the cross that he might bear it after Jesus.*

27 *And a great multitude of the people followed Him, and women
who also mourned and lamented Him.* 28 *But Jesus, turning to them,
said, "Daughters of Jerusalem, do not weep for Me, but weep for your-
selves and for your children.* 29 *For indeed the days are coming in
which they will say, 'Blessed are the barren, wombs that never bore,
and breasts which never nursed!'* 30 *Then they will begin to say to the
mountains, "Fall on us!" and to the hills, "Cover us!" '* 31 *For if they
do these things in the green wood, what will be done in the dry?"*

32 *There were also two others, criminals, led with Him to be put to
death.* 33 *And when they had come to the place called Calvary, there
they crucified Him, and the criminals, one on the right hand and the
other on the left.* 34 *Then Jesus said, "Father, forgive them, for they do
not know what they do." And they divided His garments and cast lots.*

35 *And the people stood looking on. But even the rulers with them sneered,
saying, "He saved others; let Him save Himself if He is the Christ, the
chosen of God."* 36 *The soldiers also mocked Him, coming and offering
Him sour wine,* 37 *and saying, "If You are the King of the Jews, save
Yourself."* 38 *And an inscription also was written over Him in letters
of Greek, Latin, and Hebrew: THIS IS THE KING OF THE JEWS.*

39 *Then one of the criminals who were hanged blasphemed Him,*

saying, "If You are the Christ, save Yourself and us." [40] *But the other, answering, rebuked him, saying, "Do you not even fear God, seeing you are under the same condemnation?* [41] *And we indeed justly, for we receive the due reward of our deeds; but this Man has done nothing wrong."* [42] *Then he said to Jesus, "Lord, remember me when You come into Your kingdom."* [43] *And Jesus said to him, "Assuredly, I say to you, today you will be with Me in Paradise."*

[44] *Now it was about the sixth hour, and there was darkness over all the earth until the ninth hour.* [45] *Then the sun was darkened, and the veil of the temple was torn in two.* [46] *And when Jesus had cried out with a loud voice, He said, "Father, 'into Your hands I commit My spirit.' " Having said this, He breathed His last.*

[47] *So when the centurion saw what had happened, he glorified God, saying, "Certainly this was a righteous Man!"* [48] *And the whole crowd who came together to that sight, seeing what had been done, beat their breasts and returned.* [49] *But all His acquaintances, and the women who followed Him from Galilee, stood at a distance, watching these things.* Luke 23:23-49 NKJV

[15] *But they cried out, "Away with Him, away with Him! Crucify Him!" Pilate said to them, "Shall I crucify your King?" The chief priests answered, "We have no king but Caesar!"* [16] *Then he delivered Him to them to be crucified. Then they took Jesus and led Him away.*

[17] *And He, bearing His cross, went out to a place called the Place of a Skull, which is called in Hebrew, Golgotha,* [18] *where they crucified Him, and two others with Him, one on either side, and Jesus in the center.* [19] *Now Pilate wrote a title and put it on the cross. And the writing was: JESUS OF NAZARETH, THE KING OF THE JEWS.* [20] *Then many of the Jews read this title, for the place where Jesus was crucified was near the city; and it was written in Hebrew, Greek, and Latin.* [21] *Therefore the chief priests of the Jews said to Pilate, "Do not write, 'The King of the Jews,' but, 'He said, "I am*

the King of the Jews." ' " 22 *Pilate answered, "What I have written, I have written."*

23 *Then the soldiers, when they had crucified Jesus, took His garments and made four parts, to each soldier a part, and also the tunic. Now the tunic was without seam, woven from the top in one piece.* 24 *They said therefore among themselves, "Let us not tear it, but cast lots for it, whose it shall be," that the Scripture might be fulfilled which says: "They divided My garments among them, And for My clothing they cast lots." Therefore the soldiers did these things.*

25 *Now there stood by the cross of Jesus His mother, and His mother's sister, Mary the wife of Clopas, and Mary Magdalene.* 26 *When Jesus therefore saw His mother, and the disciple whom He loved standing by, He said to His mother, "Woman, behold your son!"* 27 *Then He said to the disciple, "Behold your mother!" And from that hour that disciple took her to his own home.*

28 *After this, Jesus, knowing that all things were now accomplished, that the Scripture might be fulfilled, said, "I thirst!"* 29 *Now a vessel full of sour wine was sitting there; and they filled a sponge with sour wine, put it on hyssop, and put it to His mouth.* 30 *So when Jesus had received the sour wine, He said, "It is finished!" And bowing His head, He gave up His spirit.*

31 *Therefore, because it was the Preparation Day, that the bodies should not remain on the cross on the Sabbath (for that Sabbath was a high day), the Jews asked Pilate that their legs might be broken, and that they might be taken away.* 32 *Then the soldiers came and broke the legs of the first and of the other who was crucified with Him.* 33 *But when they came to Jesus and saw that He was already dead, they did not break His legs.* 34 *But one of the soldiers pierced His side with a spear, and immediately blood and water came out.* 35 *And he who has seen has testified, and his testimony is true; and he knows that he is telling the truth, so that you may believe.* 36 *For these things were done that the Scripture should be fulfilled, "Not one of His bones shall*

be broken." [37] *And again another Scripture says, "They shall look on Him whom they pierced."* John 19:15-37 NKJV

While picking this story back up from the trials and the verdict in the previous chapter, the timing isn't clear between Matthew 27:26 and verse 27. Nevertheless, these verses cover an unbroken time span from Thursday evening to Sunday morning. What happened at this point probably had more to do with mundane Roman Soldier work schedules than God's plan for His Son.

Earlier, I mentioned that Jesus Christ would die on the cross, but the end result we remember overlooks a very important path leading to His last breath. That of *flogging* or *scourging.* Although Scripture doesn't actually say he received forty lashes less one, many believe that is what happened, according to tradition at the time of His death. It should be noted that forty lashes less one was a Jewish tradition, not Roman. [155] He could have received twenty-six or fifty-three, or some other number. All we are sure of is that He was flogged or scourged. It is also not clear if this occurred prior to being taken to the Praetorium or when He arrived. Again, where it occurred is not nearly as important as that it happened, and what that would do to a human being.

The Medical Evidence

In an article posted in Chrisitianity.com written by Candice Lucey titled, "What is the Medical Account of the Crucifixion," we find this very graphic depiction of this method of punishment.

> *"Jesus' back was not marked with red welts; the skin was not merely bleeding. His flesh was torn. Sometimes, a victim's back was "so shredded that part of the spine was [...] exposed by the deep, deep cuts."*
>
> *Jesus was struck with a whip made of "braided leather thongs with metal balls woven into them" plus "pieces of sharp bone." "The sufferer's veins were laid bare and the very muscles, sinews and bowels of the victim were open to exposure." While the objective was to bring a person as close to death as possible without actually killing him,*

> *"many people would die from this kind of beating even before they could be crucified."*[156]

The Importance of Robes

Although it may seem trivial, there is the robe issue; how many and what color. From the beginning of the trials to His departure for the actual crucifixion site there could have been two or three different robes. Two gospels use different colors; one scarlet and the other purple. The first robe, however, would have been white, placed on Him by the Jewish guards at King Herod's residence. This robe is called "gorgeous" in Scripture which is interpreted to be white, representing purity. There is no mention of removing this garment before He is taken back to Pilate's residence.

There is mention of removing His clothes after He is returned to Pilate, which would have included the removal of the white robe, and instead placed a scarlet robe on His body. If it was scarlet, it would have been the color of a soldier preparing for battle. Later, there is an indication a purple robe, signifying royalty, being placed on Him along with a vine of thorns twisted into a crown and smashed onto His head. Also, a reed was placed into His hand. For some period of time, those present mocked Him and knocked him around.

Since the Greek words for scarlet and purple are used correctly in these two different Gospels, referring to where the color pigments originated, we must assume there were two robes, or garments of some type. One of the reasons the first scarlet robe was used could have been the bent of the Roman soldiers for inflicting maximum pain on their subjects. Remember, this incident occurred after Jesus Christ had been scourged. As in the description above, Jesus Christ would have been bleeding profusely from the back of His upper legs, shoulders, neck and back; the very places touched by such a garment. The garment could have also promoted clotting around the wounds, slowing His loss of blood.

While they may have simply slipped the purple garment over the scarlet one, they could have also ripped the first robe off, tearing away the newly formed scabs reopening many of His wounds. These openings in His flesh, along with the blood flowing down from the top of His head created by the thorny crown, would have been enough blood loss for most prisoners to collapse. Then, just before leaving for the site of the crucifixion, the soldiers removed the purple robe and placed His clothes back on Him. This would, again, have reopened most of the wounds that had begun to clot, making His ability to reach the place of execution on His own strength unlikely. In the end, these savagely tough, war-trained soldiers no doubt would have come to admire this Jew's stamina—setting aside, of course, His lunatic ideology. One soldier in particular, later at the cross, will fall before Him and proclaim that surely this man must be the Son of God.

From here to the cross takes up little space in religious text. Some of the events have been canonized, such as the route to the cross. This roughly 2,000-foot modern-day stretch replaces previous paths established prior to the 18th century when the Via Dolorosa, called *Sorrowful Way* in Latin and often translated *Way of Suffering,* was suggested. Today, this path contains the Seven Stations of the Cross observed by many believers around the world.[157] As with several relics enshrined around the Holy Land, there is no evidence this was the actual path to the cross.

As they left for the place of crucifixion, it is important to remember that Jesus Christ is very close to death. Let's review His physical condition just prior to departure. I was struck by this description found in an AZUSA Pacific University piece called, "The Science of the Crucifixion," by Cahleen Shrier, PhD, associate professor, Department of Biology and Chemistry: [158]

- He had a very light meal on Thursday evening
- As earlier described, back in another chapter, in the garden, when He was praying, He must have suffered from a condition called *hematohidrosis,* during which the capillary blood vessels that feed the sweat glands break down. Blood released from the vessels mixes with the sweat; therefore, the body sweats drops of blood. This condition results from mental anguish or high anxiety, a state Jesus expresses by praying "my soul is deeply grieved to the point of death" Matthew 26:38. *Hematohidrosis* makes the skin tender, so Jesus' physical condition worsens slightly.

- Before continuing His trials, He probably had very little, if anything, to eat while being processed.
- Traveling from Pilate to Herod and back again, Jesus walks approximately two and a half miles.
- Pilate orders Jesus to be flogged, as required by Roman law before crucifixion. Traditionally, the accused stood naked, and the flogging covered the area from the shoulders down to the upper legs. The whip consisted of several strips of leather. In the middle of the strips were metal balls that hit the skin, causing deep bruising. In addition, sheep bone was attached to the tips of each strip. When the bone makes contact with Jesus' skin, it digs into His muscles, tearing out chunks of flesh and exposing the bone beneath. The flogging leaves the skin on Jesus' back in long ribbons. By this point, He has lost a great volume of blood which causes His blood pressure to fall and makes Him prone to shock. The human body attempts to remedy imbalances such as decreased blood volume, so Jesus' thirst is His body's natural response to His suffering (John 19:28). If He would have drunk water, His blood volume would have increased.
- Roman soldiers place a crown of thorns on Jesus' head and a robe on His back (Matthew 27:28-29). The robe helps the blood clot (similar to putting a piece of tissue on a cut from shaving) to prevent Jesus from sustaining more blood loss. As they hit Jesus in the head (Matthew 27:30), the thorns from the crown push into the skin and He begins bleeding profusely. The thorns also cause damage to the nerve that supplies the face, causing intense pain down His face and neck. As they mock Him, the soldiers also belittle Jesus by spitting on Him (Matthew 27:30). They rip the robe off Jesus' back and the bleeding starts afresh.
- Jesus' physical condition becomes critical. Due to severe blood loss without replacement, Jesus is undoubtedly now in shock. As such, He is unable to carry the cross and Simon of Cyrene executes this task (Matthew 27:32).

The March to the Cross

People never knew when they were going to be suddenly thrown onto the timeline of Jesus Christ's life and ministry. Such was the case that day as the soldiers marshalled Him out onto the streets. It would not have been an unusual sight; being frightened by Roman soldiers dragging a half-dead person out of the door and into the crowd. The soldiers would expect the spectators to show approval and cheer. There were probably disciples of both Jesus Christ and John the Baptist in that crowd. It would have been interesting to see their reaction. Those standing about knew exactly where and how this parade would end. Those, however, who followed the march to the cross would witness something amazing and awe-inspiring. But to get there, the soldiers needed a volunteer. Well maybe an, "I'm right here and I will be arrested if I don't" kind of volunteer. And that is how Simon of Cyrene's day will unfold.

In Matthew 27:32 it simply says the soldiers compelled him to carry Jesus Christ's cross. In other words, he was *voluntold.* There was no choice. Simon, a traveler passing through town, depending on how you view it, was in the wrong, or, right place at the wrong or right time. The first three Gospels all mention this incident with the Gospel of John saying Jesus Christ carried the cross to Golgotha. The accounts from Mark and Luke give us three other pieces of information. Mark tells us Simon was the father of Alexander and Rufus, and he was coming out of the country and passing through Jerusalem. Luke adds, Simon was walking behind Jesus Christ in the makeshift parade after having the cross placed on his shoulders. Although, by now it is obvious why they need someone to carry His cross, and not just because it wasn't in the soldier's job description. The prisoner was almost dead, disoriented, and was probably having trouble breathing. What is not so obvious in Scripture is just what cross was being carried. Tradition, driven by Hollywood movies, usually drapes an entire, traditional-looking, plus-shaped cross with an elongated bottom resting on His shoulder. Let's look at the real history.

Crucifixion was invented by the Persians between 300 and 400 BC. Interestingly, this was at the beginning of the silent period between the Old and New Testament writings. It is quite possibly the most painful death ever invented by humankind. The English language derives the word "excruciating" from crucifixion, acknowledging it as a form of slow, painful suffering.

Jesus' cross was probably not the Latin cross, but rather a Tau cross. The vertical piece, called the *stipes*, remains in the ground permanently. The accused carries only the horizontal piece, the *patibulum*, up the hill. Atop the patibulum hangs a sign referred to as the *titulus*, indicating that a formal trial occurred for a violation of the law. In Jesus' case, this reads "This is the King of the Jews." Luke 23:38

The accused needed to be nailed to the patibulum while lying down, so Jesus is thrown to the ground, reopening His wounds, grinding in dirt, and causing more bleeding. They nail His "hands" to the patibulum. The Greek meaning of "hands" includes the wrist. It is more likely that the nails went through Jesus' wrists. If the nails were driven into the hand, the weight of the arms would cause the nail to rip through the soft flesh.

The upper body would not be held to the cross. If placed in the wrist, the bones in the lower portion of the hand support the weight of the arms and the body remains suspended on the cross. The huge nail, seven to nine inches long, damages or severs the major nerve to the hand upon impact. This causes continuous, agonizing pain up both of Jesus' arms.

Once the victim is secured, the guards lift the patibulum and place it on the stipes already in the ground. As it is lifted, Jesus' full weight pulls down on His nailed wrists and His shoulders and elbows dislocate. In this position, Jesus' arms stretch to a minimum of six inches longer than their original length.

It is highly likely that Jesus' feet were nailed through the tops, as often pictured. In this position (with the knees flexed at approximately 90 degrees), the weight of the body pushes down on the nails and the ankles support the weight. The nails would not rip through the soft tissue as would have occurred with the hands. Again, the nail would cause severe nerve damage by severing the dorsal pedal artery of the foot.

Normally, to breathe in, the diaphragm must move down. This enlarges the chest cavity and air automatically moves into the lungs. To exhale, the diaphragm rises up, which compresses the air in the lungs and forces the air out. As Jesus hangs on the cross, the weight of His body pulls down on the diaphragm and the air moves into His lungs and remains there. Jesus must push up on His nailed feet to exhale.

In order to speak, air must pass over the vocal cords during exhalation. The Gospels note that Jesus spoke seven times from the cross. It is amazing that, despite His pain, He must push up to say "Forgive them." Luke 23:34

The difficulty surrounding exhalation leads to a slow form of suffocation. Carbon dioxide builds up in the blood, resulting in a high level of carbonic acid in the system. The body responds instinctively, triggering the desire to breathe. At the same time, the heart beats faster to circulate available oxygen. The decreased oxygen causes damage to the tissues and the capillaries begin leaking watery fluid from the blood into the tissues. This results in a build-up of fluid around the heart and lungs. The collapsing lungs, failing heart, dehydration, and the inability to get sufficient oxygen to the tissues essentially suffocate the victim. The decreased oxygen also damages the heart itself, which leads to cardiac arrest. In severe cases of cardiac stress, the heart can even burst, a process known as cardiac rupture. Jesus most likely died of a heart attack.[159]

There are three times specifically mentioned during descriptions of the events of this day; the third hour, the sixth hour, and the ninth hour. The events of this day were not random or carried out whenever someone had time to perform them. This, and all other executions, were regimented and carried out with typical Roman precision. There was one complicating factor: the Jewish Day of Sabbath. That would have been the following day.

About Nine a.m.

Scripture tells us that Jesus Christ was crucified at the Third Hour of the day, or about nine a.m. Also, within that time period, two criminals were crucified, one on either side of the taunted King of the Jews. Punishment was reserved for slaves, foreigners, revolutionaries, and the vilest of criminals, not the Savior of the world! These two criminals, and others nearby, languished on both sides of the crucified one. Of the two criminals, one was repentant, the other instead belched out slurs of this king's inability to save anyone, including Himself. There were other activities within the general area of the three crosses. The Priests and other religious leaders sneered at Jesus Christ and made disrespectful and humiliating remarks about His proclamations based on their disbelief.

Soldiers had stripped Jesus Christ of all but the least of clothing to protect His modesty. Those clothes, including a seamless linen undergarment, were used as ill-gained chips as some soldiers cast lots for His meager possessions. Standing nearby is an odd couple, both likely weeping. As Jesus hung His head out of sheer lack of energy, His eyes, caked with dried blood and dirt, allowed the slightest bit of vision where His head was unintentionally pointed. Either by action of the Holy Spirit helping to move His head, or the couple physically moving into His line of sight, Jesus Christ was staring at the same mother he saw, probably through blood resulting from birth, cuddled in her arms in a cattle stall in sleepy Bethlehem. Standing beside her, apparently without shame or fear, was the disciple beloved by Jesus Christ, His young apostle, John.

It isn't clear who else might have been present; certainly not many other disciples, if any. Standing close by, the priests implored the Roman officials to make sure the Jew was dead before dark, which would mark the beginning of the Day of Sabbath. The proven method of making sure prisoners were dead was by breaking their legs, making it impossible for them to push themselves up to breath. In effect, if they were not already dead, they would die of suffocation. The religious leaders need to make sure this mess was all cleaned up by dark so their Sabbath would not be defiled. But the plan for an early death was prohibited. There was still another prophecy to be fulfilled. One about the Messiah having no broken bones when He died. "*They shall leave none of it until morning, nor break one of its bones. According to all the ordinances of the Passover they shall keep it.*" Numbers 9:12 NKJV "*He guards all his bones; Not one of them is broken.*" Psalm 34:12 NKJV

The soldiers did break the legs of the criminals on either side of Jesus Christ, but when they attempted to break His legs, the action was halted. Instead, a soldier was called up to use his spear. Apparently, to make doubly sure of death, the soldier drove his lance between the ribs, upward through the pericardium and into the heart. John 19:34 states, "*And immediately there came out blood and water.*" There was an escape of watery fluid from the sac surrounding the heart and the blood of the interior of the heart. This is rather conclusive post-mortem evidence that Jesus died, not the usual crucifixion death by suffocation, but of heart failure due to shock and constriction of the heart by fluid in the pericardium.[160] From this recorded information, Jesus is not in a coma from which He will awake. Jesus is dead.

The Seven Last Words of Christ

Famously, there were seven things Jesus Christ says from the cross. These were written as an orchestra composition by Joseph Haydn in 1786 for a Good Friday Service. He later, in 1796, wrote an oratorio using those seven famous phrases. We found an outline of these seven words (phrases) in *New Wine Magazine*, April 1982. This was originally published in Arizona Medicine, March 1965, by the Arizona Medical Association.

The first word (phrase) looking down at the Roman soldiers throwing dice for His seamless garment—"Father, forgive them for they do not know what they do."

The second word (phrase)—to the penitent thief—"Today, thou shalt be with me in Paradise."

The third word (phrase)—looking down at Mary, the mother of Jesus, He said, "Woman, behold your son." Then turning to the terrified, grief-stricken adolescent John, the beloved apostle, He said, "Behold your mother."

Twelve p.m.

The timeline within the four Gospels is a little erratic at this point. This is understandable since the sky suddenly turned dark from about the sixth hour until the ninth hour as the Earth below and around them seemed to turn into chaos. The soldiers were confused, fearful that someone might escape or they would be accused of not finishing their task with these prisoners. One soldier tries to use a sponge and give Jesus Christ a drink of old, bitter wine, which He refuses. And then, about the ninth hour, creation itself is interrupted.

The next thing He says seemed to be in close proximity to a catastrophic event inside the Temple Holy of Holies room. This was a room set aside by God Himself centuries before where only the high priest could go. This is where the high priest would speak with God and take the concerns and prayer requests of the people to a vengeful God. This one tradition completely separated every human from any direct access to their maker. That was about to change.

The Holy Spirit descended into creation and ripped the Veil of the Temple, the Holy Curtain separating that space from the rest of the entire world completely into two useless sections, from the top to the bottom! This event was closely followed by a cry of the Son of God embodied within the human flesh of a man called, Jesus. The fourth word (phrase) is a cry from the beginning of Psalm 22: *"My God, My God, why have You forsaken Me?"*

Significance of the Veil

Historians tell us that the veil separating everyone except the priestly line from God was about sixty feet tall, thirty feet wide, and four inches thick. It was said a horse tied to each end and driven away from each other could not damage the material. On that day, at the end of Christ's ministry on earth, God undid what he started. From that day forward the Temple priests would be out of a job, but more importantly… lose their control. From that day forward all mankind would have direct access to their Heavenly Father. They would be seen as perfect and clean when viewed by the Father through His Son, Christ, who died and would soon be resurrected, removing all sins—past, present, and future—for those who confess and believe.

Three p.m.

This was as close as creation had come, since its beginning, to being torn apart at the seams like the veil. This was likely one of, if not the most stressful days Almighty God ever encountered. I wonder if He, or for that matter the rest of the Trinity, really understood the toll this journey would take on all of them when Their plan finally came to fruition. This day would truly be on the level of Genesis 1:1. After the Earth calmed down, Jesus Christ finally... asked for a drink of water. His fifth word (phrase), "I thirst." Some sour wine was thrust at Him—a perfect reward for saving the world...

After a bit, realizing the end was near, it is recorded that with one of His last breaths he was heard to say his sixth word (phrase)—possibly little more than a tortured whisper: "It is finished." And then, as He always did, his seventh and last word (phrase)—"Father, into Your hands I commit My spirit"—was a complete surrender to the Trinity, Who had never left His side during the entire event. The hard part was finally over.

In closing out this scene, here are some looks from other perspectives.

Edith Stein

Edith Stein, a victim of the gas chamber at Auschwitz, wrote The Science of the Cross. *In her teaching is testimony for how to attain full personhood as she teaches others through the Science of the Cross. But though a previously self-declared atheist, she finally emphasizes that, in the confused and hungry world today, scientific answers are not enough. Rather, the way of faith provides a wisdom that is unattainable through philosophy and reason alone. In her university course, she taught on the person, and developed a method of philosophical anthropology. Here in her lecture notes, she tells us that faith has a double significance in scholarship: It is a measuring rod by which we are kept free from error. Also, revealed truth answers many questions which natural reason cannot.*

Let us remember, she died through the so-called scientific method of the gas chamber. And today, science is still killing off innocent lives, quite methodically. On the way to her own crucifixion, the gas chamber at Auschwitz, she spoke of her suffering as an offering "for the conversion of atheists, for her fellow Jews, for the Nazi persecutors, and for all who no longer had the love of God in their hearts."

How different is Edith's philosophy of life from the modern refusal to accept suffering and the crosses of life? We live in a world of illusion and escapism. As both scientist and mystic, Edith knew intimately the greatest reality there is—God. In her holy life and writings, we find God and are brought closer to him because we see an absolute manifestation of our faith. To make this great treasury of love and faith our own—St. Teresa Benedicta of the Cross—is to take a journey into holiness.[161]

From the Official Website of Jehovah's Witnesses, jw.org, "Why True Christians Do Not Use the Cross in Worship":

The cross is loved and respected by millions of people. The Encyclopedia Britannica calls the cross "the principal symbol of the Christian religion." Nevertheless, true Christians do not use the cross in worship. Why not? An important reason is that Jesus Christ did not die on a cross. The Greek word generally translated "cross" is stau·ros'. It basically means "an upright pale or stake." In several texts, Bible writers use another word for the instrument of Jesus' death. It is the Greek word xy'lon. *(Acts 5:30; 10:39; 13:29; Galatians 3:13; 1 Peter 2:24) This word simply means "timber" or "a stick, club, or tree."*

In the book Das Kreuz und die Kreuzigung *(The Cross and the Crucifixion), Hermann Fulda states: "Trees were not everywhere available at the places chosen for public execution. So a simple beam was sunk into the ground. On this the outlaws, with hands raised upward and often also with their feet, were bound or nailed." The apostle Paul says: "Christ purchased us, releasing us from the curse of the Law by becoming a curse instead of us, because it is written: 'Accursed is every man hung upon a stake ["a tree," King James Version].'" (Galatians 3:13) Here Paul quotes Deuteronomy 21:22, 23, which clearly refers to a stake, not a cross.* [162]

Christian Science

A group, Christian Science, founded in 1879 by Mary Baker Glover

Patterson Eddy (1821-1910), is one of the more sophisticated modern cults, attracting many intellectuals. The name of this group is as close as the faith community and science have ever come to unity. The problem is that the group is neither Christian nor Science. They don't believe there was a cross because there is no need for salvation, therefore no need for Jesus to have died on a cross, since man doesn't really exist, but instead is consciousness. [163]

Tomorrow's World

As the Gospel records, and as other ancient histories make clear, the crucifixion occurred during the Passover season, and that is always at the time of a full moon. A full moon occurs when the Earth is between the Moon and the Sun. But during a solar eclipse, it is the opposite—it is the moon that is between the Sun and the Earth. Thus, the darkness occurring at the time of Jesus' death could not have been caused by an eclipse!

*Notice an ancient account that precisely matches the Gospel records. Here, a 3rd-century historian named Sextus Julius Africanus quotes an earlier historian named Thallus, who wrote around 52 AD. Africanus records Thallus as writing the following: "In the time of Tiberius Caesar, at full moon, there was a full eclipse of the sun from the sixth hour to the ninth—manifestly that one of which we speak. But what has an eclipse in common with an earthquake, the rending rocks, and the resurrection of the dead, and so great a perturbation throughout the universe?... But it was a darkness induced by God, because the Lord happened then to suffer" (*The Ante-Nicene Fathers, *Translations of The Writings of the Fathers Down to AD 325, p. 137).*

But neither a "solar eclipse" event, nor the 29 AD or 33 AD dates match the biblical account. Additionally, according to NASA and NOAA, the longest duration of a solar eclipse is just over 7.5 minutes. But the darkness that fell at Jesus' death was three hours long! As Mark 15:33 reads, "Now when the sixth hour had come, there was

> *darkness over the whole land until the ninth hour." Again, that was a full three hours! While God could make an eclipse last three hours, this is not what happened when Jesus died.* [164]

In the Tomb (Scripture from All Four Gospels, and the Book of Isaiah))

> [9] *And they made His grave with the wicked—But with the rich at His death, Because He had done no violence, nor was any deceit in His mouth.* Isaiah 53:9 NKJV

The above was taken from the famous 53rd chapter of Isaiah which contains the entire prophecy regarding the last days of Jesus.

> [57] *Now when evening had come, there came a rich man from Arimathea, named Joseph, who himself had also become a disciple of Jesus.* [58] *This man went to Pilate and asked for the body of Jesus. Then Pilate commanded the body to be given to him.*
>
> [59] *When Joseph had taken the body, he wrapped it in a clean linen cloth,* [60] *and laid it in his new tomb which he had hewn out of the rock; and he rolled a large stone against the door of the tomb, and departed.* [61] *And Mary Magdalene was there, and the other Mary, sitting opposite the tomb.*
>
> [62] *On the next day, which followed the Day of Preparation, the chief priests and Pharisees gathered together to Pilate,* [63] *saying, "Sir, we remember, while He was still alive, how that deceiver said, 'After three days I will rise.'* [64] *Therefore command that the tomb be made secure until the third day, lest His disciples come by night and steal Him away, and say to the people, 'He has risen from the dead.' So the last deception will be worse than the first."* [65] *Pilate said to them, "You have a guard; go your way, make it as secure as you know how."* [66] *So they went and made the tomb secure, sealing the stone and setting the guard.* Matthew 27:57-66 NKJV

1 *Now after the Sabbath, as the first day of the week began to dawn,*
Mary Magdalene and the other Mary came to see the tomb. 2 *And*
behold, there was a great earthquake; for an angel of the Lord
descended from heaven, and came and rolled back the stone from
the door, and sat on it. 3 *His countenance was like lightning, and*
his clothing as white as snow. 4 *And the guards shook for fear of him,*
and became like dead men.

5 *But the angel answered and said to the women, "Do not be afraid,*
for I know that you seek Jesus who was crucified. 6 *He is not here; for*
He is risen, as He said. Come, see the place where the Lord lay. 7 *And*
go quickly and tell His disciples that He is risen from the dead, and
indeed He is going before you into Galilee; there you will see Him.
Behold, I have told you."

8 *So they went out quickly from the tomb with fear and great joy, and*
ran to bring His disciples word. 9 *And as they went to tell His disci-*
ples, behold, Jesus met them, saying, "Rejoice!" So they came and held
Him by the feet and worshiped Him. 10 *Then Jesus said to them, "Do*
not be afraid. Go and tell My brethren to go to Galilee, and there
they will see Me."

11 *Now while they were going, behold, some of the guard came into*
the city and reported to the chief priests all the things that had hap-
pened. 12 *When they had assembled with the elders and consulted*
together, they gave a large sum of money to the soldiers, 13 *saying,*
"Tell them, 'His disciples came at night and stole Him away while
we slept.' 14 *And if this comes to the governor's ears, we will appease*
him and make you secure." 15 *So they took the money and did as they*
were instructed; and this saying is commonly reported among the Jews
until this day. Matthew 28:1-15 NKJV

42 *Now when evening had come, because it was the Preparation Day,*
that is, the day before the Sabbath, 43 *Joseph of Arimathea, a prom-*
inent council member, who was himself waiting for the kingdom of
God, coming and taking courage, went in to Pilate and asked for the

body of Jesus. 44 *Pilate marveled that He was already dead; and summoning the centurion, he asked him if He had been dead for some time.* 45 *So when he found out from the centurion, he granted the body to Joseph.*

46 *Then he bought fine linen, took Him down, and wrapped Him in the linen. And he laid Him in a tomb which had been hewn out of the rock, and rolled a stone against the door of the tomb.* 47 *And Mary Magdalene and Mary the mother of Joses observed where He was laid.* Mark 15:42-47

1 *Now when the Sabbath was past, Mary Magdalene, Mary the mother of James, and Salome bought spices, that they might come and anoint Him.* 2 *Very early in the morning, on the first day of the week, they came to the tomb when the sun had risen.* 3 *And they said among themselves, "Who will roll away the stone from the door of the tomb for us?"* 4 *But when they looked up, they saw that the stone had been rolled away—for it was very large.*

5 *And entering the tomb, they saw a young man clothed in a long white robe sitting on the right side; and they were alarmed.* 6 *But he said to them, "Do not be alarmed. You seek Jesus of Nazareth, who was crucified. He is risen! He is not here. See the place where they laid Him.* 7 *But go, tell His disciples—and Peter—that He is going before you into Galilee; there you will see Him, as He said to you."* 8 *So they went out quickly and fled from the tomb, for they trembled and were amazed. And they said nothing to anyone, for they were afraid.* Mark 16:1-8 NKJV

50 *Now behold, there was a man named Joseph, a council member, a good and just man.* 51 *He had not consented to their decision and deed. He was from Arimathea, a city of the Jews, who himself was also waiting for the kingdom of God.* 52 *This man went to Pilate and asked for the body of Jesus.* 53 *Then he took it down, wrapped it in*

linen, and laid it in a tomb that was hewn out of the rock, where no
one had ever lain before. 54 *That day was the Preparation, and the*
Sabbath drew near.

55 *And the women who had come with Him from Galilee followed*
after, and they observed the tomb and how His body was laid. 56 *Then*
they returned and prepared spices and fragrant oils. And they rested
on the Sabbath according to the commandment. Luke 23:50-56

1 *Now on the first day of the week, very early in the morning, they,*
and certain other women with them, came to the tomb bringing the
spices which they had prepared. 2 *But they found the stone rolled*
away from the tomb. 3 *Then they went in and did not find the body*
of the Lord Jesus.

4 *And it happened, as they were greatly perplexed about this, that*
behold, two men stood by them in shining garments. 5 *Then, as they*
were afraid and bowed their faces to the earth, they said to them,
"Why do you seek the living among the dead? 6 *He is not here, but is*
risen! Remember how He spoke to you when He was still in Galilee,
7 *saying, 'The Son of Man must be delivered into the hands of sinful*
men, and be crucified, and the third day rise again.' "

8 *And they remembered His words.* 9 *Then they returned from the*
tomb and told all these things to the eleven and to all the rest. 10 *It*
was Mary Magdalene, Joanna, Mary the mother of James, and the
other women with them, who told these things to the apostles. 11 *And*
their words seemed to them like idle tales, and they did not believe
them. 12 *But Peter arose and ran to the tomb; and stooping down, he*
saw the linen cloths lying by themselves; and he departed, marveling
to himself at what had happened. Luke 24:1-12 NKJV

38 *After this, Joseph of Arimathea, being a disciple of Jesus, but*
secretly, for fear of the Jews, asked Pilate that he might take away the

body of Jesus; and Pilate gave him permission. So he came and took
the body of Jesus.

[39] *And Nicodemus, who at first came to Jesus by night, also came,*
bringing a mixture of myrrh and aloes, about a hundred pounds. [40]
Then they took the body of Jesus, and bound it in strips of linen with
the spices, as the custom of the Jews is to bury. [41] *Now in the place*
where He was crucified there was a garden, and in the garden a new
tomb in which no one had yet been laid. [42] *So there they laid Jesus,*
because of the Jews' Preparation Day, for the tomb was nearby. John
19:38-42 NKJV

[1] *Now the first day of the week Mary Magdalene went to the tomb*
early, while it was still dark, and saw that the stone had been taken
away from the tomb. [2] *Then she ran and came to Simon Peter, and*
to the other disciple, whom Jesus loved, and said to them, "They have
taken away the Lord out of the tomb, and we do not know where
they have laid Him."

[3] *Peter therefore went out, and the other disciple, and were going to*
the tomb. [4] *So they both ran together, and the other disciple outran*
Peter and came to the tomb first. [5] *And he, stooping down and look-*
ing in, saw the linen cloths lying there; yet he did not go in. [6] *Then*
Simon Peter came, following him, and went into the tomb; and he
saw the linen cloths lying there, [7] *and the handkerchief that had been*
around His head, not lying with the linen cloths, but folded together
in a place by itself. [8] *Then the other disciple, who came to the tomb*
first, went in also; and he saw and believed. [9] *For as yet they did not*
know the Scripture, that He must rise again from the dead. [10] *Then*
the disciples went away again to their own homes.

[11] *But Mary stood outside by the tomb weeping, and as she wept she*
stooped down and looked into the tomb. [12] *And she saw two angels*
in white sitting, one at the head and the other at the feet, where the
body of Jesus had lain. [13] *Then they said to her, "Woman, why are*

you weeping?" She said to them, "Because they have taken away my Lord, and I do not know where they have laid Him."

[14] *Now when she had said this, she turned around and saw Jesus standing there, and did not know that it was Jesus.* [15] *Jesus said to her, "Woman, why are you weeping? Whom are you seeking?" She, supposing Him to be the gardener, said to Him, "Sir, if You have carried Him away, tell me where You have laid Him, and I will take Him away."* [16] *Jesus said to her, "Mary!" She turned and said to Him, "Rabboni!" (which is to say, Teacher).* [17] *Jesus said to her, "Do not cling to Me, for I have not yet ascended to My Father; but go to My brethren and say to them, 'I am ascending to My Father and your Father, and to My God and your God.' "* [18] *Mary Magdalene came and told the disciples that she had seen the Lord, and that He had spoken these things to her.* John 20:1-18 NKJV

Six p.m.

Up to this point on the timeline, the events concerning the end of Jesus Christ's life and ministry have been mostly straight-forward and chronological. Beginning with the ninth hour on the day Jesus Christ approaches His last breath, things begin to fray. This was understandable, as the fabric of creation (as most knew it) was coming apart. The last thing to be taken from the frail body of Jesus was for Him to die; to no longer exist as a breathing, heart-pumping human being. Soon after the ninth hour, Jesus fulfilled His final task: He died. Near the end of Preparation Day, a Friday as it was, things begin to become quite complex.

There was certainly another figure present, Satan, lurking in the shadows. His mind must have been spinning. I see him as confused, trying to understand how this worked. And then it hit him… back in the Garden… that thing God had said about placing enmity between him and the seed of man. The bruising of a heel and a head. No! Christ was dead!! He was staring right at him. The Son of God was caught, trapped inside a human who died. There was no escape, but… he felt the Earth tremble, he heard the veil tearing, he saw something that most likely no one else saw except

God and the Holy Spirit. He saw Christ's spirit lift out of the dead body of Jesus. Satan needed to think about this. He retreated to a dark, solemn place. He would not allow God to take away his control over death. That was his and his alone... "But wait!" he may have reasoned. "It was the living spirit of Christ that came alive, not the body of a still-dead human! Why must I worry? I got this," He smirked...

Gottes Freitag

We will now cover the events from this point until the following Sunday morning. As with the series of trials preceding the crucifixion, there are several different groups and beings involved.

But first, do any of you remember back in middle school or high school always being served fish sticks at lunch on Friday? I thought a few would have that memory.

For those too young, or who grew up in very small communities, here is what I was told. It never made sense to me because I grew up as a Free Will Baptist. We only ate fish when we were lucky enough to catch them on the river bank. I was told that it was to honor the beliefs of our Catholic students and staff as they honored the significance of Good Friday. That tradition didn't bother me. What stood out in my young mind was the term itself. Although my denomination didn't observe the fish tradition on Friday, we did observe the Friday before Easter as special and called it "Good" Friday. Like the pictures on the Scripture cards I was given as a child in Sunday School, always depicting Jesus without a smile, I saw absolutely nothing *good* about what happened to my Savior on that particular day of the week. Let's follow the research.

We will begin with a reliable source, the BBC Magazine *Monitor* which quotes a few other reliable sources on April 18, 2014.[165]

> *The earliest known use of "guode friday" is found in* The South English Legendary, *a text from around 1290, according to the dictionary. According to the* Baltimore Catechism *- the standard US Catholic school text from 1885 to the 1960s, Good Friday is good*

> *because Christ "showed His great love for man, and purchased for him every blessing."*
>
> *The* Catholic Encyclopedia, *first published in 1907, states that the term's origins are not clear. It says some sources see its origins in the term "God's Friday" or* Gottes Freitag, *while others maintain that it is from the German* Gute Freitag. *It notes that the day was called Long Friday by the Anglo-Saxons and is referred to as such in modern Danish.*
>
> *It also says that the day is known as "the Holy and Great Friday" in the Greek liturgy, "Holy Friday" in Romance Languages, and* Karfreitag *(Sorrowful Friday) in German.*

There you have it. Actually, if you read the Beatitudes in the fifth chapter of Matthew, these definitions make sense. What Jesus Christ modeled for humanity on this bloody day was the essence of the definition outlined in how Jesus Christ taught as a means for being blessed. More than taught, He demonstrated.

Three Days?

Having made short work out of a misunderstood day of the week, now let's turn our attention back to the timeline and how it was measured. Jesus was quoted as using the term "three days and three nights" to describe how long He would lie in the grave. John 2:19 NKJV: *"Jesus answered and said to them, 'Destroy this temple, and in three days I will raise it up.'"* And why not? That is what was prophesied.

> 38 *Then some of the scribes and Pharisees answered, saying, "Teacher, we want to see a sign from You."* 39 *But He answered and said to them, "An evil and adulterous generation seeks after a sign, and no sign will be given to it except the sign of the prophet Jonah.* 40 *For as Jonah was three days and three nights in the belly of the great fish, so will the Son of Man be three days and three nights in the heart of the earth.*
> Matthew 12:38-40 NKJV

These Scriptures firmly state Jesus Christ would be in the tomb for three days and three nights. We know that Jesus didn't die until sometime after the ninth hour on Good Friday. Then there is this in the Gospel of John:

> 1 *Now the first day of the week Mary Magdalene went to the tomb early, while it was still dark, and saw that the stone had been taken away from the tomb.* John 20:1 NKJV

A rational mind will immediately perceive what seems to be a conflict here. Late Friday evening to before sunrise on Sunday is not three days and certainly not three nights. However, all people of that time, regardless of social standing, did not yet have a Rolex or even a Timex. This is only one of several instances in the Bible where varying methods of accounting for time and time measurements occur. While it wasn't three "full days" or seventy-two hours, the scene did take place over three days—but still one night short. Beating this point to death only overlooks what *should* be understood. Jesus' overcoming death didn't require time. As with the raising of Lazarus, it would have been immediate. The real mystery? Based on different Scriptures concerning this event, where did Jesus Christ go between death and His first encounter with people on Sunday morning? Perhaps a ramble inside the tomb and a walk in the beautiful surrounding gardens? I thought you might think that would be an interesting journey.

Journey is an excellent term for what happened. According to what we see in Scripture, it is possible that time, as we count it, had nothing to do with this journey. In fact, it may have been a lot longer than three full days and nights. It could have been that time for all humanity stood still. Science would say that it impossible on many levels. Science also doesn't believe the Trinity had anything to do with Creation. For you see, whoever created everything surely has the ability to alter what they created. Beginning about the ninth-hour on that Good Friday evening, it is possible Jesus Christ's time, related to the Earth, began to change when suddenly every sin—past, present, and future—began to accumulate on His shoulders. Time, as Jesus the fully human man would have understood it, began to slow or speed up—who knows which, but, from that moment, mystical things began to take place.

There was a point that Friday night when things seem to turn back to normal for everyone else except a lifeless, bloody body limply hanging on a Roman instrument of death. They may have taken down the body of Jesus, but is it possible Christ, the Son of God, had already departed for a predestined meeting with millions of others who had been waiting on Him for centuries? Before going there, let's talk about the sequence right after Jesus Christ exhaled His last earthly breath and committed Himself back to His Father.

There are two very distinct tracks running from Friday evening after Jesus Christ died until Sunday morning. One track includes one or two members of the Sanhedrin court, Pilate, some soldiers, the high priest, his cronies, some of the apostles, Mary Magdalene, and a few other ladies. I did not list Jesus Christ because between these two points, to mankind, He is simply a dead body; an object to be dealt with by all of these other folks. The second track, seemingly not of this Earth, occurs in parallel with the first. I will speak to that shortly.

This first track follows a normal time cycle from about sundown on Friday, which is preparation day, through the Sabbath Saturday, and up until just before dawn on Sunday. This would have been in earthly time, based on the rotation of the earth, between thirty-four to thirty-five hours. Within that period of time, it doesn't seem any of them expect the dead man to come to life. There was no planning for that event.

The Borrowed Tomb

Joseph of Arimathea is listed as a disciple of Jesus Christ. He was afraid to admit it for fear his Jewish constituents would bring reprisal to his door. Despite this, he boldly went to Pilate and asked for the body of the dead Jew from Nazareth. His motive seems to have been to follow Jewish Law forbidding such tasks to be performed on the Sabbath, which was about to occur. Pilate had no problem with this—it would free up someone else. It would also be a favor to the Jewish leaders, who wanted this man to disappear.

Accompanying Joseph was a man named Nicodemus. Both he and Joseph were wealthy and had the servants and means to pull this off. Nicodemus was the man who earlier came to Jesus one night asking for clarity concerning the phrase, "You must be born again." In the third chapter of John, we are told, or it is implied, Nicodemus was an older man and a senior teacher of Jewish Law. This is also the man to whom Jesus Christ speaks the oft-quoted Scripture, John 3:16: *"For God so loved the world that he gave his one and only Son, that whoever believes in him shall not perish but have eternal life."* We discover in another Gospel that the tomb where Jesus Christ will be laid belonged to Joseph of Arimathea.

So, two abstaining members of the same group who forced Pilate to crucify Jesus Christ have come to bury Him in a borrowed tomb owned by a rich man. Neither had a clue what was about to happen to their best-laid and well-intentioned plans. They were just doing what they thought was right under the circumstances. But Joseph was not the only visitor that Friday night.

Caiaphas, the stand-in high priest for his father-in-law, Annas, accompanied by his yes men, went to Pilate with a potential problem regarding the dead Jew. They had heard that this so-called King of the Jews had stated publicly He would rise on the third day. The Sanhedrin was concerned His disciples would turn into grave robbers, steal the body, and then claim He had risen. They wanted Pilate to post a guard around the tomb to make sure that didn't happen. Pilate, being Pilate, reminded Caiaphas of the high priest's social position and wealth that accompanied that status and told him to go take care of it himself. In other words, Rome was through throwing money at the Jew's problem. Caiaphas did what he had to do; post a guard.

Meanwhile, Joseph and Nicodemus retrieved the body and prepared it with great care and respect for burial. Since Scripture details that a very large rock was used to cover the entrance to the tomb and that it also required a few strong men to get Jesus Christ off the cross and carried to it, the two men had servants or hired men to assist them with all this preparation and burial. It seems that after they were finished, Caiaphas had the door sealed to make sure there wouldn't be any entry until at least the following Monday.

The ladies in this story happen along before sunrise on Sunday morning. That is when things change for those following this time track. What happens after this must have occurred after Christ returned. But, returned from where? Where had He been? More to the point, when did He leave? Before exploring these questions, it is worth noting that after the high priest discovers the body is missing, he pays the guards a handsome sum of money to testify that the disciples came at night and stole the body. I would have really liked to have heard that story.

The Rest of the Story – the Second Track

There are two ways to look at this part of the story. According to several Scriptures, Christ paid a visit to some folks during this period between His death and the resurrection. Even allowing for thirty-four to thirty-five hours, the places He visited according to Scripture seemingly would have taken days for a human to arrive, much less return. The fact that some Scripture seems to suggest He visited these places asks, not for time's sake, but in what form He must have taken; and when. According to Scripture describing events in Earth time while interacting with people who knew Him from His earthly ministry, Jesus Christ is already in some other form. Some of the visits He made to the disciples are clearly written to indicate He just appeared; sometimes behind closed, locked, and likely barricaded doors. In at least one instance, He simply disappears. With this kind of ability, it is very possible He also appeared in some other places not associated with the Earth or its time constraints.

This returns us to an earlier discussion regarding one or two wills possessed by Jesus Christ. If there were two separate wills between Jesus and Christ, it is possible the instant the human body of Jesus died, Christ was released to visit other unworldly places to accomplish another task, possibly built into this earlier earthly journey. That would also account for the fact that Joseph and Nicodemus had a body to embalm and place in the tomb while, at the same time, Christ visited these other places. If both Jesus and Christ made these visits together, the journey could not have begun until the tomb was sealed; further reducing the time He was actually in the tomb.

Another argument is that it was the soul or Spirit of Jesus Christ who made these visits, which would not have required a physical body. Regardless of how you interpret these Scriptures, outlining his journeys during this period must be accounted for or all Scripture is false. Failure to figure this out could be why you don't hear this preached on too often. Well… why not? Let's take a stab at it.

I am going to look at the opinions of some who have spent time searching Scripture and other historical evidence related to these questions. You are welcome to follow the footnotes and do further research to form your own opinion. I will warn you in advance to take the following precautions:

> Understand this is not a casual project.
>
> If you find something of interest, take the time to find out about the author. Any opinion you form should be based on solid Biblical understanding.
>
> Don't assume that an author's credibility on this issue is high because they are associated with a very large organization or denomination.
>
> Keep an open mind. A Trinity capable of creating the world and universe in which you live has no limits.
>
> Resist reading only select Scripture passages. Always put them in context, preferably by reading not just the entire chapter in which the passage is from, but the chapter before it and the chapter following. If you are reading something from Revelation you should really read the entire book. This book was always meant to be read in its entirety.

Following are Scriptures where you might want to begin. These are referenced in my comments below. Remember, read around these passages; previous chapter, chapter containing the reference, and the following chapter. The references below are just bookmarks!

Matthew 27:45-50 NKJV

45 *Now from the sixth hour until the ninth hour there was darkness*
over all the land. 46 *And about the ninth hour Jesus cried out with a*
loud voice, saying, "Eli, Eli, lama sabachthani?" that is, "My God,
My God, why have You forsaken Me?" 47 *Some of those who stood*
there, when they heard that, said, "This Man is calling for Elijah!"
48 *Immediately one of them ran and took a sponge, filled it with sour*
wine and put it on a reed, and offered it to Him to drink. 49 *The rest*
said, "Let Him alone; let us see if Elijah will come to save Him." 50
And Jesus cried out again with a loud voice, and yielded up His spirit.

Luke 16:22-23 NKJV

22 *So it was that the beggar died, and was carried by the angels to*
Abraham's bosom. The rich man also died and was buried. 23 *And*
being in torments in Hades, he lifted up his eyes and saw Abraham
afar off, and Lazarus in his bosom.

Eph. 4:8-10 NKJV

8 *Therefore He says: "When He ascended on high, He led captivity*
captive, And gave gifts to men." 9 *(Now this, "He ascended"—what*
does it mean but that He also first descended into the lower parts
of the earth? 10 *He who descended is also the One who ascended far*
above all the heavens, that He might fill all things.)

Colossians 2:14-15 NKJV

*14 having wiped out the handwriting of requirements that was
against us, which was contrary to us. And He has taken it out of the
way, having nailed it to the cross. 15 Having disarmed principalities
and powers, He made a public spectacle of them, triumphing over
them in it.*

Hebrews 2:10-18 NKJV

*10 For it was fitting for Him, for whom are all things and by whom
are all things, in bringing many sons to glory, to make the captain of
their salvation perfect through sufferings. 11 For both He who sancti-
fies and those who are being sanctified are all of one, for which reason
He is not ashamed to call them brethren, 12 saying: "I will declare
Your name to My brethren; In the midst of the assembly I will sing
praise to You." 13 And again: "I will put My trust in Him." And
again: "Here am I and the children whom God has given Me." 14
Inasmuch then as the children have partaken of flesh and blood, He
Himself likewise shared in the same, that through death He might
destroy him who had the power of death, that is, the devil, 15 and
release those who through fear of death were all their lifetime subject
to bondage. 16 For indeed He does not give aid to angels, but He does
give aid to the seed of Abraham. 17 Therefore, in all things He had to
be made like His brethren, that He might be a merciful and faith-
ful High Priest in things pertaining to God, to make propitiation for
the sins of the people. 18 For in that He Himself has suffered, being
tempted, He is able to aid those who are tempted.*

Heb. 9:23-28 NKJV

23 Therefore it was necessary that the copies of the things in the heav-
ens should be purified with these, but the heavenly things themselves
with better sacrifices than these. 24 For Christ has not entered the
holy places made with hands, which are copies of the true, but into
heaven itself, now to appear in the presence of God for us; 25 not that
He should offer Himself often, as the high priest enters the Most Holy
Place every year with blood of another— 26 He then would have had
to suffer often since the foundation of the world; but now, once at the
end of the ages, He has appeared to put away sin by the sacrifice of
Himself. 27 And as it is appointed for men to die once, but after this
the judgment, 28 so Christ was offered once to bear the sins of many.
To those who eagerly wait for Him He will appear a second time,
apart from sin, for salvation.

1 Peter 3:18-22 NKJV

18 For Christ also suffered once for sins, the just for the unjust, that
He might bring us to God, being put to death in the flesh but made
alive by the Spirit, 19 by whom also He went and preached to the spir-
its in prison, 20 who formerly were disobedient, when once the Divine
longsuffering waited in the days of Noah, while the ark was being pre-
pared, in which a few, that is, eight souls, were saved through water.
21 There is also an antitype which now saves us—baptism (not the
removal of the filth of the flesh, but the answer of a good conscience
toward God), through the resurrection of Jesus Christ, 22 who has gone
into heaven and is at the right hand of God, angels and authorities
and powers having been made subject to Him.

2 Pet. 2:4-11 NKJV

4 *For if God did not spare the angels who sinned, but cast them down*
to hell and delivered them into chains of darkness, to be reserved for
judgment; 5 *and did not spare the ancient world, but saved Noah,*
one of eight people, a preacher of righteousness, bringing in the flood
on the world of the ungodly; 6 *and turning the cities of Sodom and*
Gomorrah into ashes, condemned them to destruction, making them
an example to those who afterward would live ungodly; 7 *and deliv-*
ered righteous Lot, who was oppressed by the filthy conduct of the
wicked 8 *(for that righteous man, dwelling among them, tormented*
his righteous soul from day to day by seeing and hearing their law-
less deeds)— 9 *then the Lord knows how to deliver the godly out of*
temptations and to reserve the unjust under punishment for the day
of judgment, 10 *and especially those who walk according to the flesh*
in the lust of uncleanness and despise authority. They are presump-
tuous, self-willed. They are not afraid to speak evil of dignitaries, 11
whereas angels, who are greater in power and might, do not bring a
reviling accusation against them before the Lord.

Revelation 20:11–15 NKJV

11 *Then I saw a great white throne and Him who sat on it, from*
whose face the earth and the heaven fled away. And there was found
no place for them. 12 *And I saw the dead, small and great, standing*
before God, and books were opened. And another book was opened,
which is the Book of Life. And the dead were judged according to
their works, by the things which were written in the books. 13 *The sea*
gave up the dead who were in it, and Death and Hades delivered up
the dead who were in them. And they were judged, each one accord-
ing to his works. 14 *Then Death and Hades were cast into the lake of*
fire. This is the second death. 15 *And anyone not found written in the*
Book of Life was cast into the lake of fire.

I initially approached this as a simple task: "What happened from the time the tomb door was closed and sealed until the angel rolled the rock away?" It didn't take long before my research overwhelmed my initial naive curiosity. The opinions of some were, in my understanding, outlandish, even borderline heresy. What was undeniable, based on Scripture, is that something happened. While Jesus may have become a corpse, like Lazarus before being raised from the dead, something was taking place on some level, somewhere in the space occupied or managed by the Trinity.

I am going to do this in reverse order. I will begin with my conclusion, and then work backwards to demonstrate how I came to that understanding. Based on what I could find, and yes, after much prayer asking God to help me make sense out of what was before me, I believe there is a reason this span of time isn't written about like much of the rest of Jesus Christ's story. The reason, or reasons, deal with relativeness, demonstration, and hope.

We often say that Scripture is the "inspired" Word of God. The faith community looks to these words as the "infallible" history and instruction to those who accept God's promise. If we hold these statements to be our security, then we need to understand the *why*, not just the *how*. Scripture that seems to be pertinent somehow to this time in the grave should be wholly relevant to the purpose of Christ's journey with Jesus. I would point you to the ninth chapter of Hebrews for relevance.

It is here the writer compares the Old Testament ritual of blood sacrifice by the God-chosen high priest to Christ's sacrifice. The blood of an animal in the Old Testament was sprinkled within the Holy of Holies for redemption and atonement of man's sin. The blood was sprinkled on everything in the room. This ritual had to be performed often, over and over. It seems when Christ died, He entered the Holy of Holies, in the presence of God. This action could have been what caused the veil in the earthly temple to split. But the blood of an animal was not sufficient for this atonement and redemption in Heaven. It had to be the blood of a sinless human being, Jesus Christ. The beauty was that He only had to sprinkle the blood once and for all time. The relevance of Christ's trips to Hell or Hades while in the grave was to announce the coming end of the one power Satan possessed: death itself. Christ's act of sacrifice had stolen that power from the fallen angel God had, in another time, allowed him to possess.

By visiting various places during His time in the tomb, Christ demonstrated His authority over Satan and all of His minions. And with this power, He demonstrated how vast it was by freeing those prior to His death, who had been held captive since the beginning of human time.

This demonstration of vastness gives hope to all future generations for salvation when Christ casts Satan and all of His followers into the Lake of Fire, represented by eternal separation from God and all things good. A place of no return. A place without hope, love, or charity.

There is a lot I don't know. So much that, I don't know what I don't know. Know what I mean?… I think you do.

Here are a few of the places I looked in my research.

Abraham's Bosom

> Some use Ephesians 4:7-10 to teach that Jesus went to hell or to Hades to release the prisoners held there and take them to heaven or into God's presence. The idea is that before His death, all Old Testament believers were in Abraham's bosom—the paradise part of Hades. Hades or Sheol was seen as the place of the dead with three areas or compartments: (1) the abyss or tartarus, the place of confinement for those demons who sinned in the days of Noah; (2) torments, the place of suffering for all unbelievers until the time of the resurrection of the unjust and the Great White Throne Judgment when they will be cast eternally into the lake of fire, and (3) a third place separated by a great gulf (see Luke 16), called Abraham's Bosom, the place of blessing for believers. Because Christ had not yet died to pay the penalty for their sin, they were not yet able to go directly into the presence of God. After His death, however, the barrier was removed and He took them out of prison and into God's presence.[166]

Weekend in Hell

However, these traditions can get out of hand, and when traditional beliefs develop and grow to the point where they negatively impact vital Christian doctrine, they are no longer harmless, and must be dealt with. Currently, in some evangelical circles, there are two traditional doctrines which have thus mutated into a dangerous form: the "slain in the Spirit" doctrine, and the doctrine which we will call "the Weekend in Hell," (the teaching that Jesus, during the three days that His body was in the tomb, made a trip to Hell, or, in the modern version, spent the whole weekend in Hell). This paper will deal with The Weekend in Hell.

> It is not certain when or where the "weekend in Hell" teaching first originated, but it crept into the Apostle's Creed around the year 750 AD after first appearing in an Arian creed in 360 AD, and in some other creeds and doctrinal statements.
>
> In the older version, the "weekend in Hell" teaching had an air of triumph, as Christ visited the nether regions to announce His victory, and there were no elements in the old version of "weekend in Hell" which undermined vital doctrine.
>
> There are several component parts to this doctrine, and in order to understand it, we need to look at each part.

(1) Preliminary to the entire idea of the "weekend in Hell" is the idea that Old Testament believers were held captive in Hell, and could not be let loose until Jesus came and got them. Consider this:

a. The spirits of the Old Testament saints went to be with God when they departed this earth. Eccl 12:6-7: *6 Remember your Creator before the silver cord is loosed, Or the golden bowl is broken, Or the pitcher shattered at the fountain, Or the wheel broken at the well. 7 Then the dust will return to the earth as it was, And the spirit will return to God who gave it.*

b. Enoch and Elijah were bodily assumed up to heaven, and from his appearance with Elijah on the mount, Moses' spirit was also in heaven. Gen 5:24 and 2 Ki 2:11 and Luke 9:29-32. Mat 27:50-53. If Jesus was to suffer in Hell for three days to finish the atonement, what is happening here? Why are these saints rising? They are rising as a witness of the completeness of the Cross; these saints are foreshadowing what Christ Himself will do in three days. Just as we will experience in the translation of believers and resurrection of those who sleep, the spirits of these Old Testament saints came from the presence of God to collect their bodies.

(2) The second part of the "weekend in Hell" doctrine is that Satan owned "the keys" to death, hell, and the grave, and that Jesus had to go win them from him.

a. There are ABSOLUTELY NO Scriptures which even remotely suggest that Satan ever had any keys to anything.

b. Furthermore, God opens and shuts, and no one stops Him in going either direction. Isa 22:22, Isa 45:1, Rev 3:7

c. There are only two verses in the Bible where these keys are mentioned. The devil is never seen with them. Mat 16:19, Rev 1:18

Now think, did the Devil EVER have the power of life and death? Did Satan EVER have authority to sentence anyone to Hell? The answer is "no!" God alone has, and has always had, that power and authority.[167]

Two Sides of Hell

While many would question where He was during the three days His body lay in the tomb, I always questioned where His spirit was during those last few hours of His body's physical life before it was placed in the tomb.

The battle in those moments must have been the most strenuous of Jesus' entire life. He bore the sins of everyone upon his shoulders and when He took His last breath, that burden was taken from Him. Because of Jesus' willingness to "accept the cup" He was awarded a place of honor.

One of the questions that often comes up along with the issue as to where Jesus was during the three days in the tomb, is that of the sins of humanity. Wouldn't that mean that Jesus went to hell where sinful people were? To understand this, we need to first understand that there are different meanings of the word "hell."

Every time we see the word "hell" in an English translation of the Bible, it does not necessarily always have the same meaning. There are a few different Greek and Hebrew words that are all translated as simply "hell" in English. The Hebrew word *Sheol*, for example, refers to "the grave" in the Old Testament; as we see in such passages as Numbers 16:30.

Historians suggest that there are two sides of hell, one is a blessed side and the other is the suffering side, as we see in Luke 16:22-23 ESV. Some argue that Jesus went to Sheol rather than "Hades," which is what historians consider the suffering side of hell, or what we think of as hell. All of these different words mean different aspects of Satan's Kingdom, so taking the time to really think about the context of each word is vital to gaining understanding about whether Jesus did actually go to "hell."[168] [169]

Alive in the Spirit

The Bible does not specifically state what happened to Jesus immediately after He died on the cross. Because of this, there is debate surrounding the answer to the question of where He went and what He did. Therefore, I will present differing views so you might know the scope of the answer and decide for yourself which position is preferable.

When Jesus was made alive in the spirit, it is not saying that His spirit died, and then it became alive again. "Made alive in the spirit" is contrasted with "put to death in the flesh." He first lived as mortal men but "He began to live a spiritual 'resurrection' life, whereby He has the power to bring us to God."1 Furthermore, some Bibles (NIV, KJV, and NKJV) render the verse as "made alive by the Spirit" referring to the Holy Spirit's work with Christ. "By the Spirit" translates one word, *pneumati*, which could refer to the third Person of the Trinity as the agent of Christ's resurrection.

One view of where Jesus was and what He did before His resurrection is that He went to Hades (the place of the dead) and made proclamation to those who were in spiritual prison. The word "proclamation" in Greek is *kerusso*. It means to proclaim and is a different word than *euaggelizo*, which means to preach the gospel. Therefore, it is most probable that Jesus was not preaching the gospel to those in Hades/Spirit prison so they could be saved, but was instead proclaiming the truth to them. After all, the Bible says, *"And inasmuch as it is appointed for men to die once and after this comes judgment."* Heb. 9:27.

Needless to say, this passage also raises many questions and much debate can be found as to its precise meaning. Nevertheless, as far as the other option goes, that Jesus simply presented the facts concerning His work on the cross to those in spiritual prison, we can look to Eph. 4:8-10 for possible support.

So, even though we cannot precisely determine where Jesus was and what He did during those three days, it seems apparent that

He presented the gospel message (not to have them get saved) to those in spirit prison and possibly also to those in Abraham's bosom.[170]

Imprisoned Spirits

The word spirit refers to Christ's spirit. The contrast is between His flesh and spirit, and not between Christ's flesh and the Holy Spirit. Christ's flesh died, but His spirit remained alive. Jesus' body was in the tomb, of course, but His spirit, having departed at His death (Matthew 27:50), was elsewhere for those three days.

According to Peter, sometime between Jesus' death and His resurrection Jesus made a special proclamation to some imprisoned spirits.

Hades is a term that refers broadly to the realm of the dead, a temporary place where the dead await resurrection. Revelation 20:11–15 in the NASB and the NIV makes a clear distinction between Hades and the lake of fire. The lake of fire is the permanent, final place of judgment for the lost. Hades is a temporary place for both the lost and the Old Testament saints.

At some time between His death and resurrection, Jesus also visited a place where He delivered a message to spirit beings—most likely fallen angels (see Jude 1:6); these spirits were probably imprisoned because they were somehow involved in a grievous sin before the flood in Noah's time (1 Peter 3:20).

Peter does not tell us what Jesus proclaimed to the imprisoned spirits, but it could not have been a message of redemption, since angels cannot be saved (Hebrews 2:16). What Jesus proclaimed was probably a declaration of His victory over Satan and his hosts (1 Peter 3:22; Colossians 2:15).

All of this is to say that the Bible isn't entirely clear as to what exactly Christ did for the three days between His death and resurrection. From what we can tell, though, He did two things: He comforted the departed saints and brought them to their eternal home, and He proclaimed His victory over the fallen angels who are kept in prison. What we can know for sure is that Jesus was not giving anyone a second chance for salvation; we face judgment after death (Hebrews 9:27), not a second chance. Also, Jesus was not suffering in hell; His work of redemption was finished on the cross.[171]

After the Resurrection (Scripture from All Four Gospels)

9 *And as they went to tell His disciples, behold, Jesus met them, saying,*
"Rejoice!" So they came and held Him by the feet and worshiped
Him. 10 *Then Jesus said to them, "Do not be afraid. Go and tell*
My brethren to go to Galilee, and there they will see Me." Matthew
28:9-10 NKJV

9 *Now when He rose early on the first day of the week, He appeared*
first to Mary Magdalene, out of whom He had cast seven demons. 10
She went and told those who had been with Him, as they mourned
and wept. 11 *And when they heard that He was alive and had been*
seen by her, they did not believe. 12 *After that, He appeared in another*
form to two of them as they walked and went into the country. 13 *And*
they went and told it to the rest, but they did not believe them either.
14 *Later He appeared to the eleven as they sat at the table; and He*
rebuked their unbelief and hardness of heart, because they did not
believe those who had seen Him after He had risen. 15 *And He said to*
them, "Go into all the world and preach the gospel to every creature.
16 *He who believes and is baptized will be saved; but he who does not*
believe will be condemned. 17 *And these signs will follow those who*
believe: In My name they will cast out demons; they will speak with
new tongues; 18 *they will take up serpents; and if they drink anything*

deadly, it will by no means hurt them; they will lay hands on the sick, and they will recover." Mark 16:9-18 NKJV

13 Now behold, two of them were traveling that same day to a village called Emmaus, which was seven miles from Jerusalem. 14 And they talked together of all these things which had happened. 15 So it was, while they conversed and reasoned, that Jesus Himself drew near and went with them. 16 But their eyes were restrained, so that they did not know Him.17 And He said to them, "What kind of conversation is this that you have with one another as you walk and are sad?" 18 Then the one whose name was Cleopas answered and said to Him, "Are You the only stranger in Jerusalem, and have You not known the things which happened there in these days?"

19 And He said to them, "What things?" So they said to Him, "The things concerning Jesus of Nazareth, who was a Prophet mighty in deed and word before God and all the people, 20 and how the chief priests and our rulers delivered Him to be condemned to death, and crucified Him. 21 But we were hoping that it was He who was going to redeem Israel. Indeed, besides all this, today is the third day since these things happened. 22 Yes, and certain women of our company, who arrived at the tomb early, astonished us. 23 When they did not find His body, they came saying that they had also seen a vision of angels who said He was alive. 24 And certain of those who were with us went to the tomb and found it just as the women had said; but Him they did not see." 25 Then He said to them, "O foolish ones, and slow of heart to believe in all that the prophets have spoken! 26 Ought not the Christ to have suffered these things and to enter into His glory?" 27 And beginning at Moses and all the Prophets, He expounded to them in all the Scriptures the things concerning Himself.

28 Then they drew near to the village where they were going, and He indicated that He would have gone farther. 29 But they constrained Him, saying, "Abide with us, for it is toward evening, and the day is far spent." And He went in to stay with them. 30 Now it came to pass, as He sat at the table with them, that He took bread, blessed and broke it, and gave it to them. [31] *Then their eyes were opened and they knew Him; and He vanished from their sight.*

[3]*2 And they said to one another, "Did not our heart burn within us while He talked with us on the road, and while He opened the Scriptures to us?"* [33] *So they rose up that very hour and returned to Jerusalem, and found the eleven and those who were with them gathered together,* [34] *saying, "The Lord is risen indeed, and has appeared to Simon!"* [35] *And they told about the things that had happened on the road, and how He was known to them in the breaking of bread.*

[36] *Now as they said these things, Jesus Himself stood in the midst of them, and said to them, "Peace to you."* [37] *But they were terrified and frightened, and supposed they had seen a spirit.* [38] *And He said to them, "Why are you troubled? And why do doubts arise in your hearts?* [39] *Behold My hands and My feet, that it is I Myself. Handle Me and see, for a spirit does not have flesh and bones as you see I have."* [40] *When He had said this, He showed them His hands and His feet.* [41] *But while they still did not believe for joy, and marveled, He said to them, "Have you any food here?"* [4]*2 So they gave Him a piece of a broiled fish and some honeycomb.* [43] *And He took it and ate in their presence.*

[44] *Then He said to them, "These are the words which I spoke to you while I was still with you, that all things must be fulfilled which were written in the Law of Moses and the Prophets and the Psalms concerning Me."* [45] *And He opened their understanding, that they might comprehend the Scriptures.* [46] *Then He said to them, "Thus it is written, and thus it was necessary for the Christ to suffer and to rise from the dead the third day,* [47] *and that repentance and remission of sins should be preached in His name to all nations, beginning at Jerusalem.* [48] *And you are witnesses of these things.* [49] *Behold, I send the Promise of My Father upon you; but tarry in the city [m] of Jerusalem until you are endued with power from on high."* Luke 24:13-49 NKJV

[13] *Then they said to her, "Woman, why are you weeping?" She said*

to them, "Because they have taken away my Lord, and I do not
know where they have laid Him." 14 *Now when she had said this,*
she turned around and saw Jesus standing there, and did not know
that it was Jesus. 15 *Jesus said to her, "Woman, why are you weeping?*
Whom are you seeking?" She, supposing Him to be the gardener, said
to Him, "Sir, if You have carried Him away, tell me where You have
laid Him, and I will take Him away." 16 *Jesus said to her, "Mary!"*
She turned and said to Him, "Rabboni!" (which is to say, Teacher).
17 *Jesus said to her, "Do not cling to Me, for I have not yet ascended*
to My Father; but go to My brethren and say to them, 'I am ascend-
ing to My Father and your Father, and to My God and your God.' "
18 *Mary Magdalene came and told the disciples that she had seen the*
Lord, and that He had spoken these things to her.

19 *Then, the same day at evening, being the first day of the week, when*
the doors were shut where the disciples were assembled, for fear of the
Jews, Jesus came and stood in the midst, and said to them, "Peace be
with you." 20 *When He had said this, He showed them His hands and*
His side. Then the disciples were glad when they saw the Lord. 21 *So*
Jesus said to them again, "Peace to you! As the Father has sent Me, I
also send you." 22 *And when He had said this, He breathed on them,*
and said to them, "Receive the Holy Spirit. 23 *If you forgive the sins*
of any, they are forgiven them; if you retain the sins of any, they are
retained." 24 *Now Thomas, called the Twin, one of the twelve, was*
not with them when Jesus came. 25 *The other disciples therefore said*
to him, "We have seen the Lord." So he said to them, "Unless I see in
His hands the print of the nails, and put my finger into the print of
the nails, and put my hand into His side, I will not believe."

26 *And after eight days His disciples were again inside, and Thomas*
with them. Jesus came, the doors being shut, and stood in the midst,
and said, "Peace to you!" 27 *Then He said to Thomas, "Reach your*
finger here, and look at My hands; and reach your hand here, and
put it into My side. Do not be unbelieving, but believing." 28 *And*
Thomas answered and said to Him, "My Lord and my God!" 29 *Jesus*

said to him, "Thomas, because you have seen Me, you have believed. Blessed are those who have not seen and yet have believed."

30 *And truly Jesus did many other signs in the presence of His disciples, which are not written in this book;* 31 *but these are written that you may believe that Jesus is the Christ, the Son of God, and that believing you may have life in His name. John 20:13-31 NKJV*

1 *After these things Jesus showed Himself again to the disciples at the Sea of Tiberias, and in this way He showed Himself: 2 Simon Peter, Thomas called the Twin, Nathanael of Cana in Galilee, the sons of Zebedee, and two others of His disciples were together.* 3 *Simon Peter said to them, "I am going fishing." They said to him, "We are going with you also." They went out and immediately got into the boat, and that night they caught nothing.* 4 *But when the morning had now come, Jesus stood on the shore; yet the disciples did not know that it was Jesus.* 5 *Then Jesus said to them, "Children, have you any food?" They answered Him, "No."* 6 *And He said to them, "Cast the net on the right side of the boat, and you will find some." So they cast, and now they were not able to draw it in because of the multitude of fish.*

7 *Therefore that disciple whom Jesus loved said to Peter, "It is the Lord!" Now when Simon Peter heard that it was the Lord, he put on his outer garment (for he had removed it), and plunged into the sea.* 8 *But the other disciples came in the little boat (for they were not far from land, but about two hundred cubits), dragging the net with fish.* 9 *Then, as soon as they had come to land, they saw a fire of coals there, and fish laid on it, and bread.* 10 *Jesus said to them, "Bring some of the fish which you have just caught."* 11 *Simon Peter went up and dragged the net to land, full of large fish, one hundred and fifty-three; and although there were so many, the net was not broken.* 12 *Jesus said to them, "Come and eat breakfast." Yet none of the disciples dared ask Him, "Who are You?"—knowing that it was the Lord.* 13 *Jesus then came and took the bread and gave it to them, and likewise the fish.*

14 *This is now the third time Jesus showed Himself to His disciples after He was raised from the dead.* 15 *So when they had eaten*

breakfast, Jesus said to Simon Peter, "Simon, son of Jonah, do you love Me more than these?" He said to Him, "Yes, Lord; You know that I love You." He said to him, "Feed My lambs." [16] *He said to him again a second time, "Simon, son of Jonah, do you love Me?" He said to Him, "Yes, Lord; You know that I love You." He said to him, "Tend My sheep."* [17] *He said to him the third time, "Simon, son of Jonah, do you love Me?" Peter was grieved because He said to him the third time, "Do you love Me?" And he said to Him, "Lord, You know all things; You know that I love You." Jesus said to him, "Feed My sheep.* [18] *Most assuredly, I say to you, when you were younger, you girded yourself and walked where you wished; but when you are old, you will stretch out your hands, and another will gird you and carry you where you do not wish."* [19] *This He spoke, signifying by what death he would glorify God. And when He had spoken this, He said to him, "Follow Me."*

[20] *Then Peter, turning around, saw the disciple whom Jesus loved following, who also had leaned on His breast at the supper, and said, "Lord, who is the one who betrays You?"* [21] *Peter, seeing him, said to Jesus, "But Lord, what about this man?"* [22] *Jesus said to him, "If I will that he remain till I come, what is that to you? You follow Me."* [23] *Then this saying went out among the brethren that this disciple would not die. Yet Jesus did not say to him that he would not die, but, "If I will that he remain till I come, what is that to you?"* [24] *This is the disciple who testifies of these things, and wrote these things; and we know that his testimony is true.* [25] *And there are also many other things that Jesus did, which if they were written one by one, I suppose that even the world itself could not contain the books that would be written. Amen.* John 21:1-25 NKJV

Credibility

In an article written for Christianity.com by contributing writer Candice Lucey in February of 2020, she compares the conditions under which Jesus Christ died and His physical condition when He rejoined His disciples after His resurrection. This comparison clearly demonstrates that he died and His raising from the dead was not a trick, or that He just appeared to be dead.

> *"The resurrection means nothing unless Christ really died but refuting the evidence of His death and suggesting He was hidden somewhere and nursed back to health makes little sense. The gospels do not depict a weak, ailing Jesus displaying jagged, oozing wounds so that He could have recovered by the third day. 'Common complications of hemorrhagic shock include kidney damage, other organ damage, death" plus potentially "gangrene due to decreased circulation to the limbs.'"*
>
> *"The disciples saw Christ in good health, holes visible but healed (John 20:27), with energy to continue His ministry for some time. Those who believe that Christ died for their sins naturally feel a sense of guilt and pain when they realize what Jesus went through for their sake. But there is triumph here too, for God was able to rescue Christ from Sheol, and if He can do that, He is able to rescue us from our sin if we put our faith and trust in Him."*[172]

Jesus Christ continued to walk on the earth for forty days and nights after His resurrection and before His ascension. During that time there were several encounters He had with various individuals and groups prior to giving the Great Commission on the mountain, surrounded by His disciples and being drawn up into Heaven. As was indicative of His entire ministry, the first person He meets after His resurrection is a woman, Mary Magdalene, who mistook Him for the gardener. Upon hearing his voice, she realizes who He is. This initial encounter is important, among other things, because she is a female. Women during this historical period were not considered reliable witnesses. As always, God does not think like humans. In this case Jesus Christ wasn't just face-to-face with a woman, He was in the presence of a faithful, caring human being. Remember that it was a man, who was close to Jesus Christ for three years, who doubted. What He did with this encounter is even more telling. He sent this so-called unreliable female witness to tell His disciples, a bunch of men, He was alive! What a responsibility for her and a test for those He held close for three years. The disciple's reaction to this message, delivered from a woman, would be telling about what they had learned over those three years.

As if this first encounter wasn't enough, the second group He would encounter was also a group of women in Matthew 28:9! Their reaction? Worship.

Then He meets two disciples on the Road to Emmaus. These do not seem to be in His close group of twelve. One of them is identified as Cleopas, not a name associated with the twelve. This also accounts for their not recognizing Him until they sat with Him and shared a meal. It had probably been some time since they had last seen him up close. Following this, He meets a few of His original twelve disciples, who think He is a ghost. To move them away from the supernatural He shows them His scars. In Luke 24:39 He reassures them that all He had told them would happen; fulfilling what was written in prophecy in the Law of Moses, the prophets, and the Psalms.

There was one disciple not present during the previous encounter, Thomas. Thomas had been told of the miracle of seeing the risen Lord but he would not believe unless he saw it for himself and touched His wounds. Jesus Christ obliged in John 20:27. A careful reading of this passage suggests that Thomas didn't need to place his hand in the wounds or touch Him. It simply says he saw and believed.

This is also the moment where we find one of the misplaced Beatitudes. There are some fifty-one of these in Scripture, including those listed in Matthew 5.[173] After this incident with Thomas, Jesus Christ again uses the word blessed.

> 29 *Jesus said to him, "Thomas, because you have seen Me, you have believed. Blessed are those who have not seen and yet have believed."*
> John 20:29 NKJV

The next encounter includes reinstating Peter after his famous denial during Jesus' series of trials. It also includes the last recorded miracle, if you don't consider the ascension a miracle. Following Jesus' miracle of filling the disciples fishing nets with an enormous number of fish, He takes Peter off privately and asks him three unusual questions. He replies to each of Peter's answers with an instruction to feed the sheep and lambs and to follow Him. Based on Peter's life and ministry following this private conversation, it could be said that for the very first time, Peter "got it!" [174]

There were other appearances including one which is mentioned only by the Apostle Paul in 1 Corinthians. Paul, or Saul at the time, also saw Jesus on the Damascus Road. The encounter in 1 Corinthians is important because of the size of the group. "*6 After that, he appeared to more than five hundred of the brothers and sisters at the same time, most of whom are still living, though some have fallen asleep.*" 1 Corinthians 15:6 NKJV This is also important because this large group of people included people who were still alive at this time, greatly increasing the credibility of their testimony.

Not to be overlooked in these passages is John 20:22 NKJV, "*22 And when He had said this, He breathed on them, and said to them, "Receive the Holy Spirit.*" This is another instance of Scripture demonstrating the breath of God representing the Holy Spirit. It also suggests that the air God breathed into the first Adam was… the Holy Spirit. This could help fill in some other holes created by the science and faith communities. Science makes a big deal, as they should, about what the space between objects consists of. Science says molecules. In other words, the emptiness we see as air is also matter. What if that matter is actually the Holy Spirit binding all creation together? We cannot deny, based on this Scripture and others, that the final breath exhaled on the cross by Jesus Christ was different than the breath He exhaled onto these disciples after just *appearing* in their midst behind closed doors. Those doors were probably barricaded for fear of the Jewish disciples over the Temple's recent efforts to do away with everyone associated with the crucified blasphemer.

An Atheist Prospector

Credibility was a big deal to a contemporary in our 21st -century journey by the name of Lee Strobel, author of the book, *The Case for Christ.* The best-selling book was also made into a popular Christian movie by the same name. [175] Strobel was challenged by atheist author David Fitzgerald in an interview with Valerie Tarico with AlterNet.org. Fitzgerald, an atheist, challenges facts and events used by Strobel in his best-selling book. [176] As you might expect, Strobel counters with his rebuttal, clarifying and explaining away Fitzgerald's claims. [177] This is an interesting dialogue between an atheist and a former atheist. I will let you come to your own conclusions.

Why is this encounter important and why am I including it? I have stated earlier that if this event we call the resurrection didn't happen, or if any part of it isn't true, the entire Jesus story, and by association the entire Bible, is irrelevant. In other words, Christianity has no basis. Even if Fitzgerald's claim of irregularities in Strobel's book are grounded—you decide—Strobel's approach to substantiating the facts of the resurrection are point on. For Lee Strobel to move from total disbelief to accepting Jesus Christ as the Messiah and only Son of God needed a solid plan. It needed a scientific approach including testing, not just Scripture, which is in a way a biased source.

However, historical records and examining the credibility of witnesses who were close enough to the actual event to have properly recorded what they saw further clarifies the event. This would have included the first disciples. One of Strobel's most empirical tests included testimony of 20th -century medical doctors who validated, based on historical records of how Roman soldiers treated and punished prisoners during this period of time surely caused the death of Jesus of Nazareth—not a coma. If there was any doubt Jesus Christ really died, bringing Him back to life would be a worthless exercise.

With no doubt that Jesus Christ died during His torture or afterward on the cross, the only thing left for Strobel to develop was to believe that Jesus, in a very short period of time, rose from the dead, walked about, and interacted with other living people. Regarding reliability of witnesses, one of the records Strobel cited was that of Alexander the Great's actions, written 400 years after his death. In contrast, the records concerning the resurrection of Jesus Christ were written within one generation; some by people who were actually there.

Closing Thoughts on the Resurrection

In closing, there was one amazingly small, but dynamic, even seemingly subliminal message left by Jesus Christ during His time before the ascension. This one is outlined in an article on Beliefnet.com.

I often wondered why, in the rush and importance of everything we discover Jesus Christ had left to do before leaving this world, He would take the time to remove the handkerchief covering His face in the tomb. It was placed on His lifeless face by a believing Joseph or Nicodemus. It was folded neatly, and then placed on the slab where He had been lain after proper burial preparation. It is noted that the custom of the time was that if a person were to leave their seat while eating, their napkin would be placed over their plate as if through. If they were not through, the napkin would be neatly folded and laid on the table. By this tradition it would appear as if Jesus Christ was saying, "I will be back." [178]

In summarizing this forty-day period between the resurrection and the ascension, I want to close with a few verses that may bring together much of what has been said. It is important to delineate between the immaculately conceived Jesus Christ, 100 percent divine and 100 percent carnal infant, and the Christ who was reborn, or *"made alive by the Spirit"* out of the resurrection. Let's look at some Scripture.

> *For Christ also suffered once for sins, the just for the unjust, that He might bring us to God, being put to death in the flesh but made alive by the Spirit.* 1 Peter 3:18 NKJV

> *And so it is written, "The first man Adam became a living being." The last Adam became a life-giving spirit."* 1 Corinthians 15:45 NKJV

> *Therefore, from now on, we regard no one according to the flesh. Even though we have known Christ according to the flesh, yet now we know*

> *Him thus no longer.* [17] *Therefore, if anyone is in Christ, he is a new creation; old things have passed away; behold, all things have become new.* 2 Corinthians 5:16-17

It is clear that Christ gave up His flesh when He died on the cross. That flesh was embalmed and buried just like any other human. There was nothing divine about the crucifixion or the burial. However, the resurrection was supernaturally divine, just as miraculous as Mary's conception. Christ's being separating from Jesus' flesh shattered creation's core. There was an earthquake early that morning when the stone rolled away. Personally, I don't think the earthquake moved the stone an inch. I think the angels brushed it aside with their stare and a wing. The earthquake was surely creation's reaction to a divine, very rare occurrence; the Creator returning home by bursting out of nature's grasp. More Scripture.

> [51] *I am the living bread which came down from heaven. If anyone eats of this bread, he will live forever; and the bread that I shall give is My flesh, which I shall give for the life of the world.* John 6:51 NKJV

> [28] *just as the Son of Man did not come to be served, but to serve, and to give His life a ransom for many.* Matthew 20:28 NKJV

> [11] *But Christ came as High Priest of the good things to come, with the greater and more perfect tabernacle not made with hands, that is, not of this creation.* [12] *Not with the blood of goats and calves, but with His own blood He entered the Most Holy Place once for all, having obtained eternal redemption.* Hebrews 9:11-12 NKJV

... And Then There Was That

I saved Matthew 27:52-53 until now. Why? It seems out of place during the crucifixion. It appears to fit more into the period between the tomb and the ascension.

> "[52] *and the graves were opened; and many bodies of the saints who had fallen asleep were raised;* [53] *and coming out of the graves after*

His resurrection, they went into the holy city and appeared to many." Matthew 27:52-53 NKJV

Many Bible scholars do not try to force this into the visits Christ made after His resurrection. One of those who holds that these verses are right where they belong is John Piper, founder and teacher of DesiringGod.org and chancellor of Bethlehem College and Seminary. On his website he wrote on April 15, 2017:

"To be sure, Paul calls Jesus's resurrection "the firstfruits of those who have fallen asleep" (1 Corinthians 15:20)— meaning that Jesus's resurrection is part of the final resurrection of all his people. So, ours is as sure to happen as his has surely happened. It is all one harvest."

"But Matthew shows something different. Not contradictory, but different. He says that when Jesus died, and before the resurrection of Jesus, tombs were opened, and bodies of saints were raised."

"Even though it sounds very strange, Matthew locates the raising of the bodies of the saints at the death of Jesus, and their coming out of the tombs and entering Jerusalem after the resurrection."

"So, Matthew is showing that the raising of many saints' bodies at the point of Jesus's death gives glorious tribute to what was accomplished in that moment when he gave up his spirit (Matthew 27:50)."

"The greatest obstacle to our resurrection was not physical death and dissolution. The greatest obstacle was our sin and God's righteousness. God can put decayed bodies back together with the wave of his little finger. But the righteous removal of holy wrath cost the death of his Son."

"One last statement in the sequence of seven remains: "...and [they] appeared to many" (Matthew 27:53). "Many" raised saints are seen by "many." In other words, this was not a solitary visionary experience of some dreamer. This was many raised bodies appearing to many living inhabitants of Jerusalem. It was real, physical, and public. That is what it will be in the last day, when we are raised." [179]

The Ascension (Scripture from Matthew, Mark, Luke, and Acts. The ascension does not directly appear in the Gospel of John)

16 *Then the eleven disciples went away into Galilee, to the mountain*
which Jesus had appointed for them. 17 *When they saw Him, they*
worshiped Him; but some doubted. 18 *And Jesus came and spoke to*
them, saying, "All authority has been given to Me in heaven and on
earth. 19 *Go therefore and make disciples of all the nations, baptiz-*
ing them in the name of the Father and of the Son and of the Holy
Spirit, 20 *teaching them to observe all things that I have commanded*
you; and lo, I am with you always, even to the end of the age." Amen.
Matthew 28:16-20 NKJV

19 *So then, after the Lord had spoken to them, He was received up*
into heaven, and sat down at the right hand of God. 20 *And they*
went out and preached everywhere, the Lord working with them and
confirming the word through the accompanying signs. Amen. Mark 16:19-20 NKJV

50 *And He led them out as far as Bethany, and He lifted up His hands*
and blessed them. 51 *Now it came to pass, while He blessed them, that*
He was parted from them and carried up into heaven. 52 *And they*
worshiped Him, and returned to Jerusalem with great joy, 53 *and*
were continually in the temple praising and blessing God. Amen.
Luke 24:50-53 NKJV

1 *The former account I made, O Theophilus, of all that Jesus began*
both to do and teach, 2 *until the day in which He was taken up,*
after He through the Holy Spirit had given commandments to the
apostles whom He had chosen, 3 *to whom He also presented Himself*
alive after His suffering by many infallible proofs, being seen by them

during forty days and speaking of the things pertaining to the king-
dom of God.

4 *And being assembled together with them, He commanded them not*
to depart from Jerusalem, but to wait for the Promise of the Father,
"which," He said, "you have heard from Me; 5 *for John truly bap-*
tized with water, but you shall be baptized with the Holy Spirit not
many days from now." 6 *Therefore, when they had come together, they*
asked Him, saying, "Lord, will You at this time restore the kingdom
to Israel?" 7 *And He said to them, "It is not for you to know times*
or seasons which the Father has put in His own authority. 8 *But you*
shall receive power when the Holy Spirit has come upon you; and you
shall be witnesses to Me in Jerusalem, and in all Judea and Samaria,
and to the end of the earth." 9 *Now when He had spoken these things,*
while they watched, He was taken up, and a cloud received Him out
of their sight. 10 *And while they looked steadfastly toward heaven as*
He went up, behold, two men stood by them in white apparel, 11 *who*
also said, "Men of Galilee, why do you stand gazing up into heaven?
This same Jesus, who was taken up from you into heaven, will so come
in like manner as you saw Him go into heaven."

12 *Then they returned to Jerusalem from the mount called Olivet,*
which is near Jerusalem, a Sabbath day's journey. 13 *And when*
they had entered, they went up into the upper room where they
were staying: Peter, James, John, and Andrew; Philip and Thomas;
Bartholomew and Matthew; James the son of Alphaeus and Simon
the Zealot; and Judas the son of James. 14 *These all continued with*
one accord in prayer and supplication, with the women and Mary the
mother of Jesus, and with His brothers. Acts 1:1-14 NKJV

Before looking at research on this topic, let's take an individual look at what is recorded in the individual gospels and the book of Acts.

Matthew

Matthew devotes only five verses to the very end of Jesus Christ's ministry on earth. This recounting by Matthew, the former tax collector and agent of the Roman government, indicates the disciples were previously instructed by Jesus Christ to meet Him near Galilee on a mountain. In the second verse of this closing five verse summary, Matthew makes a startling comment. After over three years of following this Messianic teacher, Matthew comments, almost as an aside, "...but some doubted." Really, after what has taken place, and in their own presence? The fact that Matthew places this in his writing only confirms the statement to be true.

Verse 18 defines Heaven as a separate place than the Earth and the term Heaven doesn't seem to refer to the sky or even the universe. Christ's comment regarding *authority* is very clear that on Earth, and in Heaven, wherever that might be, was under His complete control. Christ's last words recorded in the book of Matthew is a command and a promise. The command was simple, (paraphrased) "Now go do what I taught and personally demonstrated you should do." The promise was that He would always, not sometimes, be with them until the end of the age. The term "age" as used here needs further research and discussion. Yes, we will get to that in its proper context.

Mark

Mark's Cliff Notes are consistent to the very end of his writing. Rather than the five verses dedicated by the former tax collector, Mark pens just two verses. All seem to be an afterthought or a summary of Matthew's notes. He does add, within the confirmation that Christ was taken up into Heaven, that He took a seat at the right hand of God.

Luke

Very much unlike Luke the physician, his summary is only four verses, actually less than Matthew. However, as we would expect, the brief remarks include detail not given anywhere else. Certainly, more than John, who seems to ignore the entire event. Luke names the location where this occurred. It is no longer just in the region of Galilee, but in the specific location of Bethany. Luke also records where the disciples traveled after the event. The location and what the disciples did is in contrast to what we will find later as they wait on the arrival of the Holy Spirit. Luke records they traveled to Jerusalem and, "*were continually in the temple praising and blessing God.*" The city was no surprise, but continually at the temple praising and blessing God seems odd since later they feared for their lives until the Holy Spirit fell upon them and they became bold. It is possible Luke simply skipped the fearful diagnosis and jumped to the prognosis, where they would receive the baptism of the Holy Spirit promised by Christ in the book of Acts chapter 1, verse 5.

Before summarizing this event in the Book of Acts, notice that each of the three Gospels that refer to this event end with the term, "Amen." Interestingly, so do most of the other books of the New Testament, including Revelation. Two books that do not end with *amen* are the books of John and Acts.

The pronunciation of *amen* is one of a few words that is almost always recognized internationally. It may be spelled differently, but it is generally pronounced the same way. Through both the Old and the New Testaments the word means a few different things in Hebrew and Kione Greek, including, *verily, so be it, it is so,* and *it is true.* Or, as in Deuteronomy 27:14-19, it is used as an affirmation or a response to something said. Today the affirmation comes after someone proclaims, "And all God's people said..." "Amen." Because we most often use the term at the end of a prayer, we interpret it to mean the end or "I'm done praying." Actually, according to Scripture, 1 Thessalonians 5:17, tells us to pray without ceasing. Prayers should never have an ending. Several times Jesus Christ used the term at the beginning of a sentence to set up the fact that what followed was the truth. Not just any truth, but THE Truth. He was the Truth and had every right to that phrasing. Back to why John and Acts do not end with "amen."[180]

The end of all four Gospels, including John, was the end of a period on the timeline of creation. The use of the term in the first three Gospels simply states that what we have recorded is true. John, while not using the term, in his next to last verse actually says what he wrote is true. The interesting assumption connecting John and Acts could be that John was setting up the continuing story, where the other three seem to take a pause.

Acts

Acts, as the introduction and foundation of the New Testament church, could not begin with any better opening scene than the ascension. The Book of Acts reaches back into the other four Gospels and makes a seamless transition into what would happen next. The same would be true with the end of the Book of Acts.

This book introduces humanity to many new characters. Some had been contemporaries of Jesus Christ, some witnesses to divine events, and some who never met him or heard Him speak or perform miracles.

The beginning of the Book of Acts sums up what the other Gospels recorded, adding two other details. First, we find that while the disciples were looking up into the cloud that had just swallowed up the earthly form of Jesus Christ, their reverie was interrupted by the voice of one of two men dressed in white, presumably angels, who spoke asking them why they were gazing into the heavens instead of getting on with what they had been commanded to do by the Son of God, who just departed. Second, we are given a list of names of those who gathered back in the upper room to wait for something called the anointing of the Holy Spirit. They heard Him say baptized, but what did *that* mean? It seems no one really understood. But, when it happened, the real intended ministry of the Son of God, planned before the Creation, was about to explode onto 1st-century civilization.

A Summary of the Spring of the Creation Story

Here we are at the end of one of the most celebrated periods in the history of mankind. We have followed the faithful recounting of four of the most trusted confidants of Jesus Christ, as they take us from His innocent, humble birth through a mostly unreported childhood and teenage period. They bring us inside his gathering of followers and disciples, miracles, teaching, religious confrontation, campfire talks, large-scale feeding and sermons, endless stories used to explain life choices, His betrayal, an arrest, trials, savage beatings, mocking, a cruel crucifixion, horrible death, burial, and resurrection. Then the climax with a huge Hollywood ending that includes large explosions, thousands of soldiers crashing into each other, and Jesus Christ rescuing all those in harm's way—including a lovable, sad little puppy! All the bad guys are wiped out, and Jesus establishes His Kingdom on the earth. There it is folks, the next giant picture show coming to a theater near you, or at least for rent on Netflix for $2.95.

Well, that was pretty accurate right up to the resurrection. We have discussed how it really ended. Truthfully, for a best-selling novel, the real ending was pretty lame. The four authors seem to just be tired of writing and end the story. Why is that? Shouldn't each writer have devoted at least a chapter or two to properly giving closure to everything preceding the ending? Hey, why start doubting the mind of God now? Haven't we seen, time after time, God not imitating His creation? My conclusion, after researching and thinking this through, is quite simple. This ain't the end of the story, y'all. As short and abrupt as we find the endings described by these Gospel writers, there were actually some very important points buried in their brief endings.

One of the most important was the answer to the question we found time after time inside the ministry stories within these four Gospels. It was well-known to any Jew old enough to comprehend their history that this group of people called *the Children of God* were looking for a military Messiah. A crusader on a white horse, He who would gather a mighty, divine army and take back their rightful ownership of the land and establish a *Kingdom* where He would reign and they would be free. Not to be disappointed, this subject came up… again… in the abbreviated discourse at the ascension. When the subject was raised, the thought of a Kingdom was not the issue, but, *when it would occur* became the central point. The answer, paraphrased, "It was none of their business… not important… only God knew *when.* Now, get to Jerusalem and wait for the baptism of the Holy Spirit." They did. It happened. Things changed dramatically. The New Testament Church was born!

In my research I did find something that will at least give some closure after enduring over 2,000 years of waiting. Bob Deffinbaugh, Th.M., graduate of Dallas Theological Seminary and pastor/teacher and elder of Community Bible Chapel in Richardson, Texas, wrote an extensive article in Bible.org on the subject of the ascension of Jesus Christ. In his conclusion, he lists six threads of doctrine applicable to Christians: separation, consummation, glorification, confirmation, transition, and anticipation. I encourage you to find this article and read it in its entirety. He ends with the following,

> *"It is not primarily to be viewed as the conclusion of our Lord's life and ministry, but as the introduction of a new phase of His ministry through His church, empowered by His Spirit. The assurance of His return and the measure of His presence and power in these intervening days is to be found, to a great extent, in His ascension. What a Savior!"* [181]

Due to the miraculous nature of this event, I also want to address the rejection of what happened at the ascension by some individuals, both atheist and scientific. Matt Nelson, writing for Catholic.com on May 28, 2019 accepts the challenge of various persons regarding the miracle of the ascension. He begins by stating that Catholics believe this event to be very real and profess it every Sunday morning. He quotes Episcopalian theologian John Shelby Spong, who reads the ascension as non-literal and symbolic by asserting, "A modern person knows that if you rise up off the Earth (as in the ascension), you don't go to heaven. You go into orbit."

Mr. Nelson defends his literal position with a rebuttal by C.S. Lewis:

> *After his Resurrection, it may have been that Our Lord, a being still in some mode, though not our mode, corporeal, withdrew at His own will from the Nature presented by our three dimensions and five senses, not necessarily into the non-sensuous and undimensioned but possibly into, or through, a world or worlds of super-sense and superspace. And He might choose to do it gradually. Who on earth knows what the spectators might see? If they say they saw a momentary*

movement along the vertical plane – then an indistinct mass – then nothing – who is to pronounce this improbable?

He quotes David Hume as saying, "…that a miracle is a violation of the laws of nature." The assumption is that God, if He existed, would have no right to create a supernatural effect in the natural world. Nelson's response is simply, "Why not?"[182] I would agree with him. If God is powerful enough to create something, He has the authority to change it, unless He covenants not to do so. I don't recall Him making such a covenant against nature. He did, however, make a promise dealing with nature after flooding the earth. He placed a rainbow in the heavens as a symbol He would never do that again.

Here we are at the division between the faith and science communities. These two have got to come together. My goal is to provide a platform where that can happen before the end of the book—look for it in chapter 26.

One last thing of importance I want to include here that will be covered in greater detail in the Fall section of this book. Consider Acts chapter 1, verse 11, *"Who also said, 'Men of Galilee, why do you stand gazing up into heaven? This same Jesus, who was taken up from you into heaven, will so come in like manner as you saw Him go into heaven.'"* The last half of that verse is haunting. In today's language, *He's coming back!* Remember that folded napkin on the tomb table and its meaning? He left, but He wasn't through with His meal. Don't take away His dishes! This is where the argument over Jesus Christ building a kingdom became stalled. There was no rebuttal concerning a kingdom. What remained in question was when He was coming back.

Although the term, *where will the kingdom be,* is debated to this day, the one large unknown was… when? Well, according to *bobology*, I have a theory. All through the building of the New Testament church, this phrase is thrown around in various forms, mostly to accept Jesus Christ as your savior because He is coming back soon. I think the answer is simple and quite rational. It has to do with something I have covered earlier in this book… time… Earth time. It bears repeating.

Just to remind you—once you leave the bounds of the Earth your

Timex is quite useless. Time on Earth is regulated by the relationship of the Earth's rotation to the rays of the sun. If you lived on the moon, you may use Earth time for convenience but it would have nothing to do with time on the moon. The same is true, from what I read in Scripture, concerning Earth time and time in heaven. Here is my simple explanation as to when Christ is coming back. The split-second I die... He is coming back. Why? Because the split second I die, earthly time loses all meaning to me. Time for me may speed up or slow down. Really, I don't have a clue what it will do, or if it is even at all important. So how do I prove it? Simple. The second I exhale my last breath, my ability to choose Christ as my Savior ends. It doesn't matter when He returns, my decision-making is over. So, I need to treat every second I have left as if it is happening just before He returns. I need to be packed and ready to roll. When will you exhale your last breath? Hold that thought, (not your breath!) and let's turn to the birth of the New Testament Church and the beginning of the Summer of Creation...

SUMMER

The development of the New Testament Church through Christ's third Coming (yes, there will be three)

CHAPTER 16
THE SECOND GENESIS

... What if man had never sinned, would he have remained in the Garden?
... If Jesus was the second Adam, who was the second Eve?
... Will the Summer of Creation last as long as the Winter?
... Hey, who is going to write the 'new', New Testament to record all this?

... For since by man came death, by Man also came the resurrection of the dead. For as in Adam all die, even so in Christ all shall be made alive.

1 CORINTHIANS 15:21-22 NKJV

THE BEGINNING OF the Summer of Creation starts with a recounting of the last days of Jesus Christ on Earth by His disciple Luke. His writings are addressed to Theophilus in the Book of Acts 1:1. Theophilus is also mentioned in Luke 1:3, adding the title "most excellent." This could have inferred that he was a government official or at a minimum, a man of means in Rome. Some even claim the term Theophilus, meaning *loved of God,* was simply a term of endearment Luke used to address a larger, all-Christian audience.[183]

Although many have speculated about who this individual may have been, it is clear that the Book of The Acts of the Apostles is a long introduction of the birth of the New Testament Church. More than an important resource for Luke, and therefore all of the leaders of the church, the dedication to Theophilus would have been less about his position in society and more about a compassionate Luke's effort to introduce his friend to Christ, Who could change his friend's life for eternity.

Many have written about Jesus as the second Adam. Some of these writers infer that Adam was a failure by disobeying God. The second Adam, Jesus, was perfect and overcame the *original* sin. More than the opinion of man, Scripture is clear in 1 Corinthians 15:21-22,

> *For since by man came death, by Man also came the resurrection of the dead. For as in Adam all die, even so in Christ all shall be made alive.*

Also, in Roman's 5:12-21, we find the Apostle Paul's empassioned address to those of Rome who were strict legalists,

> 12 *Therefore, just as through one man sin entered the world, and death through sin, and thus death spread to all men, because all sinned—* 13 *(For until the law sin was in the world, but sin is not imputed when there is no law.* 14 *Nevertheless death reigned from Adam to Moses, even over those who had not sinned according to the likeness of the transgression of Adam, who is a type of Him who was to come.* 15 *But the free gift is not like the offense. For if by the one man's offense many died, much more the grace of God and the gift by the grace of the one Man, Jesus Christ, abounded to many.* 16 *And the gift is not like that which came through the one who sinned. For the judgment which came from one offense resulted in condemnation, but the free gift which came from many offenses resulted in justification.*
>
> 17 *For if by the one man's offense death reigned through the one, much more those who receive abundance of grace and of the gift of*

righteousness will reign in life through the One, Jesus Christ.) [18] *Therefore, as through one man's offense judgment came to all men, resulting in condemnation, even so through one Man's righteous act the free gift came to all men, resulting in justification of life.* [19] *For as by one man's disobedience many were made sinners, so also by one Man's obedience many will be made righteous.* [20] *Moreover the law entered that the offense might abound. But where sin abounded, grace abounded much more,* [21] *so that as sin reigned in death, even so grace might reign through righteousness to eternal life through Jesus Christ our Lord.* NKJV

Before coming down too hard on Adam, perhaps we should recount what God instructed him to do. I can't find any Scripture where God specifically told Adam, or for that matter Eve, to go and be perfect. However, He did say, Go and be fruitful and multiply.

[28] *Then God blessed them, and God said to them, "Be fruitful and multiply; fill the earth and subdue it; have dominion over the fish of the sea, over the birds of the air, and over every living thing that moves on the earth."* Genesis 1:28 NKJV

Later, God told them there was one thing they should not do,

[2] *And the woman said to the serpent, "We may eat the fruit of the trees of the garden;* [3] *but of the fruit of the tree which is in the midst of the garden, God has said, 'You shall not eat it, nor shall you touch it, lest you die.'"* Genesis 3-2-3 NKJV

When God created man and woman, and many other things in the beginning, He said they were *good*, He didn't say they were perfect. In fact, many times in the Bible we are told that only God is perfect, as is the Trinity. It may be a bit unfair to blame Adam and Eve for not being perfect. That doesn't seem to have ever been the plan. In fact, there are many things in life, at least at some macro-level, that are imperfect. Things on this God-created Earth have imperfections. Some would claim that imperfections actually make them beautiful and unique.

So, why did God create Adam and Eve, and by extension—us? It is summarized in Isaiah 43:7 NKJV,

> *Everyone who is called by My name, Whom I have created for My glory; I have formed him, yes, I have made him.*

Simply put, to glorify God.

From this, one could summarize by saying that God created man to fail so His perfect Son could come to the Earth, be born, die, resurrect, and ascend back to His Father to overcome the failure God originally created. It is clear that after Adam and Eve sinned and were ejected from the Garden, it seems God's attitude toward His creation sours. Looking forward from the original sin, untold people—both children and adults—were killed by God's hand. Some examples that come to mind are Sodom, the flood, the Children of Israel in the desert, and innocent babies at the birth of Moses. To get Christ into the world, countless other babies were murdered by Herod. While God had the ability to be compassionate, sometimes He used other emotions with unexplainable consequences. To demonstrate God's ability to recognize this, Scripture tells us He placed a rainbow in the sky after the flood as a promise to never do that again. These are tough questions, but they are questions that could very well come up as you discuss your faith with an unbeliever. It is best to be prepared to express your ideas concerning these events.

I know this summary is poking at you, but I am going to leave it lying there for now. Before we address this again, we need to take another journey that is as long, if not longer, than the first one we took through the Winter and Spring of Creation. This journey will take an endless path forward just as our previous journey receded into an endless past. The previous paragraph is the axis point between these two journeys into faith, riding on the incomplete tracks of science. Both of these, faith and science, were created in Genesis 1:1. The curious thing is that when Christ returns for the last time, neither faith nor science will remain in the vocabulary of Heaven. Both will have served a divine purpose and will be permanently retired.

One last item to be addressed before we move ahead on our timeline:

If Jesus Christ was the second Adam, who was the second Eve? She was the church; the Bride of Christ. Guess what? That's where we will begin; the birth of the New Testament Church. Below is a comparison of the church to a bride.

> [22] *Wives, submit to your own husbands, as to the Lord.* [23] *For the husband is head of the wife, as also Christ is head of the church; and He is the Savior of the body.* [24] *Therefore, just as the church is subject to Christ, so let the wives be to their own husbands in everything.*[25] *Husbands, love your wives, just as Christ also loved the church and gave Himself for her,* [26] *that He might sanctify and cleanse her with the washing of water by the word,* [27] *that He might present her to Himself a glorious church, not having spot or wrinkle or any such thing, but that she should be holy and without blemish.*
>
> [28] *So husbands ought to love their own wives as their own bodies; he who loves his wife loves himself.* [29] *For no one ever hated his own flesh, but nourishes and cherishes it, just as the Lord does the church.* [30] *For we are members of His body, of His flesh and of His bones.* [31] *"For this reason a man shall leave his father and mother and be joined to his wife, and the two shall become one flesh."* [32] *This is a great mystery, but I speak concerning Christ and the church.* [33] *Nevertheless let each one of you in particular so love his own wife as himself, and let the wife see that she respects her husband.* Ephesians 5:22-33 NKJV

> [6] *And I heard, as it were, the voice of a great multitude, as the sound of many waters and as the sound of mighty thunderings, saying, "Alleluia! For the Lord God Omnipotent reigns!* [7] *Let us be glad and rejoice and give Him glory, for the marriage of the Lamb has come, and His wife has made herself ready."* [8] *And to her it was granted to be arrayed in fine linen, clean and bright, for the fine linen is the righteous acts of the saints.* Revelation 19:6-8 NKJV

What follows is Jesus Christ's metaphor about the church being prepared for His return, using a bride and bridegroom.

> 1 *“Then the kingdom of heaven shall be likened to ten virgins who took their lamps and went out to meet the bridegroom.* 2 *Now five of them were wise, and five were foolish.* 3 *Those who were foolish took their lamps and took no oil with them,* 4 *but the wise took oil in their vessels with their lamps.* 5 *But while the bridegroom was delayed, they all slumbered and slept.*
>
> 6 *“And at midnight a cry was heard: ‘Behold, the bridegroom is coming; go out to meet him!’* 7 *Then all those virgins arose and trimmed their lamps.* 8 *And the foolish said to the wise, ‘Give us some of your oil, for our lamps are going out.’* 9 *But the wise answered, saying, ‘No, lest there should not be enough for us and you; but go rather to those who sell, and buy for yourselves.’* 10 *And while they went to buy, the bridegroom came, and those who were ready went in with him to the wedding; and the door was shut.*
>
> 11 *“Afterward the other virgins came also, saying, ‘Lord, Lord, open to us!’* 12 *But he answered and said, ‘Assuredly, I say to you, I do not know you.’* 13 *“Watch therefore, for you know neither the day nor the hour in which the Son of Man is coming. Matthew 25:1-13 NKJV*

Other Scriptures highlight comparisons between Eve’s following Satan’s lead into sin and the New Testament Churches straying from God. If the Book of Acts is the introduction to the birth of the New Testament Church, who are the doctors who birthed the baby? You guessed correctly, the twelve. Well, first they had to fix a hole in the team. They needed to replace Judas. What better place to take care of this little piece of business than in the Introduction? Let’s look in on a meeting of, well, apparently all of the disciples in Acts 1:15.

> 15 *And in those days Peter stood up in the midst of the disciples (altogether the number of names was about a hundred and twenty), and said,* 16 *“Men and brethren, this Scripture had to be fulfilled, which the Holy Spirit spoke before by the mouth of David concerning Judas, who became a guide to those who arrested Jesus;* 17 *for he was numbered with us and obtained a part in this ministry.”* 18 *(Now this*

man purchased a field with the wages of iniquity; and falling headlong, he burst open in the middle and all his entrails gushed out. [19] *And it became known to all those dwelling in Jerusalem; so that field is called in their own language, Akel Dama, that is, Field of Blood.)* [20] *"For it is written in the Book of Psalms: 'Let his dwelling place be desolate, And let no one live in it'; and, 'Let another take his office.'*

[21] *"Therefore, of these men who have accompanied us all the time that the Lord Jesus went in and out among us,* [22] *beginning from the baptism of John to that day when He was taken up from us, one of these must become a witness with us of His resurrection."* [23] *And they proposed two: Joseph called Barsabas, who was surnamed Justus, and Matthias.* [24] *And they prayed and said, "You, O Lord, who know the hearts of all, show which of these two You have chosen* [25] *to take part in this ministry and apostleship from which Judas by transgression fell, that he might go to his own place."* [26] *And they cast their lots, and the lot fell on Matthias. And he was numbered with the eleven apostles.* Acts 1:15-26 NKJV

The first chapter of Acts fills in a lot of holes. Luke references his other work, the Book of Luke, and then validates the other works and teachings of Jesus Christ, including appearing to many after His resurrection and before His ascension. Luke mentions Jesus Christ's command that the eleven remain in Jerusalem until they are baptized with the Holy Spirit, but will not be told when He will return again. Luke also restates Jesus Christ's Great Commission to the eleven. After He ascends into Heaven, they all return to the borrowed upstairs room to wait. Verses 13 and 14 again name the eleven, as well as Mary the mother of Jesus and His brothers (half-brothers.) It would be assumed that His step-brother James, the writer of the Book of James, at the time a non-believer, was also present.

Verse 15 ushers in an abrupt scene change. Peter, the assumed leader of the group, is now standing in a very large group of men and brethren he is about to address. I don't have a clue what the difference is between men and brethren. We also must assume that while the eleven were back at the borrowed upper room, they must have read some of King David's writings and found a reference to what happened to their lost disciple,

Judas Iscariot. They had determined that they must choose someone to take his place so there would again be twelve. It seems they also determined that individual should come from the larger group of "about" one hundred and twenty. We find out something else that is rarely mentioned elsewhere in Scripture. The reason Judas' replacement should come from this group is that,

> 21 *"Therefore, of these men who have accompanied us all the time that the Lord Jesus went in and out among us,* 22 *beginning from the baptism of John to that day when He was taken up from us, one of these must become a witness with us of His resurrection."* Acts 1:21-22 NKJV

It has been a subject of speculation as to who was present during Jesus Christ's roaming about, preaching, teaching, and performing miracles; even during His prayer in the garden and His arrest. This narrative by Luke seems to clarify that it wasn't a small band of thirteen in the various scenes in the Gospels, but a small army. This would have added a lot more dynamics to every scene where we encountered Jesus Christ in Scripture.

Following the apparently 1st -century equivalent of drawing straws, they chose the two best candidates and then cast lots to see who would get the appointment. Out of the two candidates, Justus and Matthias, Matthias was chosen as Judas' replacement. These men were preparing for something… they just had no clue as to what. They were all still waiting on something Jesus Christ called a baptism of the Holy Spirit.

Surely all of the men, over one hundred of them, had either heard first or second-hand what John the Baptist had preached and Jesus Christ validated. They had witnessed hundreds or even thousands of water baptisms; some of them had likely performed baptisms themselves. It is disappointing but not surprising that none of them put the two terms together spoken by John the Baptist, "but He will baptize with the Spirit and with fire!" They were about to catch on.

Chapter 2 of the Book of Acts is a very big deal. The events outlined in this chapter created an entirely new celebration in the Christian Faith that is still celebrated today, Pentecost Sunday, or the Day of Pentecost. What is it? Consider the following promise:

> [25] *"These things I have spoken to you while being present with you.* [26] *But the Helper, the Holy Spirit, whom the Father will send in My name, He will teach you all things, and bring to your remembrance all things that I said to you."* John 14:25-26 NKJV

On this day, a day promised by Jesus Christ in a very intimate setting with His closest disciples, the promised Holy Spirit, the helper, the comforter, will arrive in a very unexpected way. You can imagine one of the first things the Holy Spirit will help them remember was that day written about in the 14th chapter of John.

This is a great place to use your imagination. Understanding what is happening at the moment the Holy Spirit consumes the room lets our mind explode with images. A rushing mighty wind, tongues or fire splitting and resting on the heads of each one of the chosen. Men speaking in tongues with words and sounds they had yet to learn. Jerusalem was a big city and although this event may not have originally affected the whole population, you can bet in a short period of time word would have traveled throughout the entire city.

Locally, people from all over the world would be drawn to the actions of what must have seemed chaos as the disciples proclaimed the Word of God in a language understood by everyone, regardless of where they were from in the world. Verses 12 and 13 draw a vivid comparison between the two ideologies present that day:

> [12] *So they were all amazed and perplexed, saying to one another, "Whatever could this mean?"* [13] *Others mocking said, "They are full of new wine."* Acts 2:12-13

Let your imagination stretch a little further. There were also Romans present. Word of this incident would not go unnoticed by those who crucified "the King of the Jews." Nor would it go unnoticed by King Herod, the Sanhedrin Court, Pilate, or most of all, Caiaphas. My biggest curiosity would be the effect the news had on Pilate's wife. This could have driven her mad if taken as final confirmation of the dream she described to her husband. The only answer I have for these actions by God is… He has a tremendous sense of humor.

Leaving our imaginations, (attributes of a creative God passed along to His creation), we return to the awesome but serious agenda of the newly arrived comforter. Having examined the Gospels, we know whom to call on to calm a chaotic situation. Who else but… Peter? In verse 14 through 16 of the 2nd chapter of Acts, Peter stands up and motions the other eleven to stand. With no microphone, Peter speaks a warning in his loudest voice:

> 14 *Men of Judea and all who dwell in Jerusalem, let this be known to you, and heed my words.* 15 *For these are not drunk, as you suppose, since it is only the third hour of the day.* 16 *But this is what was spoken by the prophet Joel:*

It seems Peter speaks these words from memory, since he probably didn't think to bring along a scroll. It is probably also the result of the Holy Spirit's presence. These are powerful words describing the End Times, and that only those who call on the name of the Lord will be saved. Peter's first address was to the people of Judea living within the city of Jerusalem. After quoting the prophet Joel, Peter turns his attention to the people of Israel and tells them in detail just who Jesus was, where He came from, what they did to Him, and how he overcame it by rising from the dead. He then quotes King David. He follows this with a reminder to the men and brethren that David was both dead and buried and did not ascend into the heavens as Christ did. The people were cut to the heart when Peter said to them, *"Repent, let every one of you be baptized in the name of Jesus Christ for the remission of sins; and you shall receive the Holy Spirit."*

From this impromptu revival, the New Testament Church was born in the following manner:

> 40 *And with many other words he testified and exhorted them, saying, "Be saved from this perverse generation."* 41 *Then those who gladly received his word were baptized; and that day about three thousand souls were added to them.* 42 *And they continued steadfastly in the apostles' doctrine and fellowship, in the breaking of bread, and in*

prayers. [43] *Then fear came upon every soul, and many wonders and signs were done through the apostles.*

[44] *Now all who believed were together, and had all things in common,* [45] *and sold their possessions and goods, and divided them among all, as anyone had need.* [46] *So continuing daily with one accord in the temple, and breaking bread from house to house, they ate their food with gladness and simplicity of heart,* [47] *praising God and having favor with all the people. And the Lord added to the church daily those who were being saved.* Acts 2:40-47 NKJV

From such a dynamic beginning, we will now move to a more personal look at each of the twelve and a few others, pertaining to the impact they may have had on the founding and growth of the New Testament movement.

A Personal Note to the Reader:

One of the objectives of this book is to give someone of faith a common-sense historical reference when having a discussion with a non-believer. One of the fundamentals is understanding the development and structure upon which the faith community constructs its belief. The foundation of that structure, in addition to Christ's birth, teaching, death, burial, resurrection, and ascension, is the initiation and growth of the New Testament Church.

Most of the remainder of the Summer of Creation will address in detail the people, the interactions, and establishment of what today constitutes the faith community.

For an unbeliever, reading this section may perhaps provide background that is helpful in constructive future discourse with a member of the faith community. As with previous events we have covered, there are several things to discuss that may be surprising to both communities when viewed from these perspectives.

CHAPTER 17
THE NEW BAND OF TWELVE

... I always heard there were twelve disciples
... well there were, but then there were eleven
... but they filled the vacancy by choosing
from the other seventy-two
... maybe so, but I thought there were 120 or more

After that He was seen by over five hundred
brethren at once, of whom the greater part remain
to the present, but some have fallen asleep.

1 CORINTHIANS 15:6 NKJV

THE FAITH COMMUNITY has a deep, rich history. The foundation was Jewish history stretching back thousands of years, secure in the fact that God created the world and everything in it. Thus, mankind moved forward with Divine instruction and leadership. At a specific point on that timeline, the Holy Trinity decided to turn everything upside-down and begin a new thing, led by God's Son, Christ. This adjustment period lasted for about thirty-three years. During that short time frame His Son gathered 12, 72, 120, 500—no one really knows how many disciples—to train and send out to all of the known and unknown world of the time. As pointed out earlier in this book, it is vital to remember as we peer into this specific period that there were several other significant non-Jewish civilizations all around the globe during this time frame.

Prior to this abrupt change in direction, God concentrated on what had become known as *God's chosen people*, the Jewish nation. This abrupt change in messaging would require a change in receptors. The ears of this new group of people would be called Gentiles. If you needed someone to talk to a different group of people, and I mean different in every way, you would need a very special person, someone who was respected by the Jews but believable to the Gentiles. God's choice? A Jew by birth whose sole priority at the time was seeking out Jews professing to believe in Jesus Christ, and have them put to death.

To lead the new church, God would put in place a man who was close to Jesus Christ all his life, and who had never believed Him to be who He said He was—Jesus Christ's half-brother. Okay, that's a start. Now there is the small issue of the rag-tag band of brothers—literally, in some cases. Such were those Jesus Christ pieced together during His short miracle tour, whom He would leave behind after He returned to the comforts of Heaven.

Looking back on what was to be the start-up team for the New Testament Church, one would wonder how they survived their first year in business. That's the catch; it wasn't a start-up. *It was the continuation of creation.* Read that last six-word sentence again. It is huge.

In Matthew Chapter 10, we are reminded of the leadership training session held by Jesus. Following that session, He then led the trainees out into the real world and demonstrated what He taught. Did they get it the first time? Nope. New trainees rarely do until they are given the opportunity to fail; and fail they did... but they had an eternally patient teacher. Thank God! (literally).

Combined with these seasoned evangelists, if you could have called them that, God mixed in some other interesting folks to establish His church. Before we take a look at the many others who made up the New Testament church—the Bride of Christ—let's examine a number of that 500 or so disciples chosen and attracted by Jesus Christ during His ministry. Several of these would slowly move from the title of disciple or student, to apostle or teacher.

Of the original twelve apostles, it would make sense that those closest to Jesus Christ during His ministry would become the first leaders of this new thing. Of course not. Not the way the Lord works, His wonders to perform. What about Simon Peter, the Rock upon which Jesus Christ said He would build His church? Perfect choice… didn't happen! At least, not at first. What about John, the disciple beloved by Jesus Christ during most of His ministry and a member of the Divine trio—Peter, James, and John? No way. Then the last member of the trio, James, the oldest of the trio? Not the choice. In fact, we don't hear a lot out of Him after the ascension and he is the first to die a martyr's death in 44 AD. No, and it wasn't Paul. He isn't a Christian yet. How about one of the other men by the name of James, the half-brother of Jesus Christ.

As the story goes, believe it or not, boisterous Peter, always the first to jump up and proclaim Jesus Christ as Savior and Lord, and the two brothers renamed for their fiery dispositions, the "Sons of Thunder," the Zebedee boys, all out of humility, eventually invited a new convert to Christianity to take charge of the new church in Jerusalem. Oh, this story is just beginning. There is still plenty of time to mess this up and of course… all with the best intentions and reasons. Let's pause and take a quick look at the other candidates, including what they actually did after the ascension, according to some recorded religious and secular records, and some unproven, passed-down myths.

The First Trail of Tears

Andrew

Born in Bethsaida. He was Peter's older brother and originally a disciple of John the Baptist. Andrew believed Jesus to be the Messiah before his brother Peter, and is also known as the *Protocletus*, or the *first called.* Andrew was known to introduce others to Jesus, beginning with his brother. He also brought the little boy with five loaves and five fishes to Jesus, who then fed the five thousand. He traveled far from the Holy Land during his post-ascension mission. It is said that he preached around the Black Sea arriving as far north as the Ukraine, Romania, and Russia. He established the See of Byzantium (later Constantinople) in 38 AD. When martyred, at his request, he was crucified on an X-shaped cross in Patras, in Greece, on November 30, 60 AD after being flogged. Like his brother, Peter, he did not feel that he should be crucified on the same type of cross as the Messiah.

The X-shaped cross became known as the St. Andrew's Cross. Many country flags bear this symbol. Notably, Scotland's flag, where Andrew is the Patron Saint of that country. Stories say he continued to preach for two days before expiring on the instrument of his destruction. His remains have been divided among churches in Patras, Amalfi, Edinburgh, and Warsaw as well as many other places.

Because of his patronage of Constantinople, Saint Andrew's relics were transferred there around the year 357 AD. Tradition holds that some relics of Saint Andrew were taken to Scotland in the 8th century, to the place where the town of St. Andrews stands today. In the wake of the Sack of Constantinople during the Fourth Crusade, the remaining relics were brought to the Cathedral of Saint Andrew in Amalfi, Italy. In 1964, in an attempt to strengthen relations with the Ecumenical Patriarch in Constantinople, Pope Paul VI returned all relics of Saint Andrew that were then in Rome to the Greek Orthodox Church.[184]

Peter

Peter was the younger brother of Andrew, both of whom came from Bethsaida in Galilee. He and Andrew were Jewish fishermen living at the northwest end of the Sea of Galilee when they were called by Jesus to follow Him. According to Luke 5:10, it is suggested that he and Andrew were in a fishing business with James and John, the sons of Zebedee. Peter was probably married.

Peter's original name was Simeon (Hebrew) or Simon (Greek). He was later renamed Cephas (Rock) by Jesus, Who proclaimed that He would build His church on this rock. He is credited with writing 1st and 2nd Peter.

Peter was chosen early on by Christ as leader of the twelve. Upon the betrayal and death of Judas, one of his first tasks after the ascension was to lead the apostles in the selection of a replacement for their disgraced colleague. The group chose Matthias. Then, he was the first to raise his voice and preach at Pentecost following the anointing of the Holy Spirit promised by Jesus Christ at the ascension. In these first years, the yet-to-be-established church looked to Peter for guidance. His leadership would dominate the faith community for fifteen years after the ascension, and up until his arrest and imprisonment. These early years of the new church fulfilled Jesus' assertion that it would be built on the foundation laid by Peter. Even so, he would not become the long-term leader and voice of the church.

It was during this time that the official church of Jerusalem began to take shape and James, the half-brother of Jesus Christ, who had refused to believe anything his brother proclaimed to be true, was encouraged to become the leader of this new church. This transition was not without dissent and disagreement over, of all things, tradition. Imagine that. Let's review.

Peter served as an advocate for the apostles before the Jewish religious court in Jerusalem in Acts 4:5-22. He acted as judge in disciplining those who erred within the church, Acts 5:1-10. Peter went first to Samaritans in Acts 8:4-17, who then received the Holy Spirit and encountered the magician and faith healer Simon Magus in Samaria. Later, he led the other eleven in extending the church in Acts 9:32 including healing the paralyzed Aeneas in Lydda in the Plain of Sharon in Acts 9:32-35. In Joppa, a Mediterranean Sea Port, Acts 9:36-43, he cured Tabitha (Dorcas) in the name of Christ.

Most importantly, in Acts 10:1 thru 11:18, he introduced the Gentiles into the church through Cornelius, a centurion of the Italian Cohort in Caesarea. This became a huge issue, since at that time the Gentiles were required to be circumcised before becoming accepted by the Church. Peter instead had them baptized in the Name of Jesus Christ in Acts 10:48. This ensured opposition by Jewish Christians and others. Although blessed by the Spirit in Acts 10:10-15, this could have been what led to the beheading of James the brother of John and the arrest of Peter by Herod in Acts 12:2-3.

Peter had his chains removed by an angel in prison and he escaped in Acts 12:1-8. He immediately went to the house of John Mark's mother.

He asked them to report his escape to James and the brethren, and then went to another place in Acts 12:17. It is this incident where we lose track of Peter for a bit. It is also during this time that James, the half-brother of Jesus, is led into the position as head of the Jerusalem church. It is also the time when the converted Apostle Paul becomes involved in the circumcision debate. We will back up and address Paul's involvement later when we review his contributions.

Up until his imprisonment by Herod, sanctioned by the Sanhedrin, Peter enjoyed an unchallenged leadership role in the Jerusalem church. As an example, Paul went to see Peter in Jerusalem three years after he was converted on the Road to Damascus and spent two weeks with Peter, (Galatians 1:18-19). When he left Jerusalem, in something of a missionary journey, leadership finally fell to James. Peter was still not over his problem concerning eating with the Gentile Christians at Antioch (Galatians, 2:11-14). This was addressed by the so-called Apostolic Council of Jerusalem in Acts 15:7.

What at first was thought to be a humble transition of leadership in the Jerusalem Church turned into growing dissention between Peter and James over treatment of Gentile Christians.

When Paul goes to see Peter after his conversion, Peter's name is listed first. Fourteen years later, in Galatians 2:9, James' name is listed first, indicating a change of leadership.

Paul had tried very hard to get the Jews and Gentiles of the Antioch church to eat together. In Paul's absence, Peter visited Antioch as a missionary, often accompanied by Mark, and ate with the combined group. Later some folks came from James and opposed their eating together. The two leaders, James and Peter, drew away from each other and the unity of the church was destroyed. From this the Jerusalem Council, in 49 or 50 AD, settled that Paul would be entrusted with the gospel to the uncircumcised, and Peter would tend to the mission of the circumcised (Galatians 2:8).

Later, Paul refers to a party of Cephas (Peter) in 1 Corinthians 1:12 that suggests a group in the church of Corinth was especially devoted to Peter (leading some to assume a residence of Peter in Corinth) and refers to Peter in 1 Corinthians 9:5 as carrying on missionary activity accompanied by his wife. A missionary journey to Asia Minor may be suggested in the First Letter of Peter 1:1.

Peter was not originally a cheerleader for the Gentiles. The opening of Acts chapter 10 introduces what might be the human holding the key to the door locked between Jews and Gentiles for centuries,

> 1 *There was a certain man in Caesarea called Cornelius, a centurion of what was called the Italian Regiment,* 2 *a devout man and one who feared God with all his household, who gave alms generously to the people, and prayed to God always.* 3 *About the ninth hour of the day he saw clearly in a vision an angel of God coming in and saying to him, "Cornelius!"* NKJV

God, realizing how difficult it was going to be to get His message of salvation to the world through the traditions of Jewish centuries, created a well-timed incident utilizing someone who, compared to others in Scripture, would make only a cameo appearance. That appearance would remove centuries of traditional bias allowing Peter and other Jews to work alongside the newly converted Apostle Paul to reach both Jews and Gentiles, the name then associated with every non-Jew. The tradition in question was circumcision.

The short story revealed in Acts chapter 10 can be summarized as follows. I would encourage you to read this chapter in its entirety, in several versions of the Bible.

While Peter already understood the Gospel was to be given to everyone according to the Great Commission, spoken by Jesus Christ just before the ascension, he was convinced Jews and Gentiles were not to mix, especially eating together. To become *real* Christians, they must, in addition to accepting the message of salvation, be circumcised according to Jewish traditional Law.

In this 10th chapter Peter becomes hungry and goes up on the roof while someone is making him something to eat. It even says he gets angry. Not sure if that was because he was so hungry or not, but he calms down and falls asleep. He has a dream involving a sheet coming down from Heaven with all four corners tied to something. The sheet is covered with all sorts of unclean animals according to Jewish tradition. He doesn't understand the dream, so, as normal with Peter, he has the same dream three times. Time will pass and he will get to eat. Prior to this dream someone else, Cornelius, is having a vision in another town.

Cornelius is visited by angels who tell him about a man he needs to send for who will bring him a message from God. He obeys and sends for the man with specific instructions about where to find him. When Peter comes down off the roof, he finds out there are some people looking for him. Guess who? They tell Peter about Cornelius's visit from angels and Peter decides to go home with them. Peter takes a few of his friends with him.

When he arrives, he hesitates to enter because he has understood for some time he is not to be associated with their kind. However, he remembers being told in the dream that God did not create anything unclean. Unlike the old Peter, he gets it. He enters the home, meets the people, preaches to them the Gospel of salvation, they accept Jesus Christ, and the Holy Spirit descends on them like it did on the disciples on the Day of Pentecost. Instead of insisting they be circumcised, he baptizes them in the name of the Father, Son, and Holy Spirit. This new baptism, as proclaimed by John the Baptist, did away with the need to be circumcised. After returning to Jerusalem and telling the others, including how he baptized them instead of insisting on circumcision, some accepted. This one change almost destroyed the New Testament Church movement. Kind of like the church music wars of the early 2000s. Humans are quite predictable. Back to the story of Peter.

In modern times Peter is the recognized leader of the twelve by the Roman Catholic Church, and as their first Pope. Tradition says Peter founded the church in Rome with Paul and was later crucified there in 64 AD. It is said he insisted he be crucified upside down as he felt unworthy to perish in the same way as his Savior. His remains are said to be beneath the high altar of St. Peter's Basilica in Rome.

When considering Peter as being named the first Pope of the Roman Catholic Church, much confusion arises from the fact there is no reference in Acts or Romans to any residence in the city of Rome, other than a small mention of Babylon in 1 Peter 5:13. Babylon is a cryptic term for

Rome as used in Revelations 14:8, 16:19, 17:5-6. Although there are other non-biblical writings (Letter to the Romans by St. Ignatius) which claim that the early church of Rome was founded by Peter or served as the first bishop are in dispute relying on evidence no closer to the event than the middle or late 2nd century.

Peter's death is referenced in John 21:18-19 as prophecy. Strongest evidence that he was martyred in Rome under the reign of Nero around 66 AD, not 64 AD, is found in the Letter to the Corinthians of St. Clement of Rome C. 96 AD; 5:1-6:4.

> *Peter, who by reason of wicked jealousy, not only once or twice but frequently endured suffering and thus, bearing his witness, went to the glorious place which he merited (5:4)... To these men [Peter and Paul] who lived such holy lives there was joined a great multitude of the elect who by reason of rivalry were victims of many outrages and tortures and who became outstanding examples among us (6:1).*

The actual burial site of Peter is still very controversial to this day. There appears to be no absolute archeological evidence.[185]

James the Greater

The son of Zebedee and Salome, this James was the older brother of John and the first of the apostles to be martyred. He was from the town of Capernaum and was thought to be, not only older than his apostle brother, but the oldest of the original twelve. Not much is attributed directly to James, sometimes called James the Greater. The added "Greater" was perhaps because of his age or his stature, not because of power. He and his brother's mother tried to convince Jesus Christ to give her sons elite positions in His kingdom. This was likely due to an ongoing hope that Jesus Christ would establish and rule over an earthy Kingdom. James and his brother earned the title, "Boanerges," or "sons of thunder" from Jesus Christ after a Samaritan village rejected Jesus and they wanted to call down fire from heaven upon the place. Tradition says James was decapitated by King Herod Agrippa in 44 AD.[186] It is said he preached in Iberia

(Spain) as well as the Holy Land. Upon his death his followers took his remains back to Spain where he was buried at Santiago de Campostella. The Catholic Church does not necessarily ascribe to the notion that James ever actually worked in Spain, but the legend has persisted since 700 AD.[187]

John

John was the younger brother of James, both sons of Zebedee and Salome. Salome was one of the women in the inner circle who closely ministered to Jesus Christ and the disciples regularly. Often referred to as the "disciple whom Jesus loved," John, also later to be called Saint John the Evangelist and Saint John the Divine, was thought to be the youngest of the apostles. He is the only one of the apostles to die of old age, though stories abound of the many times he was persecuted and tortured. It is believed that it is he who authored several books of the New Testament, though there are those that say some were written by another John. Most credit him with writing the Gospel of John, 1st – 2nd and 3rd John, and the Book of Revelations. John was banished by the Romans to the island of Patmos during the persecution in the middle 90s by Domitian. An early Latin tradition has him escaping unhurt after being cast into boiling oil in Rome. John's tomb is in the remains of the Basilica of St. John in Ephesus (near modern Selçuk, Turkey).[188]

Philip

Philip was born in the same little town of Bethsaida as were Andrew and Peter. This Philip should not be confused with the Philip who was chosen as a deacon to help with food distribution in the New Testament Church in Acts 6:5. He went on to preach in Asia Minor, Greece, Syria, Phrygia, and possibly Carthage in North Africa. Having converted the wife of the proconsul of Rome in Hierapolis (an ancient city in Turkey), the angry official condemned Phillip to torture and crucifixion (upside down). Another story says he was beheaded. Though it is agreed his relics are in the crypt of the church of Santi Apostoli in Rome, a 2011 archeological dig in Hierapolis unearthed his original tomb.[189]

Bartholomew (Nathanael)

Also known as Nathanael, Bartholomew, originally from Cana in Galilee, is said to have gone to India, Ethiopia, Mesopotamia, Parthia, Lycaonia, and Armenia. He is considered to be one of the patron saints of the Armenian Apostolic Church. Tradition says he was flayed alive and beheaded in Albanopolis, Armenia, but another account places his martyrdom in India. His relics are found in the Cathedral of St. Bartholomew in Benevento, Italy. Another relic is in San Bartolomeo All'Isola in Rome, the site of a hospital called Bartholomew whose name means "protector of medicine and hospitals."[190]

Matthew (Levi)

As a tax collector and a Publican, Matthew was literate in Aramaic and Greek. He was also the son of Alphaeus which could make him the brother of James the Just who was also the son of someone known as Alphaeus. Though the Gospel that bears his name is anonymous, his authorship has been ascribed since the 2nd century. According to Hippolytus, Matthew wrote his Gospel first in Hebrew and published it in Jerusalem.[191] Most scholars agree that he preached in the Holy Land but cannot come to consensus on where else in the world he traveled. It is interesting to note that Muslim texts place his missionary work around the Caspian Sea. Others also suggest he went to Persia and Ethiopia where he is thought to have been stabbed to death. There are conflicting stories about his martyrdom, but his relics are found in the Cathedral of Salerno. Some say he died by a spear. He possibly died at Hierees, a town of Parthia (Iran).[192]

James the Son of Alpheus (James the Less and James the Younger)

The son of Alpheus, this James is only mentioned four times in the Gospels. He is believed to be the brother of the Apostle Matthew since their father's name is the same.[193] One account of his martyrdom has him beaten to death with a fuller's club. Another story has him crucified in Egypt. His mother stood with Mary, the mother of Jesus Christ at His crucifixion.[194]

Thomas (Didymus)

Thomas was also a fisherman and remembered as the "doubter." He traveled to India about twenty years after Christ's death. He established seven churches in Kerala (southwest India) and descendants of those he baptized there proudly trace their lineage to those first converts. He is known as the "Apostle of India." He was martyred during his mission in southwestern Calamene, India, by the thrust of a pine spear through the four members of his body.[195] Some scholars believe he might have traveled to Indonesia. His remains were brought to Edessa in Mesopotamia but have been divided and moved several times. In 1258 AD, a portion made its way to the city of Ortona in Italy.[196]

Simon the Zealot

To distinguish him from Simon Peter, he often is referred to as Simon the Zealot. Zealots were a Jewish revolutionary group who opposed Rome. He was originally from Cana and was there called Simon the Canaanite. After working in Egypt, he joined Jude in Persia and Armenia. Different legends say he was crucified as the second Bishop of Jerusalem following the death of James the Just[197], sawn in half in Persia for refusing to sacrifice to the sun god, or he died peacefully in Edessa at the age of 120. Others say he was martyred with Jude.[198]

Jude Thaddaeus, Judas, Son of James

Some historians say he changed His name from Judas to Thaddaeus, meaning "warm-hearted," due to the lingering connection between the name and the other Judas' misdeeds.[199] Usually pictured with the flame of Pentecost over his head, Jude preached in Judea, Samaria, Idumaea, Syria, Mesopotamia, Beirut, and Libya. This Jude is not to be confused with the author of the Book of Jude who was the half-brother of Jesus and the brother of James the Just. He was martyred around 65 AD in Beirut with Simon the Zealot. His body was later brought to Rome and lies in St. Peter's Basilica. He is known as the patron of lost causes and, of course, is the patron of the famous children's hospital and research center in Memphis, Tennessee.[200]

Judas Iscariot

Judas was obviously not involved in the New Testament Church.

Matthias

He is unique because he wasn't personally chosen by Jesus, but by the other apostles upon the death of Judas Iscariot. He went on to preach in Syria, Cappadocia, and the area of the Caspian Sea. He was stoned to death in what today is the country of Georgia. Others say he was burned to death.[201]

It is often assumed these first chosen remained together to start the New Testament Church. That likely did not happen. As it turns out, the Apostle Paul was not the only one spreading the message throughout the world. After receiving the Holy Spirit on the Day of Pentecost, this band of brothers were changed. They received the ability to fully understand, perhaps for the very first time, what had just happened over the past three years or so. In addition, the words of Jesus Christ just prior to His ascension took on new meaning. All of them, in varying degrees, left Jerusalem for all parts of the civilized world known at that time, to spread the Salvation Gospel. Ten of them would die a noble death, just as Jesus Christ had predicted they would on multiple occasions over the years. They each took up *their* cross, not Jesus Christ's cross, and for the first time really followed Him into the world. These men, and several others contributing in varying degrees, sowed seeds on fertile ground and tended it. These efforts would continue to develop an amazing story of faith until present time. It will continue until Christ returns… Just as He said He would…

CHAPTER 18
WHY THE SECOND ADAM SUCCEEDED

... the night shift makes the day shift look really good!
... no team is successful without a strong bench.
... they didn't create it, they made it work.
... curing is the secret spice of every fine cuisine.

"From his obscure birth in Tarsus of Cecilia, through his remarkable conversion in route to Damascus, and along all those journeys from Antioch of Syria to the very presence of Nero enthroned in Rome, including his brutal martyrdom beside the Ostian Way, the man leaves his mark on all who take the time to pause and ponder. Left in the wake of its significance, you're unable to remain the same."

Chuck Swindoll
(On the impact of the Apostle Paul)

AS IN BASEBALL, my favorite sport, a few superstars often overshadow the small, but strategically important outcome of a game. Bottom of the ninth, two outs, the home team is behind by two runs with runners on first and second base. The superstar, twisting his knee three innings back, would have been due up next. The manager looks at his available subs and none seem to match the enormity of the moment. A second glance narrows to the name of a young kid, just drafted from Double-A ball. The kid has played baseball all his life. He has been in this very situation multiple times, but never in the big league, the Show, the game played at the level of Ruth, DiMaggio, Robinson, or Mays. *Why not?* the manager thinks. *Tomorrow is another day; we have 106 games left to play. What's another loss? The kid has to get some experience sometime, somewhere.* The manager looks at the kid and gives a little gesture toward the bat rack.

What happens next doesn't really matter. The manager was giving the kid experience. That was going to happen regardless of a hit or an out. In reality, in as many games as I have watched, it would not be out of the ordinary if the rookie stepped in and swung at the first pitch depositing the ball in the upper deck of right-center field. Home runs by a rookie during their first at-bat happen more often than you would think. The core of the New Testament Church's first days begins much the same way as the above illustration.

While we need to remember the actions and enormous contributions of Paul, Peter, and James the half-brother of Jesus Christ at the beginning of what was early on considered just another fledgling Jewish Sect, we would be overlooking Cornelius and others who appear to have very minor parts in this immortal story that will dominate history for the next 2,000 years and beyond. One other such minor character was a young man by the name of Stephen.

The Early Years

The New Testament Church had its formal introduction in the second chapter of Acts with Peter's first fiery sermon. Following this scene attention is diverted from many of the original disciples who begin to start ministries of their own, some leaving Jerusalem and entering surrounding cities, villages, towns, and eventually other countries and geographic regions. They would lay the groundwork for missionary trips that would follow, designed and implemented by some new minor and major characters. One of the new, larger-than-life characters entering the story who will play the largest roll in the New Testament, has an off-hand encounter with a minor character creating what will probably haunt him until he dies. That minor charter's name was Stephen.

Stephen

Beginning in Acts chapter 5, we find that Peter and others have been healing people brought to them from many different parts of the territory. They were preaching both Jewish history and the salvation offered through the resurrection of Jesus Christ. Their refusal to stop these efforts, when told to by a sect of the Sanhedrin Court, resulted in being placed in prison. The first night, angels appeared and released the disciples. They immediately returned to the Temple area and again began to preach. The next morning, the Court called for the prisoners to be brought before them to answer for their disobedience.

The guards returned without the prisoners and told the Court that they had somehow been able to return to the Temple and pick up where they left off. The Court officers, although furious, called for the disciples to be returned to the Court in a civil manner. Waiting for their return, the Court realized that they must be careful not to upset the general populace, who had grown to admire these brazen blasphemers. They consulted with Gamaliel, a respected scholar. This was the same scholar who was day-schooling a young man by the name of Saul.

Gamaliel reminded the Court of a few other Jewish saviors who had made similar claims. After attracting a following, they were killed or disappeared without any evidence that they were who they claimed to be. His wise advice was to let them go. If they were not sent by God, the same thing would happen that always happens when these illegitimate claims are made about being the Messiah. He also opined that if they are sent by God and this Jesus Christ is Who He says He is… well… better safe than sorry, as we would say in the 21st century. These guys didn't get where they were by being stupid. So, they did what they always did… the best thing for themselves. They ordered the prisoners to be whipped and released.

After being removed from the royal court, you can guess what they did and where they went immediately. The church grew tremendously and so did the demands on the disciples.

The sixth chapter of Acts relates a story of just one of the issues dividing the new church, again with the food! The Hellenists of this period were Greek-speaking people from outside Jerusalem who converted to Christianity and remained in the Jerusalem Church. The problem was that the Hellenistic widows of the Jerusalem church were "being overlooked in the daily distribution of food" (6:1). That is, the church apparently had an organized charity, like a daily "soup kitchen" for the needy, including widows. But the immigrant widows were not getting an equal share. This was considered a shortcoming of the church and by association, its leaders.[202] Something had to be done. As usual, they appointed a committee. Wisely, most members were from the Hellenistic group. Among the seven who were appointed and given authority were Stephen and a man by the name of Philip – not the one from the original twelve. Some agree this was the beginning of the term *deacon.*

The motive for appointing these seven resulted from the disciple's opinion that they were very busy proclaiming the gospel and didn't have time to wait on tables. While it was important for them to be spreading the gospel, it could also have been misconstrued that serving tables was beneath them. We will never know. The seven, however, were required to have three attributes: a good reputation, full of the Holy Spirit, and exhibiting wisdom in their decisions. Seven were chosen and placed before the apostles who laid hands on them and prayed. Problem solved… for now.

Acts 6:8 starts with,

> 8 *And Stephen, full of faith and power, did great wonders and signs among the people.* 9 *Then there arose some from what is called the Synagogue of the Freedmen (Cyrenians, Alexandrians, and those from Cilicia and Asia), disputing with Stephen.* 10 *And they were not able to resist the wisdom and the Spirit by which he spoke.* NKJV

Sounds like Stephen was doing more than waiting tables! Perhaps 21st-century leaders, regardless of title, should sit up and take notice. Stephen would not have made a good deacon or elder in the 21st century, which includes attending a board meeting and running the business... *the* business of the church. Then again, it would be difficult to recruit new folks to serve as deacons and elders if it included stoning. Beginning in Acts chapter 6:11 we find some very familiar actions taken by the elite of the Jewish court.

> 11 *Then they secretly induced men to say, "We have heard him speak blasphemous words against Moses and God."* 12 *And they stirred up the people, the elders, and the scribes; and they came upon him, seized him, and brought him to the council.* 13 *They also set up false witnesses who said, "This man does not cease to speak blasphemous words against this holy place and the law;* 14 *for we have heard him say that this Jesus of Nazareth will destroy this place and change the customs which Moses delivered to us."* 15 *And all who sat in the council, looking steadfastly at him, saw his face as the face of an angel.* NKJV

Acts chapter 7 is a summary by Stephen of many of the most important events in Jewish history including the words and actions of two key characters, Abraham and Moses. This summary follows a simple question from the court in chapter 7 verse 1, *Then the high priest said, "Are these things so?"* NKJV From verse 2 to verse 50, Stephen is uninterrupted and gives a very intellectual and eloquent response... without answering the question. Then, in three verses, sounding more like an angered Peter, he says what he really thinks!

> 51 *"You stiff-necked and uncircumcised in heart and ears! You always resist the Holy Spirit; as your fathers did, so do you.* 52 *Which of the prophets did your fathers not persecute? And they killed those who foretold the coming of the Just One, of whom you now have become the betrayers and murderers,* 53 *who have received the law by the direction of angels and have not kept it."* NKJV

As quickly as Stephen is introduced in Scripture, he exits from the building of the New Testament Church. His exit is witnessed by many that day, but one young man in particular, a man by the name of Saul. This Saul, taught by that same Gamaliel whose advice freed Peter and others after the angel of the Lord broke them out of prison, would watch over the expensive coats of the elite court and other spectators as they circled Stephen and stoned him to death. Saul, a Jewish Roman citizen, would later go to the authorities and gain papers authorizing him to travel to Damascus to locate other blaspheming Jews and Gentile converts and bring them to the same Sanhedrin justice met by Stephen. Saul, later called Paul, will spend many nights camped out on lonely stretches of road. What he witnessed that day from the gallery may have been the "thorn in his side" he could never heal.

This brief appearance is just another day in the life of another human being who greatly aided in the opening of the gates that had been locked between Jews and the rest of humanity for centuries. There would be others.

James the Just

Earlier, I made reference to a decision God made to place a very unlikely individual in charge of this brand-new thing called the church. This individual doesn't receive a lot of scriptural facetime, but he is still a very important part of God's plan. I will cover a few other minor, strategic characters later, but first we need to fit this individual into the early story of the church.

In Matthew 13:55-56 we find that Mary and Joseph had four other male children and two females. This doesn't mean they had no future children. It is a snapshot in time, but it does address the individual who Jesus Christ chose to eventually be in charge of the New Jerusalem congregation.

The half-sibling male's names are recorded as: James, Joseph, Simon, and Judas. The girl's names are not mentioned, likely because of the attitude toward women during that period of history. We are also not told in what order the girls and boys were born. Consistent with other lists in Scripture, names are usually record first to last for order, either by age or status. Based on this traditional criterion, we can assume he probably was Mary and Joseph's first normal birth.

Based on Scripture that outlines the union of Mary and Joseph, we can draw the conclusion that Jesus Christ was their first child. Some speculate Joseph was previously married, and that some or all of these other children were from that previous marriage. One reason given for this previous marriage was the protection of the virgin status of Mary, that no other humans would have come from her womb. This idea has no scriptural or secular support I could find.

To determine the approximate age of Mary and Joseph's children, if Jesus Christ died around thirty-three years of age, the next oldest child would likely have been around thirty-two or younger when that occurred. In any event, James, the half-brother or Jesus Christ, would have been in his early thirties or late twenties when his half-brother was crucified.

Scriptural evidence that Jesus Christ had half-brothers is striking, since one of those unbelieving brothers was eventually chosen to lead the first church in Jerusalem. This evidence can be found in the comments contained in John 7:3-5:

> [3] *His brothers therefore said to Him, "Depart from here and go into Judea, that Your disciples also may see the works that You are doing.* [4] *For no one does anything in secret while he himself seeks to be known openly. If You do these things, show Yourself to the world."* [5] *For even His brothers did not believe in Him.* NKJV

... and then Mark 3:31-35, following healing the man on the Sabbath and running into disbelieving, large crowds:

> [31] *Then His brothers and His mother came, and standing outside they sent to Him, calling Him.* [32] *And a multitude was sitting around Him; and they said to Him, "Look, Your mother and Your brothers are outside seeking You."* [33] *But He answered them, saying, "Who is My mother, or My brothers?"* [34] *And He looked around in a circle at those who sat about Him, and said, "Here are My mother and My brothers!* [35] *For whoever does the will of God is My brother and My sister and mother."* NKJV

... and Mark 6:1-4, while traveling into His original homeland among people that have known Him since He was a boy:

> 1 *Then He went out from there and came to His own country, and His*
> *disciples followed Him.* 2 *And when the Sabbath had come, He began*
> *to teach in the synagogue. And many hearing Him were astonished,*
> *saying, "Where did this Man get these things? And what wisdom is*
> *this which is given to Him, that such mighty works are performed by*
> *His hands!* 3 *Is this not the carpenter, the Son of Mary, and brother of*
> *James, Joses, Judas, and Simon? And are not His sisters here with us?"*
> *So they were offended at Him.* 4 *But Jesus said to them, "A prophet*
> *is not without honor except in his own country, among his own rela-*
> *tives, and in his own house."* NKJV

Think about your siblings. Think about growing up with them daily, experiencing their ups and downs, listening to their discussions, arguments, and hearing what they want to be when they grow up. Then, as they reach their teenage years, watching them go through what you are going through as your body and mind changes and matures. Adulthood sets in and the oldest children begin to pull away, to begin their own family, occupation, education, achievements, and disappointments. That dynamic was no different in the case of Jesus Christ's family, except in this case the oldest child was different… very different. For one thing, there was no decree issued for Mary's other children to be put to death as infants. The parents didn't have to flee their home in order to avoid being discovered and having those children massacred. There were other unsettling facts we do know.

Jesus Christ, at the age of twelve, decisively rejected His parents while on a trip to Jerusalem. He wasn't an only child, so his brothers and sisters were most likely on that same trip. Perhaps they were even aware that Jesus Christ was not with them when they departed the city, but did not tell their parents. What twelve-year old tells His parents to leave Him alone, *that He must be about His Father's business?*

We are not told expressly how all of Jesus Christ's siblings felt about Him, but there is an indication, for whatever reason, they and many of His neighbors and friends also doubted His actions and words. One of these half-brothers, later called James the Just, was chosen to step in, during a critical time in the founding and growth of the New Testament Christian church as the recognized leader. Why? What happened?

Again, thinking about your siblings and parents, in most cases regardless of what may have happened between family, there are moments when that family comes together... because they are family. Imagine word had come to you that one of your brothers or sisters had been arrested and was going to probably be executed. Would you, if at all possible, try to be near them and lend support? Even if things were really bad between you and your sibling, you would likely be there at least to support your mother. All of these emotions were probably present after the execution had taken place and the family had received the impossible news that their half-brother was reported to be alive!

Regardless of the reason, Acts 1:14 tells us that the remaining eleven as well as His mother Mary and His brothers, following an earlier instruction for them all to remain in Jerusalem until they received baptism by the Holy Spirit, were gathered waiting in the upper room. They were present when the Holy Spirit appeared to them like fire in a rushing, mighty wind. For one of these family members there had been an earlier encounter with the risen Lord.

That encounter shows up in Scripture several years later. The Apostle Paul, writing to the Corinthian church for the first time after establishing that church, while listing a sequence of events, and just before telling the church that the risen Lord had appeared to him, we find these words in 1 Corinthians 15:7, "After that, he was seen of James; then of all the apostles."

Back to family: If your mother had told you, who was an eyewitness, that your half-brother was brutally beaten and died while hanging on a Roman cross, and that He then rose from the dead, perhaps you would doubt. But add to that having that same dead, half-brother appear personally to you... and then to be present at an event like Pentecost. This would probably make a believer out of the most unbelieving human.

Proof of James' involvement in the Jerusalem church can be found in Acts 12:17 and Galatians 1:19, when Peter instructed those with him in prison to report to James regarding his miraculous escape. Later in Acts 15:13-21, it is James who makes the final declaration during an early ministerial conference. After his conversion, the Apostle Paul first met with James and Peter before speaking with any of the other disciples. In Acts 21:18-26, James became an advisor to the Apostle Paul, who later acted on that advice.

James the Just, the writer of the Book (Letter) of James in the New Testament, begins the writing by calling himself the "servant" of God and the Lord Jesus Christ, not the brother or half-brother of the Lord Jesus Christ. I take this as total acceptance of Who his brother really was; a 180-degree change in his heart. The Book, or Letter of James, thought to be one of the first books written during the beginning of the movement, is often associated with the doctrine of *works* (James 2:24) as opposed to *faith* alone. This debate has separated the faith community on several occasions, and created entire denominations that place works above faith. However, while the majority of the faith community hold to accepting Christ through faith, they still contend that an apple tree should produce apples, not grapes.

> *The second-century writer and historian Hegesippus referred to Jesus' brother as James the Just and characterized him as zealous for the law of God. Many statements from James' letter prove Hegesippus was right; it represents a book of Christian proverbs that cover subjects that touch many aspects of Christian life.*
>
> *Not long after writing his epistle, James was martyred in Jerusalem in AD 62. According to the first-century Jewish historian Josephus, James was accused by the high priest and condemned to death by stoning. Eusebius, a fourth-century church historian, adds details of James' death. He states that the scribes and Pharisees took James to a public place, the top of a wing of the temple, and "demanded that he should renounce the faith of Christ before all the people . . ." But, rather than deny Jesus, James "declared himself fully before the whole multitude, and confessed that Jesus Christ was the Son of God, our Savior and Lord"*
>
> *Hegesippus writes "they went up and threw down the just man [from the temple height], and said to each other, 'Let us stone James the*

Just.' And they began to stone him, for he was not killed by the fall, but he knelt down and said, 'I entreat thee, Lord God our Father, forgive them, for they know not what they do' [thus following his brother's example to the last]. One of them, who was a fuller, took the club with which he beat out clothes and struck the just man on the head. And thus he suffered martyrdom"[203]

Along with Peter and Paul, James the Just was also instrumental in managing an early conflict within the Jerusalem Church regarding the Jewish condemnation of Gentiles. The conflict stemmed from Gentiles eating unclean food in the presence of Christian Jews. What is more, the Gentiles were also being excused from the ceremony of circumcision, replacing the tradition with baptism performed in the name of the Father, the Son, and the Holy Spirit. This conflict almost destroyed the efforts of the early church to become the Bride of Christ to the world. Only God would use a man who denied Him, a man who set out to kill as many believers as possible, and a half-brother who refused to believe for almost three decades, to build a church that would last for thousands of years. A church through which hundreds of millions of humans would come to know and accept His sacrifice for their sins.

Barnabas

One of the Apostolic Fathers was Barnabas, later called St. Barnabas. His original name was Joseph or Joses the Levite. At a time when the Apostles were preaching the resurrection of the Messiah with great power and success, the Apostles were given charge over everything owned by this group of Christians in Jerusalem to be fairly distributed to those in need. And, through this socially accepted method, no one seemed to be in need during this period. Believers were selling what they had and laying it at the feet of the Apostles. Shortly, after the resurrection of the Savior, Barnabas, from the country of Cyprus, became a convert after moving to Jerusalem. This man, whose translated name is "Son of Encouragement," owned land which he sold and also lay at the feet of the Apostles. We learn most of this background by reading Acts 4:36-37.

Barnabas becomes a Hellenized Jew when he joins the Church in Jerusalem, since he is not a hometown boy. He and others become founders of the Church in Antioch, where Scripture records these believers are first called Christians.

Barnabas' first encounter with Paul, then called Saul, was in the ninth chapter of Acts, or the conversion chapter of Saul of Tarsus. After this great ordeal by Saul on the road to Damascus and the eventual return of his eyesight, he finds himself in Jerusalem. His new-found journey with Christ drives him to attempt a meeting with the disciples. After being threatened with death by an angry mob, he finds his way to the disciples, who also reject him. It is only after Barnabas steps in and recounts what Saul has recently gone through, and his encounter with Jesus Christ on the Road to Damascus, that Saul finds acceptance among the disciples.[204]

Although Paul eventually becomes the most prolific writer of the New Testament, he begins his new life called as an assistant to Barnabas, who is preaching in Antioch. Barnabas would become deeply involved in Paul's first missionary journey, but they would eventually part company during the planning of a future trip. Both eventually return to Jerusalem.

Barnabas and Paul spent a year preaching in the Antioch Church, during which time they became better acquainted and probably had several discussions about the new church movement and how it might spread. At the end of the year, they were asked to travel to the Jerusalem Church and deliver contributions collected by the Antioch Church to be used as relief for the poor in Judea. After that they returned to Antioch, taking Barnabas' cousin (or perhaps nephew) John Mark with them. Later they would travel to Cyprus, the homeland of Barnabas, as well as some of the principal cities of Pamphylia, Pisidia, and Lycaonia. This would later be labeled as Paul's first missionary journey. This journey was revealed to them by the Holy Spirit in Acts 13:1-3.

Before Paul was called to put together a second missionary journey, there was work for them both in the Jerusalem church. Earlier I mentioned the Jerusalem Counsel and the dispute between these early leaders over the subject of circumcision regarding Gentile converts. It is at this point in the timeline that this event occurs. It seems Barnabas initially joined Peter in the refusal to accept Gentile Christians without following

Jewish Law. This upset Paul, who gave a scathing address to the Church, accusing them of not walking upright in the truth of the Gospel.

It is not uncommon in the 21st century for a church to split, and split over something as frivolous as the color of new carpet to be installed in the sanctuary. What happens next in the 1st century will be the first split in the very young Christian Church. This split was confined mostly to leadership. Paul asks Barnabas to accompany him on a second missionary journey into Syria and Cilicia. Barnabas was okay with the trip but insisted on taking his nephew John Mark. Whether intentional or not, Barnabas' request to take John Mark was met by Paul with an absolute negative response because John Mark had abandoned them during the first missionary trip.

John Mark had traveled with them before to Cyprus, but did not continue and returned home. Apparently, Paul felt the young man was not reliable and didn't want to bother with him. It is quite possible that the lingering resentment created by the Gentile circumcision dispute as well as Paul's public humiliation of him contributed to Barnabas' decision not to accompany him. And so, they split up. Paul left for Syria and Cilicia with another brother, Silas, and Barnabas returned to familiar country in Cyprus with his relative John Mark. Barnabas is not mentioned again in the Book of The Acts of the Apostles. There is Scripture later in Paul's writings that implies the two eventually forgave each other. [205] In 2 Timothy 4:11, as Paul nears the end of his life, he will describe John Mark as useful to him, also indicating their differences were settled.[206]

There is no certainty about the last years of Barnabas' life but there are different stories and traditions associated with his life and death. Among these is an alleged martyrdom and burial in Cyprus recorded by a questionable author in the 5th century. Another finds him in Alexandria, Egypt where he writes the Letter of Barnabas, a treatise on the Old Testament. He is also pictured in Rome, where some attribute to him the authorship of the New Testament Book of Hebrews. Barnabas' reputed tomb, discovered in 488 AD, is near the Monastery of St. Barnabas in the Cypriot city of Salamis, whose Christian community was founded by Paul and Barnabas.[207]

Here again is another instance of God using a single individual who was living a very normal, successful life away from the heartbeat of the resurrection story. Barnabas is led to Jerusalem where he hears the story of Jesus Christ, accepts Him, sells everything he owns and gives it to the new movement. He then hears God's call to preach in a small community, and ends up introducing the man God chose to cement the connections of the newly established church, the Bride of Christ. Barnabas enters the story for a brief period, accomplishes his task, and then exits.

This bench player was put in the game by the team manager during one of those critical ninth innings. He simply stood at the plate, waited for the right pitch, hit a grand slam home run, and then calmly returned to the bench and had a seat. God knows how to assemble a winning team.

John Mark—Cousin or Nephew of Barnabas

Many folks probably think John Mark was one of the original disciples because of the Gospel of Mark. It is easy to assume he was one of the disciples since the Gospel of Mark is placed second in a list of two other very prominent disciples who were part of the original twelve. But he was not. In fact, he is instead considered a disciple of the Apostle Peter.[208] Mark's writing is also considered by some to be simply the life and preaching of Peter dictated to Mark as a scribe or recorded as Mark heard Peter preach. Luke was another author of one of the first four Gospels who does not have the distinction of being one of the original twelve. However, he was a major player, and we will discuss him in another section.

We first hear of Mark and his widowed mother in Acts 12:12-17. In this passage the disciples were meeting in the widow's home. It is here to which Peter flees on his escape from Herod's prison after being released by angels. This home must have been a regular meeting place, since it seems Peter knew someone would be there when he arrived. Colossians 4:10 discloses John Mark as the cousin of Barnabas. This Scripture also records that later in Paul's life, John Mark would become useful to the missionary close to the time of his death. Other sources claim he was Barnabas' nephew, but there is a possible misunderstanding regarding the translation of the mother as Barnabas's sister. We also know that John Mark accompanied Paul and

Barnabas for part of their first missionary journey, and he traveled with Barnabas on his return to his home country in Cyprus.

As I covered earlier in this book, there is a reference in Mark 14:51-52 to a young man who is captured when Jesus is arrested, who then escapes from the grip of the soldiers, causing him to have his cloths torn away, leaving him naked as he escapes. Many think this is a reference by John Mark of himself. This incident appears nowhere else in the Bible.

External proof of John Mark's legitimate claim to authorship of the Gospel of Mark can be found in the Papias of Hierapolis (AD 60-130) who writes,

> *And the presbyter said this. Mark having become the interpreter of Peter, wrote down accurately whatsoever he remembered. It was not, however, in exact order that he related the sayings or deeds of Christ. For he neither heard the Lord nor accompanied Him. But afterwards, as I said, he accompanied Peter, who accommodated his instructions to the necessities [of his hearers], but with no intention of giving a regular narrative of the Lord's sayings. Wherefore Mark made no mistake in thus writing some things as he remembered them. For of one thing he took especial care, not to omit anything he had heard, and not to put anything fictitious into the statements.*

Further, Irenaeus (AD 130-200) also writes,

> *After their departure, Mark, the disciple and interpreter of Peter, did also hand down to us in writing what had been preached by Peter.*

The early church unanimously concurred that John Mark was the author of the Second Gospel. [209]

The last mention in Scripture of John Mark is in 1 Peter 5:13. Here, Peter calls Mark his "son," no doubt a sentimental reference because Mark had been so helpful to him. According to Coptic Church tradition, John Mark is the founder of the Coptic Church in Egypt. They believe Mark was tied to a horse and dragged to his death by a mob of pagans on Easter, 68 AD, in Alexandria. The Coptic church also count him as the first of their line of 118 patriarchs (popes). Later legend suggests that in the early

9th century, John Mark's remains were moved from Alexandria to Venice and buried under the church of St. Mark.[210]

John Mark, in addition to being an intelligent, skilled writer with a sharp memory, had a servant's heart. He was a humble servant-helper and is referred to by several major writers throughout the New Testament. Nowhere in the Book of Mark does he draw attention to himself, but only points to Jesus Christ; with maybe the exception of Mark 14:51-52.

Silas

I have mentioned the Jerusalem Council several times and the impact it had on the early church. There were several people involved in that meeting other than Paul, Barnabas, Peter, and James the Just. Most of these others received little mention, if mentioned at all in Scripture. One stands out: Silas or Silvanus as he is called in other parts of Scripture.

Silas is another person who receives little attention in Scripture, but who undoubtedly made a large impact in the lives of many. When we are first introduced to Silas, he is one of the outstanding leaders and teachers in the Jerusalem Church (Acts 15:22,32). In addition to being a leader and teacher, and a missionary partner of Paul on his second journey, in Acts 16:37 it is implied that Silas, like Paul, was also a Roman citizen. And, like many in the early church in Jerusalem, he was also a Hellenistic Jew, meaning he was a converted Christian from outside the main Jerusalem district.

After the Jerusalem Council concluded, leadership felt it was important to send a delegation of respected representatives to the up-and-coming church at Antioch to explain what was determined by the Council to be the message given to newly converted Gentiles. Silas was the one who was chosen.

After the message was delivered to the Antioch Church, most of the representatives returned to Jerusalem. Silas, however, remained behind with Paul and Barnabas. This was at the point where the two men had their disagreement regarding John Mark. After the matter was settled and the two went their separate ways, Paul chose Silas to accompany him to Greece. Their welcome at Philippi involved an arrest, a beating, and a good old hymn sing-along. It appears they had a captive audience in Acts 16:25. When they left town, they left behind a body of believers—the first Christian church in Europe.

Later we find Silas ministering with Timothy in Berea, Acts 17:14. After Paul left the city of Corinth, Silas remained behind for a while to minister and teach. 1 Peter 5:12 validates Silas' involvement with Peter, and records that Silas was the one who delivered the letter later known as 1 Peter to its intended recipients.[211]

It is clear, with the few written passages in Scripture about his involvement, that Silas had a servant's heart, was a faithful brother, and a critical piece of God's plan for the beginning of the birth of the Bride of Christ.

Timothy

Yet another person popping up at various times throughout the New Testament was a man by the name of Timothy. The name Timothy means either *honoring God* or *precious to God,* both of which define the life of this man. We learn that he was born in Lystra in Lycaonia. We also know that he was of mixed race with a Jewish mother, Eunice, and a Gentile father. Eunice's mother's name was Lois. Timothy's intimate knowledge of the background, beliefs, and traditions of both Jewish and Greco-Roman culture would prove invaluable to Paul as he worked to unite these two civilizations in Christ. The Apostle Paul found Timothy to be personable, trustworthy, and efficient, evidence of Timothy's educational background. Coming from a highly educated background himself, this trait would have attracted Paul's attention when searching for an assistant. Paul's instinct and the testimony of the brethren in Derbe and Lystra in 51 or 52 AD placed Timothy permanently by Paul's side almost until Paul's death.

Man of God, his dearly beloved son, companion of his labors, and brother are terms Paul used to describe Timothy. Their travels would cover a lot of ground for persons walking or possibly catching a ride on a mule or wagon. Some of the places they visited included Macedonia, Thessalonica, and Corinth. First, and sometimes second letters were sent to these cities that find their way into the New Testament.

To head off a petty controversy over circumcision, which nearly destroyed the Jerusalem Church, Paul suggested to Timothy that he should be circumcised in Acts 16:3. Although Timothy's mother was Jewish, apparently his Gentile father objected to the ritual, leaving Timothy uncircumcised when he first met Paul. Timothy agreed and went through the painful adult procedure even though he was assured by Paul it was not relevant to his salvation, yet another statement of Timothy's character.

Little is said of it, but Paul also took a Third Missionary Journey. Paul sent Timothy, and another briefly mentioned man, Erastus, to Macedonia and then to Corinth to remind them of the truths he had taught them. Later Paul writes to the Corinthians requesting they take care of Timothy and then send him back to him. As it turns out Timothy returns to Paul who was then in Asia. Paul associates Timothy's name with his in second Corinthians in about 57 AD.

Paul returned from Rome in 64 AD, and on the way left Timothy in Ephesus to care for the church. Recognized by the council of Chalcedon, Timothy is that church's first bishop. Apparently, Timothy was a bit anemic because Paul, in the first letter to Timothy urges him to drink very little wine due to the weakness of his stomach and frequent infirmities. He would eventually live the rest of his life and die a martyr serving this church. But first, he had to make one more trip to Rome.

Paul wrote a second letter to Timothy around 65 AD near the time of his death. The letter contains many salutations of kindness and praise for Timothy, and his assistance with Paul's work. This second letter to Timothy is sometimes called Paul's last will and testament. He requests Timothy to bring several of the things he had left in Troas with a brother whose name was Carpus. What did he ask for Timothy to bring to him? First, his cloak. Did you ever have a coat or vest that seemed to just fit perfectly and somehow you felt safe and comfortable when you put it on? I think that was the reason for the cloak. Remember Paul is in prison and he is older, also probably not treated well by the Roman soldiers who would eventually execute him. Next, Paul requested the letters, or writings, or in some interpretations, the parchments. Paul's notes over the years must have needed one more review before his life on Earth ended. The Apostle Paul *was* his writings.

It isn't indicated anywhere but I believe Paul kept a diary. He had covered thousands of miles and encountered innumerable different people and situations. I just believe there is a book that didn't make it to the Bible, or it was so personal it didn't make sense to include those writings.

When Paul wrote to Timothy the last time and asked him to come to Rome, he pleaded for him to arrive before winter. We don't know if he made it. We can assume, due to their close relationship, that if it was at all possible, Timothy made the trip and spoke with Paul one last time. Because of the proximity of dates, if he did make it to Rome, he likely witnessed Paul's execution. That would have been devastating to Timothy.

Those visions of his mentor must have flashed before his eyes at the time of his own violent death. He had returned to Ephesus and continued as Bishop for several years, confronting his most ardent pagan opposition. At one point, this group held a feast and worshiped images of idols while wearing masks and waving clubs. Tradition says Timothy rushed the group to stop their sinful worship, but was stopped when the group grabbed him and began beating him with their clubs. His injuries quickly led to his death. It is said they left him where he died in the street. He was eventually removed by his disciples and given a proper burial.

The above history is paraphrased and adapted from John Kitto's 1870 History of the Bible and represents the commonly accepted views about this apostle among rank-and-file believers in the late 19th century. [212]

Many of the characters who played lesser roles in Scripture contributed vastly to the establishment and growth of the New Testament Church in the 1st century. Timothy holds a special place among all of these other heroes of Christianity, since two of the apostle Paul's thirteen traditionally recognized books of the New Testament bear his name.

Titus

Similar to Timothy, Titus was a young man, but unlike Timothy, he was all Greek with no Jewish heritage. Also, unlike Timothy, Paul did not insist he be circumcised upon accepting Christ. He instead became a Christian by way of the decision made by the Jerusalem Council by being baptized in the name of the Father, Son, and Holy Spirit and abstaining from certain activities.

Titus was converted after hearing Paul preach during his first missionary journey. Through the years Paul would come to recognize Titus as one of his spiritual sons since he led him to trust in Christ. In Galatians 2:1-4, we find Paul returning to Jerusalem with Titus to demonstrate to the church and its leaders that a Greek Gentile could accept Christ and love God. He would come to represent all non-Jewish people who would accept Christ through faith.

Titus would travel with Paul on parts of all three missionary journeys and play an important role in a few of the ministries started by Paul. One such place was Corinth where Paul sent him to ease the tensions growing in that church and to collect money for the poor.

After being released from prison, Paul and Titus traveled to the island of Crete. As a result of their preaching several people accepted Christ, and soon there were a few churches scattered throughout the region. Paul felt a need to return to Corinth, and due to his confidence in Titus' ability, he left him in charge in Crete. While there, Titus grew the church and appointed leadership. Paul eventually sent someone to replace him and asked Titus to join him in Macedonia where they departed northward into what is now Albania (2 Timothy 4:10). This and other actions would increase the presence of Christianity in Europe. Titus would eventually return to Crete. He would receive a letter from Paul, later included in the New Testament as the Book of Titus, with hopes that he could again join Titus for more travels (Titus 3:12). Sadly, they would never see each other again.

There is little written about the latter part of Titus' life, but tradition records his return to Crete where he would live out the remainder of his life, closing out another chapter that made the novel of the Apostle Paul's life and legacy so very impactful in the history of humanity.[213]

CHAPTER 19

THE END OF THE APOSTOLIC ERA

... a timeline has no end, just footnotes
... a timeline has no beginning,
it just fades into the past
... a timeline has no middle, just
occasional transitions
... excuse me, where would you suggest I
place eternity and infinity on this thing?

The closing years of the apostolic age (70-100 AD) remain one of the most obscure periods of church history due to the scantiness of information relating to this period. It seems that all of the New Testament books, except for John's writings (95-98 AD), had been written by the time Jerusalem was destroyed (70 AD). Consequently, there is little information provided in the Scriptures as to what occurred among Christians during this period. Information from sources outside the Scriptures is likewise sparse. Yet, the close of the apostolic age is one of the more important periods of church history because of its transitional nature.[214]

> *The first twenty years of Christianity can be summed up by two themes: explosive growth and continual conflict. There is no such thing (as) an idyllic pristine early Church. What we find is a fugitive church that experiences attacks, murders, famine, and heresy. We find a church marred by internal conflict and schism.*
>
> *Yet we see in the midst of this Jesus' promise: I will build my church. The gospel spread to Europe (Acts 13–14) and Africa during this time (see Acts 8) despite the severe problems. And the next twenty years were no less chaotic, culminating in the invasion of Israel by Rome and the sacking of Jerusalem. But that's a story for another time.* [215]

Following the death of John the Revelator just before the end of the 1st century, the newly established almost 100-year-old church began a slow, painful transition from apostolic, divine leadership, to hometown, sometimes individual, untrained guidance. Prior to this grand awkward transition, there were two individuals, neither of which were part of the original twelve, who held undeniable, irreplaceable positions within the structure and foundation of the New Testament Church.

Their work, beginning after the ascension of Jesus Christ and transformed through future centuries, would be self-evident proof of a Holy God Who set the entire universe and the humanity it contained in motion. Their names were simple, but other than the commonality of four letters in each name, they were two very complicated, uniquely educated individuals with totally different talents, gifts, and abilities. However, under God's guidance and in His eyes, both were very similar. Remember, that's how God thinks and works. Their names were Paul and Luke.

The Apostle Paul

If you thought God's choice of Moses, a murderer, was odd to lead the Children of Israel out of Egypt, how about His choosing a man who was responsible for the death of many Jews whose only fatal mistake was converting to Christianity? That is where the story of this young man, a Jewish Roman citizen, home-schooled by one of the greatest scholars of his time, begins... standing beside a crowd watching over the robes and personal belongings of those stoning another young Jewish man to death for the simple act of accepting Jesus Christ as his Messiah and converting to Christianity. At the time, the keeper of the cloakroom's name was Saul. Let's take a look at the man who is credited by many scriptural scholars with the authorship of most of the New Testament—Saul, later to be called Paul the Apostle.

He was not chosen by Jesus Christ to be one of the original twelve—he wasn't ready at the time, nor was he chosen as Judas' replacement after the ascension by the remaining eleven. Instead, it was Matthias, one of the other 120 or so disciples also following Jesus. The Apostle Paul, according to Scripture and testified as true by Paul, was chosen by Jesus Christ on the Road to Damascus. Let's start at the beginning.

Early on Paul was known as Saul of Tarsus, solely because that was the place of his birth. Tarsus was, at the time, a major city in eastern Cilicia, a region in Asia Minor. Rome, by the time Paul reached adulthood, would have made that area part of their occupied territory, called Syria. Saul would have been born about the same time as Jesus Christ and lived until martyred in Rome around 64 AD.

Saul was not raised in an elite Jewish family, and according to Scripture was a tradesman earning a living in the making of tents. These skills would have been useful to him in the later journeys he would take by simply carrying with him a small pouch of tools designed to work leather. If needed, he could have sustained himself and others with this background and skill. Though he was born into a working-class family, he nonetheless received a sterling education. In Acts 22:3 it is told where Saul came from and that he was, "*...educated at the feet of Gamaliel according to the strict manner of the law of our fathers, being zealous for God as all of you are this day.*" This passage is taken from a scene where his life hangs in the balance. To understand the importance of his defense we must understand who Gamaliel was within the Jewish community.

> *Rabbi Gamaliel I, son of Simon and grandson (according to the Talmud) of Rabbi Hillel (founder of the more liberal [i.e., more tolerant] of the two main schools of the Pharisees, Shammai being the other). Although an alternate tradition makes Gamaliel the son of*

Hillel, the Talmud is surely to be preferred on this point. A member of the Sanhedrin and a teacher of the law (Acts 5:34), he was known in rabbinical writings as Gamaliel the Elder to distinguish him from his grandson, Gamaliel II. He was the first of seven successive leaders of the school of Hillel to be honored with the title Rabban ("Our Rabbi/Master").

Scholarly, urbane, a man of great intellect, he studied Greek literature avidly. What we know of his tolerance and cautious spirit is entirely in keeping with the account of his appeal in the Sanhedrin to spare the lives of Peter and his companions (Acts 5:33–39).

Gamaliel's reputation as one of the greatest teachers in the annals of Judaism, however, remains untarnished and is perhaps best exemplified in Mish Sotah ix.15: "Since Rabban Gamaliel the Elder died there has been no more reverence for the law, and purity and abstinence [perîšûṯ, cf. "Pharisee"] died out at the same time."[216]

The social rank of Saul's family would not have given him the opportunity to study at the feet of someone this well-known and admired. His opportunity likely is attributed to the Rabban's acute desire to teach students hungry for knowledge, despite not being the child of rich parents. Early on it was evident that Saul had a gift for learning and was indeed thirsty to understand Jewish Law. More than just the Rabbinical Law, Gamaliel taught philosophy and conversational debate, or sociology. Saul, later named Paul, will use what he learned from this teacher several times both in defense of the Gospel and in defense of his very life.

Road to Damascus

Now that we have taken a look at where Saul came from, let's turn our attention to what happened to this determined, well-learned young Pharisee. We have touched on Saul's bent toward cleansing the Jewish Nation of newly converted Christians who seemed to be set on destroying the history and traditions he had come to appreciate and love. We first meet Saul in Scripture when he is assisting those who are stoning a young man to death merely because he converted to Christianity and was trying to introduce his new Savior to other Jews. Following that incident, Saul goes to the authorities and requests legal papers allowing him to travel to the town of Damascus for the intended purpose of seeking out others who were promoting this Christian idea and bringing them to justice. Saul was about to meet someone way above Gamaliel's pay grade.

While on the road to Damascus with a few traveling companions, only Saul is blinded by a brilliant light. He hears a voice asking him why he is persecuting the one speaking. Here are several different replies taken on different occasions, including the actual event and other times when Paul later recounts the incident.

> Actual event—*And he said, "Who are You, Lord?" Then the Lord said, "I am Jesus, whom you are persecuting. It is hard for you to kick against the goads."* Acts 9:5 NKJV
>
> Paul's defense in Jerusalem—*So I answered, 'Who are You, Lord?' And He said to me, 'I am Jesus of Nazareth, whom you are persecuting.'* Acts 22:8 NKJV
>
> Paul's defense before King Agrippa—*And when we all had fallen to the ground, I heard a voice speaking to me and saying in the Hebrew language, 'Saul, Saul, why are you persecuting Me? It is hard for you to kick against the goads.' So I said, 'Who are You, Lord?' And He said, 'I am Jesus, whom you are persecuting.'* Acts 26:14-15 NKJV
>
> Paul's letter to the Galatians—[15] *But when it pleased God, who separated me from my mother's womb and called me through His grace,* [16] *to reveal His Son in me, that I might preach Him among the Gentiles, I did not immediately confer with flesh and blood,* [17] *nor did I go up to Jerusalem to those who were apostles before me; but I went to Arabia, and returned again to Damascus.* Galatians 1:15-17 NKJV
>
> First letter to the Corinthian Church—[1] *Am I not an apostle? Am I not free? Have I not seen Jesus Christ our Lord? Are you not my work in the Lord?* [2] *If I am not an apostle to others, yet doubtless I am to*

you. For you are the seal of my apostleship in the Lord. 1 Corinthians 9:1-2 NKJV

Following this encounter Saul is instructed to go into the town of Damascus, led by those who were with him, and wait there for three days. That was it... go... and wait. He did. He could have just returned to Jerusalem, but... blind? The cure was apparently in Damascus, not where Saul might have wanted to go. Again, another minor player is used here in a mighty way by exhibiting an unwavering faith in the face of an unrealistic Heavenly command.

That command found in Acts 9:11-12 reads:

> 11 *So the Lord said to him, "Arise and go to the street called Straight, and inquire at the house of Judas for one called Saul of Tarsus, for behold, he is praying.* 12 *And in a vision he has seen a man named Ananias coming in and putting his hand on him, so that he might receive his sight."* NKJV

Ananias was a disciple and therefore a true believer. He was probably one of the folks Saul would have searched out and arrested if he had arrived as originally planned. It is very likely that the human grapevine in Jerusalem had already communicated to their brothers and sisters in Damascus. And what did God want Ananias to do? To go where? To see whom? What would you have done, Mr. or Ms. 21st -century Christian?

The Lord is patient and the Lord is kind. He did not get upset when Ananias responded with a paraphrased, "Are You sure?" God took a deep breath, I am sure, and then explained it to the probably now, pale messenger. So, what did Ananias do? He went. What did he call Saul when he arrived? Brother Saul, what else. He trusted the Lord.

The story has a couple of different turns at this point based on who was writing and what details they may have had. The author of the Book of Acts says Saul remained in Damascus for some time and the congregation and disciples were amazed at the transformation of the vile Pharisee from Christian bounty hunter to Christian preacher. Toward the end of his time in Damascus, Saul discovered a plot by some Jews to kill him. He shared this concern with the disciples, who arranged to help him escape at night by rigging up a manual elevator and letting him down from the top of the city wall. From there, the Book of Acts takes Saul back to Jerusalem to search for the disciples in charge of the New Jerusalem Christian Church. He wants to meet them and become part of what they are doing. He will get an Ananias reaction from them, and, as earlier mentioned, another one of the minor characters will come to his rescue. [217]

I will outline Paul's domination of the Biblical timeline in broad strokes, not to minimize its importance, but to emphasize its impact within the first one hundred years of the infant Christian movement. However, before we go there, we should spend a few minutes prospecting information related to yet another major name change in an already four-thousand year lineage.

Saul? Paul?

If you ask most folks when Saul's name was changed to Paul and then to the Apostle Paul, they will say it was when he was converted on the road to Damascus. Ask them for Scripture to validate that idea and you will probably get a blank look. Well, it's just how God and Jesus Christ did things. Right? Yes, and... no. It is the way two-thirds of the Trinity did things occasionally, but they had nothing to do with this one. While it works well into lessons and sermons, Peter changed to *the rock on which I will build my church*, James and John to the Sons of Thunder due to their temperament, Abram to Abraham and Jacob to Israel for lineage and prophecy. But changing Saul the persecutor to Paul the Apostle just isn't scriptural, regardless of how powerful it sounds.

Neither God nor Jesus Christ changed his name. Note the following scriptural references: [218]

1. Jesus addresses him as "Saul, Saul" on the Road to Damascus. Acts 9:4 And, in Acts 26:14, Paul draws attention to the fact that during Jesus Christ's address to him on the Road to Damascus He speaks to him in Hebrew.
2. Ananias addresses him as Saul after his conversion. Acts 9:17

3. The Holy Spirit calls him Saul before his first missionary trip. Acts 13:2
4. After the conversion experience, he is called Saul eleven more times.
5. The decisive shift from Saul to Paul happens only once, in Acts 13:13, as he and his companions set sail.
6. Acts 13:9 gives a big hint to why the name change took place: *"But Saul, who was also called Paul, was filled with the Holy Spirit."*

A little research will fill in a lot of holes in this mystery and offer a much better explanation relative to his mission and character. Saul is his Hebrew name, probably derived from the famous first King of Israel who was, as this Saul, from the Tribe if Benjamin, one of the twelve Tribes of Israel. Paul is a *koine* or Greek form of the name Saul derived from the Latin surname Paulus. Remembering Paul's early teaching under the direction and influence of Gamaliel and Paul's early dedication to the Pharisee belief, he would have greatly preferred the Hebrew title of Saul.

During Saul's conversion, Jesus Christ's directions to him while he was blind was that he would be the messenger to the Gentiles. The insertion of this one-time mention of his name change in the 13th chapter of Acts is very telling. It is only coincidental, but very memorable, that the change is mentioned in connection with the conversion of the intelligent proconsul, Segius Paulus, at the beginning of the First Missionary Journey while blinding Elymas the sorcerer for trying to turn the proconsul away from the faith. Although we are not given a specific reason for this name change, it is logical that the Greek form of his name will be used for the rest of his life since his primary mission was to Gentiles. Who changed Saul's name? No one. It seems Paul made the decision to use the Greek form to better minister to his new Gentile (mostly Greek) congregations.

The 13th chapter of Acts occurs at about 47 to 49 AD. Saul, soon to change his name to Paul, and his companion Barnabas receive a vision about the need to take a trip. This trip will later be branded the First Missionary Journey. This trip will occur about fourteen to fifteen years after Saul's conversion. During that time, he has several opportunities to talk with eyewitnesses to the crucifixion, resurrection, and ascension. He also has ample time to meet and get to know several Hellenist and traditionalist Jews. This is all preparation for this first venture into the hostile world in surrounding areas outside Jerusalem. It has also given him time to assemble in his head how he will approach the thousands of Gentiles he is about to attempt to convert to Christianity. The best-laid plans are always met with the unexpected. That is precisely why God chose this particular human, at this particular time on the timeline. How does the phrase go, "...when in Rome... within limits?"

Who made you an Apostle, Paul? Or was it Apostle Saul?

We now know when Saul changed his name to Paul. Now we need to know when it was changed to Apostle Paul. The answer comes pretty quickly in Acts 13:1-3. In verse 1 we learn that both Barnabas and Saul are already called teachers and prophets, along with several others in the Antioch Church. Verse 2 describes a time of worship and fasting that both were participating in when the Holy Spirit shows up with a Heavenly message. These two are to be set apart from the rest for a special project. This verse is interesting because it doesn't reference God or Christ, nor does it indicate the instructions are in any way from the Trinity. It specifically says He wants them set apart for the work which He has called them to do.

This group has encountered the Holy Spirit before and fully understand the importance of following through with divine instructions. Without hesitation, verse 3 simply says they fasted and prayed, laid hands on the two and sent them on their way. This setting apart would have been the act giving these two teacher/prophets the designation of Apostle. Notice however, his name would have been Apostle Saul. He has not yet changed his name. Rather, he gains the apostle designation before deciding to use the Greek form of his name.

The name change however comes fairly soon after they are sent on their way. As described earlier, the name change happened during an encounter with one of the first groups they meet after sailing to Salamis on the island of Cyrus, the home area of Barnabas. When reaching the other side of the island, along with John Mark, they convert a proconsul and it is during that incident where Saul first becomes the Apostle Paul. They eventually sail to Perga where John Mark soon takes his leave and returns to Jerusalem—an incident Paul will not easily forget.

First Missionary Journey

Their next stop is Pisidian Antioch where they are at first received well, but eventually encounter a group of Jews who do not accept Jesus Christ as the Messiah and throw the two traveling evangelists out of town. Next stop, Iconium. Same result, except these folks' plot is to kill them rather than simply toss them out of town. Discovering this, they are off to Lystra. It will not, however, rid them of the band of Jews who are amassing to kill them. Lystra turns out to be an interesting visit.

It is here where Paul heals a crippled individual, amazing the people who then try to worship Paul the man. Again, Jews from other areas, probably some who have followed them from their other stops, incite the people to rise up against these two, ultimately stoning Paul to death. The angry crowd drags his seemingly dead body out of the city and leaves it to rot. Paul regains consciousness, and just to prove his point, walks back through the city on his way to another location. I see humor here, and take Paul's return trip through town as his way of saying, "You could have saved me some time and drug my body to the side of town closer to my next travel destination." At any rate, his bloody face must have been resolute and with a smile.

The pair travel to Derbe, preach the gospel, then retrace their steps back through the previously visited locations. And yes, they walked back through Lystra one… more… time. Ultimately, they are back where they started in Syrian Antioch. The Gospel has officially been opened to the world, although they have yet to leave the area adjacent to the Mediterranean Sea. This first trip took about two years to complete. Much of it was by ship, but when on land almost entirely on foot; and uphill both ways… literally—check the topographical map.

Dissention... the Jerusalem Council

After returning, the pair would remain in Antioch for about three years. It would be a turbulent time in the life of the church with dissention and disagreement that almost tore the church apart in Jerusalem and then Antioch. The meeting where this issue was debated was called the Jerusalem Council or the Conference at Jerusalem.[219] All in the meeting, which included Paul, Barnabas, Peter, and James, the half-brother of Jesus Christ, agreed that circumcision was not required of a Gentile to become a believing Christian convert. As with the issue Paul and Barnabas encountered on their First Missionary Journey, this was a hard thing for many of the traditionalist Jews to accept. The leadership split their duties, accepting that a quick resolution was not about to occur. Some were assigned to minister to the Gentiles and the others to the traditionalist Jews. James gives the final disposition of the Jerusalem Council indicating that he, not Peter, has become the head of the Jerusalem congregation.

Second Missionary Journey

In about 49 AD, after a three-year sabbatical, Paul again feels the urge to travel, but when he approaches Barnabas, he encounters an insurmountable conflict. Barnabas insists that John Mark must again accompany them. Paul adamantly refuses. The resolution will be two separate journeys, one led by Paul with a new companion, Silas, and the other led by Barnabas accompanied by John Mark. Paul and Silas will head off toward uncharted territory after first traveling through his home town of Tarsus while Barnabas will return to Cyprus, his home country.

Paul and Silas return to Derbe and Lystra again, where Paul meets Timothy. They turn toward Pisidian Antioch and preach in western Asia. The Holy Spirit then turns them toward Troas. Receiving further guidance from the Holy Spirit, they are directed to Neapolis then Philippi, where they are thrown in prison for casting a demon out of a slave girl, thereby removing income from the slave's owner. An earthquake allows them to escape prison, sending them to Thessalonica. There, many believe after hearing Paul preach, but the pair is expelled from the city due to riots and unrest by the unbelieving Jews and they have to leave. The next part of this journey leads them to several cities to which Paul will later send letters of encouragement and instruction. Many of these letters will be incorporated into the New Testament.

After Paul's famous sermon and interaction on the Areopagus in Athens, they leave for Corinth in 50 AD where he meets back up with Silas and Timothy. He remains in the city for eighteen months. In late 51 AD he is brought before Gallio in Acts chapter 18, but soon after he travels to Cenchrea in early 52 AD, taking a Nazarene vow and shaving his head. [220] He and other companions sail to Ephesus where he then travels back to Jerusalem in time for the Feast. During this period there is a bad scene between Paul and Peter over, you guessed it, circumcision and eating with Gentile converts. As he did years before to Barnabas, Paul publicly rebukes Peter over his actions in the matter. Not long after, Paul sets out on his third journey in 53 AD.

Third Missionary Journey

Paul spends the year 53 AD revisiting the Galatian churches he earlier established, followed by a visit to the Phrygia province to strengthen the believers (Acts 18:23). From there he will travel to Ephesus, where he will actually live for a little over three years. In late 57 AD, Paul has to leave the city of Macedonia after upsetting the idol-makers trade union by converting Gentiles away from their use. Again, Paul is saved by a minor character, a city clerk who steps in to calm the crowd. (Taking the message to the Gentiles at first sounded like such a simple assignment!). Now in his third missionary journey covering some fifteen to sixteen years, anyone with less commitment and calling would have probably long ago returned to Tarsus and opened a used tent business.

58 AD is a whirlwind of visits to several locations including Assos and Miletus, before sailing to Patara and then Tyre where he calls for the elders in Ephesus to come to Miletus for instruction concerning false teaching and conditions that cause church splits. Paul was well prepared to give directions concerning both. He remained in Tyre for only a week

before sailing to Caesarea as noted in Acts 21:5-6. After spending time with Philip the Evangelist for a few days he leaves for Jerusalem. This visit to Jerusalem in late 58 AD would not be as pleasant as previous visits.

One Too Many Trips to Jerusalem

After visiting with James, the new head of that church, he decides to take four Jewish converts to the Temple, an unusual move on his part. Their arrival causes quite a riot after an angry Jewish mob seizes him, drags him outside, and severely beats him. Nearby Roman soldiers hear the ruckus and do what they are paid to do: keep the peace among these low-class Jews. More than any other reason, the soldiers arrest Paul to protect him. This is another instance when Paul used his Roman citizenship to be released, or so he thought. Instead, he is taken by the soldiers out of Jerusalem to Governor Felix in Caesarea who would then decide what to do with him.

Hoping to make a little money from someone who may try to bribe him for Paul's release, Felix keeps Paul in prison for the next two years. Felix is replaced by Porcius Festus in 60 AD. Unfortunately for Paul, the new governor attempts to repeat the entire trial. Paul, being a Roman citizen, has rights. One of which was to refuse to have the trial in Jerusalem. Instead, he requests the trial be heard in front of Caesar himself. Left unsaid, the probable reason for the request is to allow Paul to witness to the son of *a* god who at the time was ruler of all the Roman Empire. Be careful what you wish for. Paul found himself, along with other prisoners, on a ship headed for Rome.

Fourth Missionary Journey, Voyage to Rome

This voyage began in 60 AD and is considered by some as Paul's fourth missionary journey. The seafaring trials of this voyage are noted in Acts 27:37-44. The ship and its manifest of 276 passengers abandon ship when it runs aground on its original course to the Port of Phoenix in Crete. It had hit bad weather and the winds and current moved the ship out into the open Mediterranean Sea for about two weeks. But who was keeping up with the days when trying to survive? The ship ran aground off the island of Malta, Acts 27:37-44. As promised by Paul the prisoner, all 276 passengers survived… unharmed.

After spending some time in Malta, the manifest again sets sail for Rome through the Italian Port of Puteoli and then follow the infamous Appian Way, Acts 28:12-16. Possibly, due to his Roman citizenship, Paul is allowed to live by himself in the City of Rome with only one guard for the next three years or so. Oddly, he is allowed a great deal of freedom… and he continues to preach the Gospel. Then in 63 AD Paul's charges are presented to the Roman authorities. They are dropped, and he is set free. It would not be the last time Paul would find his home in a Roman prison.

Fifth Missionary Journey, Back to Rome

For the next four years, 63 to 67 AD, Paul would take his last trip, considered to be his fifth and final missionary journey. It is not clear how far away from Rome he traveled, but he did have access to international ships from the Roman ports. It is said in Titus 1:5 that he sailed again to Crete and then in Titus 3:12 to Nicopolis. Paul had a dream of preaching the Gospel in Spain (Romans 15:22-28). He may have even made it as far as England and other areas. We are not sure. For reasons unknown, Paul is again arrested and thrown into prison in Rome in 67 AD.

From his cell, guarded with great care and security, Paul is allowed to write letters. One of those letters, considered his last will and testament, is sent to his beloved son in the Lord, Timothy. This letter will be included in the New Testament as 2nd Timothy. He will request that Timothy come to visit him and bring with him some of Paul's most cherished treasures: his cloak and his parchments – his writings.

Paul, at the hands of the same Roman Government as Jesus Christ, loses his life sometime in the months of either May or June, 68 AD. [221]

It is commonly accepted that much of the New Testament was either written by, or influenced by, the Apostle Paul. E. P. Sanders, Arts and Science Professor of Religion at Duke University and author of *Jesus and*

Judaism and *The Historical Figure of Jesus* wrote an article titled, "St. Paul the Apostle," entered into the *Britannica* data base July 20, 1998, explaining one accepted outline of Paul's New Testament writings.

> *Of the 27 books in the New Testament, 13 are attributed to Paul, and approximately half of another, Acts of the Apostles, deals with Paul's life and works. Thus, about half of the New Testament stems from Paul and the people whom he influenced. Only 7 of the 13 letters, however, can be accepted as being entirely authentic (dictated by Paul himself). The others come from followers writing in his name, who often used material from his surviving letters and who may have had access to letters written by Paul that no longer survive. Although frequently useful, the information in Acts is secondhand, and it is sometimes in direct conflict with the letters. The seven undoubted letters constitute the best source of information on Paul's life and especially his thought; in the order in which they appear in the New Testament, they are Romans, 1 Corinthians, 2 Corinthians, Galatians, Philippians, 1 Thessalonians, and Philemon. The probable chronological order (leaving aside Philemon, which cannot be dated) is 1 Thessalonians, 1 Corinthians, 2 Corinthians, Galatians, Philippians, and Romans. Letters considered "Deutero-Pauline" (probably written by Paul's followers after his death) are Ephesians, Colossians, and 2 Thessalonians; 1 and 2 Timothy and Titus are "Trito-Pauline" (probably written by members of the Pauline school a generation after his death).* [222]

Even if authorship is slightly different than listed above, it cannot be denied that the Apostle Paul was the single most important individual, with the expected exception of Jesus Christ, in the formation, development, and execution of the Great Commission given by Jesus Christ just prior to the ascension.

Saul, from his first encounter with Christ, was rescued by some unnamed traveling companions. These included a single disciple minding his own business and a team of dedicated and quite inventive church disciples. All of this initial personal interaction was before ever meeting those in Jerusalem who had already began the task of fulfilling the command of Jesus Christ to go into all the world and preach His message. All the way to Paul's last breath, many barely mentioned individuals were constantly sent by God to care for him, encourage him, test him, teach him, and love him. What about you? Do you look for these people in your life? Do you expect them to show up… at just the right time? If someone, another human being who breathes the same air as you do and was used to establish a church, the Bride of Christ, that would last for over 2,000 years could accept their help and support, who are we to refuse it or proclaim God doesn't care about us? The problem might just be that we still have on the scales that fell off Paul's eyes when he accepted the very first gift of love sent to him in the form of a man who, scared to death to be in Saul's presence, followed God's lead.

As we finish Paul's story, keep these questions in the back of your mind. It might just help you figure out some of your most passionate problems and just as important, prepare you to have that discussion with someone who needs to be introduced to Christ, but needs you to listen to them with an understanding of their heart *and* their mind.

Luke

First, let's state what we know about Luke. It won't take long. Luke was not one of the original twelve disciples. Luke was not considered to be an Apostle in 1st -century Christian circles. That's it. So, what else might we know about Luke from studying Scripture? Many scholars say he wrote two of the most influential books of the New Testament: The Gospel of Luke and the Book, the Acts of the Apostles. Although not by the number of books, letters, or epistles, the two aforementioned books he authored comprise, by volume, 27.5 percent of the New Testament. There is also some speculation that he also either wrote the Book of Hebrews or had some editorial contribution based on its literary style. He is considered the most literary writer of the New Testament.

It also seems Luke was an advisor to the Apostle Paul and from close association, a friend and fellow small businessman. Scripture also seems to brand Luke as a Gentile, with reference to the distinction drawn between he and other colleagues, "of the circumcision," Colossians 4:11-14. This distinction would also make him a perfect traveling companion for Paul during many of his missionary journeys to Gentile regions. Luke's Gentile heritage is questioned, however, by several scholars. If Luke was, in fact, a Gentile he would be the only Gentile author contributing to the New Testament. The rest of what we know about the man is tradition. [223]

Luke 1:1-4 validates, unlike the Gospel of John, the writer was not an eyewitness to the ministry of Jesus Christ. Instead, Luke becomes the historian of the New Testament, observing and listening to others who were present during Jesus Christ's daily ministry and, like an observing scientist, recording everything. Regarding Luke's source material, some speculate that the accuracy of the birth of Jesus Christ's story may have been through Luke speaking directly with Mary, the mother of Jesus Christ. Who better to recount the details?[224] Only Matthew and Luke contain the story of the infancy of Jesus Christ.

Luke is referred to as Luke the Evangelist, as are Matthew, Mark, and John; because they composed the four Gospels contributing to the teachings and background of Jesus Christ. He was thought to be a physician, whatever that meant in the 1st century. It probably indicated he was literate and read everything he could get his hands on. His credentials as a physician are borne out in Colossians 4:14 where he is specifically called... a doctor. Luke also uses very specific medical terms when describing the condition of a person. In particular, in Luke 14:2, Jesus encounters a man with dropsy, or "abnormal swelling of his body," where he uses the word *hudropikos*—a common term in Greek medical literature that's found nowhere else in the Bible.[225]

Interestingly, although never called one, Luke could have been considered a scientist, if indeed he was a physician. This point should be drawn into discussions between the faith and scientific communities.

Before Scriptures link Luke and Paul at Troas, he is speculated to have begun his studying to be a physician in Antioch, in Syria. In the ancient world, Egyptians were the most skilled in medicine, having taken centuries to perfect their art. First-century doctors like Luke could perform minor surgery, treat wounds, and administer herbal remedies for everything from indigestion to insomnia.

Further speculation is that when Scripture speaks of Luke joining Paul in Troas on Paul's missionary journey, that is where Luke first encounters Paul. Some also think Luke was converted during these early encounters in that region. Luke travels with Paul to Philippi where he is left to the administration and teaching of that church. He will later join Paul during his third missionary journey in Miletus, Tyre and Caesarea, eventually ending in Jerusalem. Later, Luke apparently accompanied Paul to Rome, where Paul will eventually be martyred. Luke is last mentioned in 2 Timothy 4:11.[226]

According to ancient sources, St. Luke was martyred at age eighty-four in the Greek city of Thebes. His remains were taken to Constantinople about 338 AD and later moved to Padua, Italy, where they are kept in the Basilica of Santa Giustina. A rib is interred at his original burial place in Thebes.[227]

Who Was Paul?

Since he wrote most of the New Testament, you would think that determining just who the Apostle Paul was would be self-evident. As wise people understand, it isn't what you know, it is what you don't know you know. Reading Paul's letters and the history of the New Testament Church as recorded in Acts by Luke, it is easy to get lost in infinite details and character development of those who came into contact with the Apostle. In other words, it's quite easy to miss the point. Why? Because it is so simple.

Before you read another word from the New Testament, I invite you to cement in your mind the following question: "Not *who* is Paul, but *what* is Paul? We covered the who of Paul earlier in this chapter. Let's look at the *what* question. As with any discussion, much can be learned from questions. Here are a few:

- Why was Paul so controversial during his life, and why is he still controversial today?

- What is the justification doctrine outlined by Paul?
- What was Paul's central idea or concept?
- Not who, but what did God desire Paul to be?

Maybe you have never thought about Paul being controversial; especially today in the 21st century. The Apostle Paul, to the Jews, began as a bounty hunter basing his entire life on Jewish Pharisaical tradition and law. Like Jesus late in His ministry, the Apostle Paul seemed to never answer the question he was asked, but instead gave a broken-record response. And like Jesus Christ, he wasn't called to define the Law but to broadcast saving grace through the death and resurrection of Jesus Christ, through faith, not through following the Law of Jewish tradition or the dictates of an emperor. Based on his background and intellect Jews expected him to be the same person when converted. This expectation led to a lot of controversy over circumcision and baptism in the early church. This same argument exists today.

What Paul was appeared foolish to the Greek culture and scandalous to the Jewish community. [228] Why? It was not what either culture expected. As today, Paul's message did not fit into everyone's box. This is where many Gentiles and Jews underestimated Paul's intellect and his training under Gamaliel.

One of the things I encourage you to take away from this book is the dramatic impact that this brief period of time—from the beginning of Jesus' ministry until the turn of the 2nd century—had on the course of Western civilization. So profound an impact in fact, that events in our world are dated from this period, hence our use of BC and AD. One could say that for the Christian community it is the Genesis of the second creation.

Not only was the Genesis Creation split open when Jesus Christ died on the cross and His Spirit came alive, ripping from top to bottom the veil in the earthly Holy of Holies, but His resurrection was the transition to a new creation and a new covenant. This period on the faith community timeline ushered in the opportunity for me, for you, to become a new creation and… with all things becoming new. This same period on the faith community timeline ushered in a new covenant, not between Israel and

God but between all mankind and God. Jesus coming to Earth in human form was God launching His new creation to make all things right in the end. That is why Jesus Christ will be coming back.

Long before Jesus Christ was born, God made a covenant with His friend, Abraham. Abraham was promised a lineage that would outnumber the stars in the sky. To accomplish that goal several things needed to occur, beginning with the twelve tribes of Israel. God surely understood that to make good on His promise, he would need to broaden the scope of, compared to the whole world, God's chosen people. This new creation included:

- A second Adam—Jesus Christ
- A second Eve—The Bride of Christ, the Church
- A new baptism—by water
- The elimination of circumcision and becoming a Jew as a requirement for becoming one of God's people
- A once-and-for-all sacrifice in the Holy of Holies in Heaven using the unblemished blood of Jesus Christ instead of blood from countless animals created in the first Creation
- And… salvation was opened to every human on the Earth past, present, and future.

And what was in the middle of all of this divine chaos? The unlikely evangelist and church planter, the Apostle Paul and his literary companion, Luke. Yes, there were others, many others who assisted but, according to the bulk of the writings of the New Testament these two were appointed by God to be squarely and perfectly positioned to build the foundation on which we live and accept Christ by faith today.

Paul is recorded to have taken three or five missionary journeys, according to the way you choose to interpret Scripture. We have covered a lot of ground in a relatively short period of time. Even though many of the New Testament books remind us of the ground covered by Paul, it was always Paul's end goal to go to Rome and witness by telling his story to Caesar, considered the ruler of the entire world at that time. He almost accomplished that goal.

As you read the New Testament, now looking for what Paul was, notice in most every book touched by Paul's passion and intellect a central theme of justification and unity. The Apostle Paul was well-versed in both subjects and empathized with both concepts. He endured through many battles within the Jerusalem Church and the Antioch Church over circumcision, an issue that almost split both churches. Paul witnessed the unity of the Trinity and eventually that of the church and her leadership.

Paul knew well the Jewish requirements of being a Jew and the tradition and Laws that must be followed to be considered faithful to Jewish culture. He came to understand that all of those requirements were the same as works addressed in the Book of James. He eventually came to understand that justification came through Jesus Christ; it was a gift, not earned or purchased.

As said by Michael F. Bird in his book co-written with N.T. Wright, *The New Testament You Never Knew*, we must understand Rome did not feed Christians to the lions because they believed in Jesus. People in the Roman Empire believed in a lot of different things. Rome fed Christians to the lions because they believed He was Lord of all. That He was God, not the emperor (paraphrased).

So, what was Paul? Read the New Testament with these fresh eyes. It will jump out at you.

CHAPTER 20

PAGE TWO OF AN INFINITE HISTORY

... has religion always been around?
... when did science become a thing?
... Why do we rarely discuss science and religion in the same breath?
... what is the actual difference between science and religion?

Who Do Men Say That I Am?

The study of Jesus has been an extraordinarily active enterprise in recent decades. Though rooted in the past, it is among the least antiquarian of historical or theological pursuits.

This is one of the most resonant questions in the whole of the New Testament. It is the question, it seems, of a man who wishes to disturb but who is also himself disturbed; of a man who has somehow found himself in deeper waters than anticipated; of a man at once baffled and intrigued by a destiny that he may have begun to glimpse but of which he is not fully aware. And thus, seeking guidance, seeking perhaps to ken the range of possibilities, Jesus put the question to his followers. It is an affecting and very human moment. Some of the disciples

> ***replied, "John the Baptist." Others said "Elijah" or "One of the prophets." Unmentioned by the disciples on this occasion were the many choicer and less admiring epithets for Jesus then current in Palestine—blasphemer, false prophet, madman. Jesus persisted: "But who do you say that I am?" Only Simon Peter ventured a reply: "Thou art the Christ." In the end Jesus declined to enlighten his followers. He simply bade them to hold their tongues.*** **(Cullen Murphy, The Atlantic, December 1986)** [229]

I HAVE NOT addressed science as it relates to faith for several chapters because the 1st century so dominates a defining moment in both eternity (faith) and infinity (science). This chapter, in particular, will pair the science community and the faith community in an effort to demonstrate why these two vital entities must accept and therefore gain credibility from each other. Discussions between these two communities should never end with "either or" but instead, "together we." Let's review some writings together to determine how that might be possible.

We have spent a great deal of time discussing when things began and humans became aware. We have not specifically looked at the possible events surrounding and possibly securing the births of what we today call religion and science. This will be an interesting, informative detour.

When did they finally call it Religion and Science?

The relationship between religion and science is the subject of continued debate in philosophy and theology. To what extent are religion and science compatible? Are religious beliefs sometimes conducive to science, or do they inevitably pose obstacles to scientific inquiry?

Before looking at these two communities as stand-alone monoliths, each denying the other ownership to the real estate of truth on which each stand, it will be helpful to understand the meaningful and rightful place of science in religion and vice versa. The area of religious studies is highly dependent on the sciences. Without Anthropologists, Archeologists, Biologists, Ethnologist, Geologists, Geochronologists, and several other scientific disciplines, the faith community would be left with little other than Scripture and limited references in secular historical documents to establish a credible religious timeline.

The systematic study of science and religion started in the 1960s, with authors such as Ian Barbour (1966) and Thomas F. Torrance (1969) who challenged the prevailing view that science and religion were either at war or indifferent to each other. Peter Harrison (1998) challenged the warfare model by arguing that Protestant theological conceptions of nature and humanity helped to give rise to science in the 17th century. Peter Bowler (2001, 2009) drew attention to a broad movement of liberal Christians and evolutionists in the 19th and 20th centuries who aimed to reconcile evolutionary theory with religious belief. [230]

Today the largest interaction between science and religion concerns evolutionary theory and creationism/Intelligent Design. Past legal battles and legislative lobbying over teaching evolution and creationism in public schools would suggest science and religion were in conflict. As far back as 1959, Karl Raimund Hopper, an Austrian-British philosopher, academic, and social commentator claimed that scientific hypotheses (unlike religious ones) are in principle, falsifiable. One way to distinguish between science and religion is the claim that science concerns the natural world, whereas religion concerns both the natural and the supernatural. Scientific explanations do not appeal to supernatural entities such as gods or angels (fallen or not), or to non-natural forces (like miracles). Some hold that taking the results of science seriously entails negative answers to such persistent questions as free will or moral knowledge.

Some would argue that a professional lobbying group called, the X Club, formed by Thomas Huxley in 1864 aimed to promote a science free from religious dogmas to remove competition by amateur-clergymen scientists, opening up the field of science to full-time professionals. This could have also released the flood of applications for grants to fund those full-time science professionals.

Over time three major types of interaction between science and religion have surfaced; the independence view—no overlap between science and religion, the contact view—some overlap between science and religion, and a union of the domains of science and religion. Underneath these classifications are numerous sub-headings combining various elements of one or more of the three main classes, one of which is called *conflict*. The definition of conflict as used here is that science and religion are in perpetual and principal conflict. The basis for this belief can be found in two historical narratives; the trial of Galileo (see Dawes 2016 for a contemporary re-examination) and the reception of Darwinism (see Bowler 2001). The conflict model was developed and defended in the 19th century by the following two publications: John Draper's (1874) *History of the Conflict between Religion and Science* and White's (1896) two-volume opus *A History of the Warfare of Science with Theology in Christendom.* [231]

Religion as We Define It Today

Let's begin with the book widely recognized, in several translations, by the faith community, the Bible. In one version, the ESV, there are five references to the word *religion:* [232]

> *Rather they had certain points of dispute with him about their own religion and about a certain Jesus, who was dead, but whom Paul asserted to be alive.* Acts 25:19 ESV
>
> *They have known for a long time, if they are willing to testify, that according to the strictest party of our religion I have lived as a Pharisee.* Acts 26:5 ESV
>
> *These have indeed an appearance of wisdom in promoting self-made religion and asceticism and severity to the body, but they are of no value in stopping the indulgence of the flesh.* Colossians 2:23 ESV
>
> *If anyone thinks he is religious and does not bridle his tongue but deceives his heart, this person's religion is worthless.* James 1:26 ESV
>
> *Religion that is pure and undefiled before God the Father is this: to visit orphans and widows in their affliction, and to keep oneself unstained from the world.* James 1:27 ASV

These five references, very few to support a vehicle as massive as our modern-day interpretation, are not consistent in their inference within their context. Each one refers to specific topics but never to how we define the term today. In these specific Bible references it might be concluded that early pioneers of the New Testament Church understood the word as a general term, not the label to sit along-side atheist or agnostic.

Today, the Oxford English Dictionary defines *religion* as follows:

> *Belief in or acknowledgement of some superhuman power or powers (esp. a god or gods) which is typically manifested in obedience, reverence, and worship; such a belief as part of a system defining a code of living, esp. as a means of achieving spiritual or material improvement.* [233]

After fifty-two years of close association with the modern concept of church, I have developed a layperson's perception of religion. While somewhat close to the above definition, I view certain religious aspects that clearly separate it from what others conceive as faith. To me personally, religion is mankind's efforts to define and establish the beliefs of like-minded individuals concerning the elements and tenets of written Scripture. Others who wish to become or remain within a particular religion must accept and pledge to honor those stated elements and tenets.

According to a former Senior Fellow at the Intellectual Takeout and Doctor of Systematic Theology from Duquesne University, Daniel Lattier:

> *The etymological source of the term "religion"—which comes from the Latin noun* religio *(there isn't really a precisely corresponding term in the Greek or Hebrew)—has been an ongoing subject of debate in the West since ancient times, and we're still no closer to a resolution.*

He goes on to say:

> *Many authors today will confidently claim that the word religion comes from the Latin verb religare, which means to bind. This interpretation can be traced back to the Christian philosopher Lactantius (240-320 AD), who wrote that religion gets its name from the ongoing act of a person binding himself to God.*

Having called the previous interpretations theories, he also quotes other interpretations from Cicero (106-43 BC), in his *De Natura Deorum*, Christian author Isidore (560-636 AD), and points out that St. Thomas Aquinas, derives from Augustine (354-430 AD) in the City of God, where he claims that "religion" comes from the Latin verb *religere* meaning "to recover." (Augustine later retracted this etymological explanation of "religion" in favor of Lactantius.) [234]

Surveys such as those conducted by the Pew forum (Masci and Smith, 2016) find that nearly nine in ten adults in the US say they believe in God or a universal spirit, a number that has only slightly declined in recent decades. Among younger adults, the percentage of theists is about 80 percent.

Atheism and agnosticism are widespread among academics, especially among those working in elite institutions. A survey among National Academy of Sciences members (all senior academics, overwhelmingly from elite faculties) found that the majority disbelieved in God's existence (72.2 percent), with 20.8 percent being agnostic, and only 7 percent theists (Larson and Witham, 1998). [235]

In spite of this negative association between science and Western modernity, there is an emerging literature on science and religion by Muslim scholars (mostly scientists). The physicist Nidhal Guessoum (2009) holds that science and religion are not only compatible, but in harmony. He rejects the idea of treating the Qur'ān as a scientific encyclopedia, something other Muslim authors in the debate on science and religion tend to do, and he adheres to the no-possible-conflict principle outlined by Ibn Rushd (Averroes): There can be no conflict between God's word (properly understood) and God's work (properly understood). If an apparent conflict arises, the Qur'ān may not have been interpreted correctly.

While the Qur'ān asserts a creation in six days (like the Hebrew Bible), "day" is often interpreted as a very long span of time, rather than a twenty-four-hour period. As a result, Old Earth creationism is more influential in Islam than Young Earth creationism. [236]

Today, with such a large mass of people in Christian and Moslem Communities, much of the emphasis on the religious timeline is based on their development and growth. However, when looking at religion globally it would be wrong to omit other groups who can backdate their origins long before those who are the majority. Two of the recognized oldest religions are Hinduism and the religion(s) of ancient China. Hinduism, known as *Sanatan Dharma,* or Eternal Order, is thought to have been founded as early as 5500 BC and those in ancient China, documented as early as 4500 BC, as evidenced by designs on ceramics. [237]

The First Religion of Mankind

On a site called ewtn.com (Eternal Word Television Network), the largest religious media network in the world broadcasting Catholic programming in more than 140 different countries and territories since 1981, appeared a reprint of an article written by David Rooney in 1993 titled, "The First Religion of Mankind."[238] The author lays out a detailed analysis of several scientists' groundwork and ultimate conclusions regarding the origin of religion in the world. The article is worth reading. I do, however, suggest brewing a stiff pot of coffee before diving into the text.

Mr. Rooney's formal essay conveys some ideas that make a lot of sense and help define the perimeters of religious development over time. His first point is that humankind was not created or even evolved with the initial ability to communicate verbally and especially not through writing. It would have been thousands of years before formal language and writing developed beyond basic tones and words, and those would have been locally understood, not universal.

His second point deals with the first Creation. Adam probably did not think of it as religion but he was very aware of an all-powerful God, and also a deceiving evil force. There is no mention in Scripture of language or writing being created on any of the six days of Genesis Creation. Apparently, Adam could speak because he was given the task of naming all the animals. Scripture also records a conversation he had with God which would indicate some sort of language skills. Likewise, Eve spoke with the Evil One.

This essay suggests that the concept of God was probably diluted after Adam and Eve were exiled from the Garden and sent out into the world to make it on their own. By the time of Noah there was apparently some concept of mathematics since Noah was given specific dimensions for a very large house-boat construction project. The main thrust of the essay seemed to address the other mass of people not mentioned in Scripture; those whose bones date back thousands of years, a time questioned by many in the faith community.

Rooney's research indicates pre-historic mankind, even without language or writing skills would have still had a reasoning brain. That ability to reason would have been a driving force to identify in some way man's desire to attribute certain realities, as they knew them, to a higher power… a God. Unlike Adam, who would have an innate sense of Monotheism (one God), since he interacted with his creator, earlier man would have had a tendency to attribute God status to several things as they became associated with various aspects of mankind's daily life. Just as resistant to attributing a time or place for the origin of religion, the author also believes assigning the use of discovered artifacts and bones is arbitrary and should not be used as a factual basis for defining the habits and daily routines of these early inhabitants of the Earth.

An article on probe.com, 2004, an outreach of Probe Ministries founded in 1973 by James F. Williams, Jr. and Jon Buell to bridge the frontier between the agonizing questions man asks and the profound answers the Gospels offer, puts forward a laymen's observation by Dr. Patrick Zukeran on the question regarding the beginning of religion. Dr. Zukeran suggests three primary theories: subjective, evolutionary, and the theory of original monotheism. [239]

Subjective—teaches that religion originates with man. Humans have a psychological need for a transcendent being that provides meaning and hope to their existence in this vast, impersonal universe. Supporting this theory, he sites Friedrich Schleiermacher, Ludwig Feuerbach, and Sigmund Freud all who support some version of an idealized person. He further states that while the subjective theory may teach us about human nature, it does nothing to explain the origin of religion or where this desire comes from.

Evolution—religion is the result of an evolutionary process in human culture. Primitive man begins with an innate feeling that a spiritual force exists called *Mana*. The next step is animism where the force is visualized as personal spirits. These spirits are contained in two categories: nature—plants, rocks, or lakes and ancestral—spirits of ancestors. These stages progress to polytheism, the worship of more than one god. The author holds that there is no record or evidence of a culture moving from a mana/animism stage to a monotheistic stage.

Original Monotheism—teaches that religion originates with God disclosing Himself to man. He gives nine qualifications or characteristics of man's first form of religion. He states that, "Cultures that are very primitive provide some of the strongest proof of original monotheism." This theory fit most closely with the Genesis description of Adam and Eve.

Religion, Science, the Catholic Church, and the Printing Press

The concept and acceptance of religion and its broad meaning had a significant collision on our timeline. This could be overlooked with a casual glance at occurrences and relative dates. During the 15th century, next to the wheel and the spoon, one the most impactful inventions of all time appeared. Johann Gutenberg revolutionized the society of his time, dramatically affecting the Renaissance, religion, and other areas associated with that period of history. This invention, the printing press, became one of the pioneering points in a scientific revolution. Regarding the religious discourse at the time, the authority of the church was strengthened, laying the ground work for the Protestant movement.

During this period the Catholic Church had been going through difficult times, rocked by the 1378—1417 Western Schism. At one point during this schism, three Christian rival popes from different countries asserted their claims to ascend to the throne of Christian spiritual authority. Among Gutenberg's other business printing applications, including involvement in printing pamphlets calling for a Crusade against the Turks as retaliation for their overthrow of Constantinople, he also printed indulgences. These were special forms issued by Catholic authorities offering liberation from one's sins in exchange for acts of penance *and* a monetary payment. These publications, as well as other activities, caught the eye of a man who would turn the religious world upside down.

About fifty years after the invention of the printing press, and 100 years after the end of the Western Schism, a Catholic Priest by the name of Martin Luther announced his now famous *Ninety-Five Theses.* These where proposed major changes to the Catholic Church, including a call for the abolition of the indulgences being printed by the Gutenberg press. While accusing Gutenberg of using his invention to promote the sins of the Catholic Church, Luther also wisely envisioned its use in the promotion and spread of Scripture to the masses. The ability of the printing press to manufacture enormous amounts of printed literature which were then widely distributed, so unlike centuries past when word of mouth took years, helped Luther's writings spread all over Europe like wildfire.

At the time, about one-third of all books printed in Europe, amounting to about one million, were written by Martin Luther. Although not a thing at the time, that would have made him the first-ever best-selling author. The result? By the year 1600, nearly half of Europe was Protestant. And it wasn't just Luther's writings. The printing press also began to print the Bible… in other languages than Latin. The everyday citizen learned to read and now had access to a Bible. For the first time, the printing press eliminated the need for a priest to read and interpret Scripture. The P*riesthood of the Believer* was finally possible: a direct link between man and God through Jesus Christ's death, burial, and resurrection. [240]

After reading this small sample of religious history, you should have no difficulty understanding why I have been and will continue to refer to the *faith community* rather than the *religious community.* With that, I will change gears and shift to the other community—science. As you read on, keep in mind that religion, as broadly defined, is static for the most part. Interpretations, practices, and application may differ, but the sources do not. Science, however, never stops changing.

Science as We Define It Today

> *Exclude God from the definition of science and, in one fell definitional swoop, you exclude the greatest natural philosophers of the so-called scientific revolution—Kepler, Copernicus, Galileo, Boyle, and Newton (to name just a few).* [241]

Like religion, science, or more correctly the concept of science, has been around since mankind discovered they could think. Even though they may not have known how to form the word "why," their brains contained a compartment called *reason.* Since that time, the human race has tried to use many different words and phrases to define this process. Let's take a look backward.

In 1833-1834 AD, Cambridge University historian and philosopher William Whewell coined the term "scientist" to replace terms like "cultivators of science" and others. Professor Laura Synder, a Fulbright Scholar and professor of Philosophy at St. John's University, was asked to give a TED Talk on this subject a few years ago. The results of her research into this topic is fascinating, informative, and entertaining. [242]

She gave credit for the acceptance of the science community's decision to adopt the universal term *scientist* to four individuals who, in 1812—the same year the British again attacked the United States—formed the "Philosophical Breakfast Club." The purpose of this group was to discuss the state of science in Britain and the world. Professor Snyder gives these men credit for ushering in a new scientific revolution that even reached Charles Darwin.

According to her research, The British Association for the Advancement of Science was holding its third meeting on the campus of Cambridge on June 24, 1833. At the organization's first meeting, held at night, an elderly, white-haired gentleman stands and addresses the gathered group. Most are surprised the man is out and about, because he rarely had left his house in years. The elder philosopher and poet, Samuel Taylor Coleridge, spoke: "You must stop calling yourselves natural philosophers." Professor Snyder's interpretation was that Coleridge felt none of the members of this association, who were advancing science like himself, pondered the cosmos from their armchairs. Instead, they were mucking around in fossil pits or conducting messy experiments with electrical piles.

The crowd was not accepting of this. Attempting to control the boisterous reaction of the audience, a young Cambridge scholar by the name of William Whewell stood and addressed the members. "If 'philosophers' is taken to be too wide and lofty a term," he said, "then, by analogy with 'artist,' we may form 'scientist.'" This was the first time the word "scientist" was uttered in public, only 188 years ago.

Following this meeting, these talented amateurs with inquisitive minds, people like Charles Darwin and Joseph Priestley, were forever after known and remembered as scientists—professionals with a particular scientific method, goals, societies, and most important… funding.

The four men from Cambridge University who met in 1812 and formed the Breakfast Club were: [243]

- *Charles Babbage—invented the first mechanical calculator and the first prototype of a modern computer.*
- *John Herschel—mapped the stars of the southern hemisphere and, in his spare time, co-invented photography.*
- *Richard Jones—became an important economist who later influenced Karl Marx.*
- *William Whewell—not only coined the term* scientist, *as well as the words* anode, cathode *and* ion, *but spearheaded international big science with his global research on ocean tides.*

After pointing out the amazing attributes of these four individuals, Professor Snyder mentions the "flip-side" of these events. Although she obviously didn't know about my book project, nonetheless she directly addressed the issues I have been discussing. [244]

> *So, the philosophical breakfast club helped invent the modern scientist. That's the heroic part of their story. There's a flip side as well. They did not foresee at least one consequence of their revolution. They would have been deeply dismayed by today's disjunction between science and the rest of culture. It's shocking to realize that only 28*

percent of American adults have even a very basic level of science literacy, and this was tested by asking simple questions like, "Did humans and dinosaurs inhabit the Earth at the same time?" and "What proportion of the Earth is covered in water?" Once scientists became members of a professional group, they were slowly walled off from the rest of us. This is the unintended consequence of the revolution that started with our four friends.

A quote from one of the most notable scientists to adopt this new platform, Charles Darwin:

I sometimes think that general and popular treatises are almost as important for the progress of science as original work.

As we can see, the contemporary battle between science and religion doesn't go back very far on the eternal, infinity timeline. In fact, centuries ago it was difficult to tell them apart, except when an elite group needed to emphasize a particular point or aspect favoring their societal or financial position. Both the scientists (or whatever they were called at the time), and religious prognosticators were tried and banished or worse—put to death.

The net result of scientific findings since the 17th century has been that God was increasingly pushed into the margins. This encroachment of science on the territory of religion happened in two ways: first, scientific findings—in particular from geology and evolutionary theory—challenged and replaced biblical accounts of Creation. While the doctrine of Creation does not contain details of the mode and timing of creation, the Bible was regarded as authoritative. Second, the emerging concept of scientific laws in 17th- and 18th-century physics seemed to leave no room for special divine action. [245]

Advances in 20th-century physics, including the theories of general and special relativity, chaos theory, and quantum theory, overturned the mechanical clockwork view of Creation. In the latter half of the 20th century, chaos theory and quantum physics have been explored as possible avenues to reinterpret divine action. [246]

Today, it appears the boundaries between faith and science are shrinking as the West grows more and more secular and the discoveries of evolutionary biology and cosmology increase. With this changing dynamic, claims that science and religion are compatible grow louder. There are those (obviously, I am among them) who believe that the relationship between science and religion is "not one of conflict, but of accommodation, a coexistence of knowledge systems." Those who hold to this belief are known as "accommodationists," who believe in "accommodationism." [247]

The concept of accommodationism can be illustrated as follows:

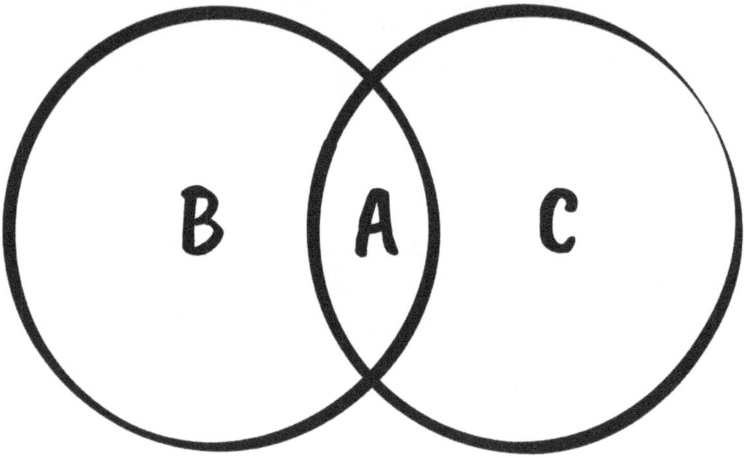

A = Accommodationism
B = Science Community
C = Faith Community

The elliptical area "A" within this diagram suggests a third fascinating arena where science and faith, working together, can properly define their important contributions to the future of our world.

Jerry Coyne, Professor of Ecology and Evolution at the University of Chicago, argues that accommodationism is misguided. He asserts that not only are science and religion in conflict, but at war with each other, since both have totally different world views. [248] The argument against the science and faith community ever helping each other, much less coming together, rests on how the two communities develop their individual positions and express them to each other.

Unpacking Differences

Many (but not all) in the science community start with an assumption that the Earth and everything in it began from a single source that was either brought to this specific planet, developed from something that was already on this planet, or after some cosmological event, e.g., the Big Bang. This assumption (theory) does not include any higher power that created everything we see from nothing; or almost nothing.

Many (but not all) in the faith community start with a higher power creating everything we see from nothing; or almost nothing at a single beginning point. This assumption does not include the concept of something growing from something into any life form we recognize today.

In the referenced on-line commentary by Professor Coyne above you can find his definition of science and religion:

> *I'll construe "science" as the set of tools we use to find truth about the universe, with the understanding that these truths are provisional rather than absolute.*
>
> *And I'll define religion as does philosopher Daniel Dennett: "Social systems whose participants avow belief in a supernatural agent or agents whose approval is to be sought."*

Coyne's conflict between science and religion rests on "…the methods they use to decide what is true, and what truths result…" My view on the difference between these two communities is similar, but from another viewpoint.

While it is true that religion is based on faith, the basis for this faith comes from written records, the Bible, and secular history recording the lives and experiences of characters and incidents mentioned in the Bible, and the individual, even personal, experience of "a" believer. While interpretation of these written records and experiences can change, the source doesn't change, or does it? For instance, finding the Dead Sea Scrolls?

Faith also has something science cannot have… absolute truths. These are tenets upon which the faith community cannot change or alter. Some of these are: the virgin birth of Jesus Christ, the resurrection of Jesus

Christ, and the ascension of Jesus Christ. If any of these absolute truths are not true, then none of them are valid.

Science is based on the latest finding or discovery. A long-held position can change overnight upon a new, startling, replicable discovery. There are no absolute truths in science. The faith community's position is based entirely on everything being created from nothing, or almost nothing, by a higher power, at a beginning point.

Actually, the science community does not have a similar position, or even conflicting position. All science can do is investigate and discover what is already there. The faith community holds to a creator who created everything from nothing, or almost nothing. The science community has discovered, invented, and combined things that already exist. So, what's the problem? How did a "world view" get brought into this discussion? Using this term is divisive. Once applied, the daggers come out, lines are drawn in the sand, and no one gets invited to dinner.

Above I inserted the phrase "or almost nothing" numerous times. This is based on different understandings of the use of, "without form" in the first part of Genesis.

The above information was rather recent, so let's go back further to an article published in the American Humanist Association's paper appropriately branded, the *Humanist* in their May/June issue, 1986. The article was written collectively by Dr. Norman F. Hall and Lucia K. B. Hall. Like Professor Coyne, the Halls have a very militant view of any hope for bringing the science and faith communities together. Here is a quote from this article's opening paragraphs:

> *Let us be blunt. While it may appear open-minded, modest, and comforting to many, this conciliatory view is nonsense. Science and religion are diametrically opposed at their deepest philosophical levels. And, because the two worldviews make claims to the same intellectual territory — that of the origin of the universe and humankind's relationship to it — conflict is inevitable.* [249]

As I also find in faith community writings, this article makes a statement and then later contradicts itself based on its ultimate goal of proving its point. As an example:

> *It is possible, of course, to define a nonsupernatural "religious" worldview that is not in conflict with science. But in all of its traditional Western forms, the supernatural religious worldview makes the assumption that the universe and its inhabitants have been designed and created—and in many cases, are guided—by "forces" or beings which transcend the material world. The material world is postulated to reflect a mysterious plan originating in these forces or beings, a plan which is knowable by humans only to the extent that it has been revealed to an exclusive few. Criticizing or questioning any part of this plan is strongly discouraged, especially where it touches on questions of morals or ethics.*

A casual glance at the opening sentence in the second insert above could lead you to believe the author is saying that he was wrong in his opening assumption about unavoidable conflict. That isn't at all what the sentence says. It says, *nonsupernatural "religious" worldview,* which is the expected view of a humanist. In reality, it is impossible to have a nonsupernatural spiritual faith. Again, I want to emphasize the faith community, at times, uses this same banter to prove its credibility.

The article goes to some length to emphasize the absence of emotion and the inability of nature to care about anything. Rather, it states that nature just does what it does without any plan or agenda. No one with a basic understanding of the universe will disagree with that position. The remainder of the article suggests several other humanistic thoughts which clarify its position as a nonsupernatural religion as contained in these few examples:

> *A naturalistic system of ethics is not likely to be popular, however, until science can overcome the currently evident public attitude of ignorance and hostility.*

> *Accept the supernatural and the hard work of making and testing theories becomes a pointless enterprise, along with all human-made*

> *explanations and meaning. But if we allow such myths to limit the scope and uses of science, we will do so to our own peril and shame.*

The authors, perhaps without intent, do properly define the role of science in their writings; *"Rather, a nonmysterious, understandable, material universe is the basic assumption behind all of science."* Such an assignment does not in any way mandate an atheistic view of God. Neither science nor faith's goal should be to investigate God or even try to figure out who He is. Science is simply looking at things that already exist, trying to know more about them, and also to find ways to use the things they study for the betterment of human life. At the same time, the study of faith should not reject science, nor its findings solely because it doesn't fit on their developed timeline, which is invariably based on incomplete, uncorroborated historical writing.

The article's closing paragraph makes an attempt to say just that, but stops short to reserve a naturalist position:

> *In an article in the October 4, 1985, issue of* Science, *cosmologist Steven Weinberg said that, even if science manages to trace the materialist explanation back to the first ten-billionth of a second of the existence of the universe, we still don't know what started the clock. "It may be that we shall never know," he wrote, "just as we may never learn the ultimate laws of nature. But I wouldn't bet on it."*

Although I disagree with Professor Coyne's theory and the position of Dr. Hall that science and religion are at war and can never come together, I still see gigantic matters yet to be resolved. These unresolved differences are tragic when you consider how easily they can be resolved. Such a resolution could eventually allow more progress on things both sides would probably agree are more important.

Resolution

Let's look at what each community is missing by refusing to accept the other. The faith community is missing an opportunity to assist the science community in solving many of their greatest challenges. What are those?

Where did the first atom come from? How big is space? What really *is* all that stuff between you and I that looks transparent? How can calming one's soul, combined with medicine, heal quicker and more completely? Providing a more plausible answer to why the universe works like a perfect machine? A better understanding of what an alien is and that they do exist? Can a person of faith be a scientist?

The science community is missing an opportunity to assist the faith community in solving many of their greatest challenges. What are those? Why is our timeline so messed up? How do we explain dinosaurs to our children? Is it possible that there was life before Adam? Can a scientist have faith in God?

A person of faith can be a person of science.

A person of science can be a person of faith.

What stands in the way?

A person of faith can be a person of science if they ask God to handle their insecurities.

A person of science can be a person of faith if they ask God to handle their insecurities.

What stands in the way? Absolutely… nothing!… not even Dark Matter.

The Birth of Chronology

Science Direct published an article on September 22, 2017 titled, "The Creation of the World and the Birth of Chronology" handled by Vincent Courtillot. I would encourage you to read his entire work. The publication begins with an abstract outlining the cyclical nature of time as agreed upon by all Greek philosophical schools of the time except Platonists. The schools are in agreement that stated time was eternal and therefore the idea that something could be made from nothing was absurd. It is worth mentioning, apart from this publication, most modern-day scientists still hold to this position. [250]

We have covered in detail, earlier in this book, that the Genesis recorded in the first book of the Old Testament is in direct opposition to this long-held position and has become the basis for the main conflict between science and the faith community. The author of this publication states, "Unambiguously, the Creation became the reference point for the world's history." On his basis, Scriptural analysis places the beginning of all things roughly 5,500 years in the past. The new science of Geology, using the study of the stratus of the earth's core, claimed that the Earth's beginning goes back millions of years. This is far from a minor discrepancy.

This modern chronology, set from fossil records beginning in the 18th century and then forward to the late 19th century and beyond, posted a thesis of an expanding universe, making a definitive beginning of everything very uncertain. This new thesis demanded a new science, as it had become known, causing Hutton to utter his famous quote in 1788, "The result, therefore, of our present enquiry is, that we find no vestige of a beginning, no prospect for an end." [251] This change in thought blasted the age of the earth from 5,500 years to more than 4.55 billion years suggesting that the chronology of the Earth must be completely rethought. To fully understand this article will require your reading the entire well-documented publication.

More than the arguments presented by Courtillot, our purpose is to motivate the faith community to properly determine how to merge the faith communities' use of Scripture into the chronology of this new science. Before addressing a few possible scenarios, it is vital that we look at a linear chronology including more than those in Scripture. In truth, the faith community cannot just ignore science, but instead find its appropriate place in the entirety of eternity and infinity.

An Aggregated View of Everything

The following chronology was developed by combining dates and incidents from the Holy Bible (mostly NKJV), websites; Explainthatstuff.com, Sciencedirect.com, Historyworld.net, Sciencetimeline.net, Christianity.com, and Wikipedia.org (to some extent). [252] The purpose of this aggregated chronology is to allow the reader to better imagine the development of both the faith community and science community over a very long period of time. In doing so, certain things should stand out. A few of these are:

- Late entry of both communities, as we understand them today, into this timeline
- The flattening of faith contributions against the incredible acceleration of science
- Based on contributions and purpose, neither originally set out to compete with the other
- It was only when each side decided to dig their heels into the origin of humankind and the age of the Earth that complications arose
- The faith community has always held that our existence began from nothing; or almost nothing. Science, never creating anything original, can only build on the discovery and invention of elements already in existence.

The end of this aggregated chronology moves civilization toward the Fall of Creation and much closer to the last exhaled breath by humankind. It is probable that the science community will continue to make even greater discoveries and invent exciting new ways to reorganize the elements of our existence for the betterment and destruction of civilization as we all move forward toward the end of what we currently understand. In contrast, the faith community will continue to discuss and disagree over traditions and text written as far back as man has been able to communicate verbally and through written and electronic record.

As you read through this astonishing list of people and incidents, I have highlighted those times and incidents that I personally feel were very significant. I encourage you to take a pencil, not a pen—you may change your mind—and circle those that you feel were most significant. Enjoy!

4—5 Billion Years BC
The Sun starts to produce energy.

3.5 Billion Years BC
Humans make the first tools from stone, wood, antlers, and bones.

1—2 Million Years BC
Humans discover fire.

10000 BC
Earliest boats are constructed.

9000—8000 BC
Beginning of human settlements and agriculture.

7000—6000 BC
Hand-made bricks first used for construction in the Middle-East.

4000 BC
Iron used for the first time in decorative ornaments.

5000—3500 BC
Glass is made by people for the first time.

3500 BC
Humans invent the wheel.

3000 BC
First written languages are developed by the Sumerian people of southern Mesopotamia (part of modern Iraq).

2500 BC
Ancient Egyptians produce papyrus, a crude early version of paper.

3000—600 BC
Bronze Age: Widespread use of copper and its important alloy, bronze.

2000 BC
Water-raising and irrigation devices like the *shaduf* (shadoof), invented

by the ancient Egyptians, introduce the idea of lifting things using counterweights.

1700 BC
Semites of the Mediterranean develop the alphabet.

1000 BC
Iron Age begins; iron is widely used for making tools and weapons in many parts of the world.

900—500 BC
Nazca people of Peru are believed to have experimented with balloon flight.

600 BC
Thales of Miletus discovers static electricity.

400—300 BC
Chinese experiment with flying kites.

300—200 BC
Chinese invent early magnetic direction-finders.

250 BC
Ancient Egyptians invent lighthouses, including the huge Lighthouse of Alexandria.

250 BC
Archimedes invents the screw pump for moving water and other materials.

150—100 BC
Gear-driven, precision clockwork machines (such as the Antikythera mechanism) are in existence.

50 BC
Roman engineer Vitruvius perfects the modern, vertical water wheel.

62 AD
Hero of Alexandria, a Greek scientist, pioneers steam power.

70 AD
The Romans under Titus destroy Jerusalem, after a long siege; 1.5 million Jews die. [253]

(Additional Resources: *First Persecutions* written by John Foxe)

71 AD
Center of Christianity moves from Jerusalem to Antioch, Alexandria, and Rome.

90 AD
Book of Revelation and Gospel of John is written.

105 AD
Ts'ai Lun makes the first paper in China.

27 BC—395 AD
Romans develop the first, basic concrete called *pozzolana.*

100—590 AD
Post Apostolic Age. [254]

131 AD
Jerusalem rebuilt.

161—180 AD
Widespread persecution of Christians under Roman emperor Marcus Aurelius.

200—300 AD

Appearance of Earliest Bible Translations. [255]

249—305 AD

Further Christian persecution under the Roman reign of emperors Decius Diocletian.

301 AD

Armenia is the world's first country to adopt Christianity as the state religion.

312—337 AD

Constantine – Edict of Milan – 313 AD.

It is important to pause at this juncture and assess what has happened since the birth of the One called Jesus Christ, the Son of the Almighty God, the Messiah. In a short synopsis: [256]

1st century—The end of thousands of years of traditional Jewish law through the birth of a Child, Jesus Christ. This transition occupied about one-third of the 1st century. During most of Jesus Christ's life and ministry He still pointed to the heritage of His earthly parents, the Jews springing from the Twelve Tribes of Israel. However, toward the end of that ministry, and at His death, He began to speak of spreading His message to the Gentiles, meaning… the rest of the entire world.

Moving further into the 1st century a man named Saul (Hebrew) or Paul (Greek) was called by God to lead the development of something different, something new, something disruptive… Christianity. Its roots were located in Jerusalem but quickly spread to other surrounding regions like Antioch. In addition, Saul/Paul engineered three to five missionary journeys in several parts of the surrounding regions and countries to establish even more places that would start groups following this new, disruptive theology.

2nd century—With limited ways to mass record literature and communicate, leadership of these organizations changed from eyewitnesses and founding members to more uninformed and diverse leadership. However, the movement continued to spread despite increased oppression from the Roman Government. The thing that set the new movement apart was the opportunity to totally disconnect from the Jewish nation, their history, their traditions, and their laws. Christianity was quickly becoming its own thing, not a mixture or Jewish and Gentile backgrounds. With these broken ties, the Romans were also losing long-established alliances with top Jewish leaders and Priests. The Romans found their grip on a lower-class society slipping away. To maintain control, persecution of the Christians and Jews escalated. The very foundation of one of the longest surviving civilizations yet in history was at stake.

3rd century—What started at the end of the 2nd century dominated the 3rd century.

4th century—What takes place in the 4th century is to many historians a mystery. With the passing of four short centuries a sect, group, religious rebellion, or however you wish to label it, moved from non-existence to being the State-Religion of the Roman Empire. By the end of the 4th century, it becomes illegal to participate in any public worship except Christianity. Rome has moved from arresting and executing Jesus Christ as a criminal and threat to social order, to proclaiming Him as God. No longer were Roman Emperors called gods. While the state of the Roman Empire is hard to define at the close of the 4th century, one thing is for certain… Christianity is nothing like it was in the 1st, or even the 2nd century.

This would be the foundation of a new church and new church leaders that would evolve over the next thousand years while still holding tightly to the original proclamation that Jesus Christ was born of the Virgin Mary, lived, was crucified, dead, buried, rose from the dead, and ascended to His Father in Heaven… and that He came to take away the sins of the world. But the problem was not the message, it became the messengers and how it was used.

So, why did we pause here? The individual who would take center stage in this religious shift was Constantine the Great, who became partial emperor in 306 AD, and emerged victorious as sole emperor of the Roman Empire by 324 AD. He had begun to favor Christianity around 312 AD, leading to his eventual baptism by Eusebius on Nicodemia. Most importantly, he played an influential role in the proclamation of the Edict of Milan in 313 AD, which declared tolerance for Christianity throughout the Roman Empire. Constantine was also a central force in calling the Council of Nicea in 325 AD, an attempt to settle a divisive dispute in the church over the divinity of Jesus Christ. The disagreement centered largely around who was considered the father based on interpretation of the words in Greek used to describe the person. [257]

325 AD
Council of Nicaea—Called by the Roman Emperor Constantine, it was the first ecumenical conference of bishops of the church.

367 AD
Saint Athanasius is the first to list all twenty-seven New Testament Books in his festal letter.

381 AD
Ecumenical Council of Constantinople revises the Nicene Creed to its current form.

397 AD
Synod at Carthage ratifies the twenty-seven books of the New Testament as sacred Scripture.

400 AD
Vulgate.

431—451 AD
Councils of Ephesus and Chalcedon.

481 AD
Horse shoes made of iron.

590—early 1300s AD
The Medieval Church.

600 AD
Windmills are invented in the Middle East.

622 AD Muslim AH (Anno Hegirae)
The Muslim chronology is the only one to be established at the same period as the history which it sets out to record. In 622 Muhammad moves from Mecca to Medina, an event taken as the beginning of Islam. It is known as the Hegira*—a word traditionally translated in western texts as "flight" but usually now described as "emigration."*

Within twenty years of the event, by the time of the caliphate of Omar, the Hegira is being officially used as the starting date of the new Muslim era. By analogy with (Anno Domini), dates are given as AH (Anno Hegirae). Being based on a lunar calendar, the years of Muslim chronology do not tally with the Christian solar years.

The Muslim calendar is the only one in widespread use to be based uncompromisingly on lunar months, with no adjustments to bring the years into balance with the solar cycle. The twelve months are alternately 29 and 30 days long (the lunar cycle is approximately 29.5 days), giving a year of 354 days. There are two significant results. Muslim months bear no relation to the seasons, and Muslim years do not coincide with those of other chronologies. There are about 103 lunar years in a solar century. By the millennium there will have been 1421 lunar years but only 1378 solar years from the start of Muslim chronology in AH 1 or 622. The year AH 1421 will be 2000. [258]

625 AD
Mohammad begins the Koran.

635 AD
Pens made of quill.

666 AD
Organs used in church.

700—900 AD
Chinese invent gunpowder and fireworks.

787 AD
Ecumenical Council at Nicaea ends the controversy over the use of icons in the church.

800—1300 AD
The Islamic "Golden Age" sees the development of a wide range of technologies, including ingenious clocks and feedback mechanisms that are the ancestors of modern automated factory machines.

800 AD
Charlemagne is crowned emperor of the Holy Roman Empire by Pope Leo III.

988 AD
With the conversion of Prince Vladimir in Kiev, Christianity begins to grow in Russia.

1000 AD
Chinese develop eyeglasses by fixing lenses to frames that fit onto people's faces.

1054 AD
Great Schism—Eastern Orthodox and Western Catholic churches separate.

1095—13th century AD
The Crusades—A series of several military campaigns. Originally, they were Roman Catholic endeavors to capture the Holy Land from the Muslims, but some were directed against other Europeans.

1157 AD
Bank in Venice established.

1163 AD
Notre Dame de Paris—beginning of construction.

1180 AD
Glass windows first used for light.

1206 AD
Arabic engineer al-Jazari invents a flushing hand-washing machine, one of the ancestors of the

modern toilet.

1215 AD
King John of England and his nobles sign the Magna Carta.

1232 AD
Chinese repel Mongol invaders using early rockets.

1234 AD
Coal dug for fuel.

1240 AD
Spectacles invented by an Italian.

1258 AD
The first English House of Commons.

1266—73 AD
Thomas Aquinas writes his great work of systematic Theology: *Summa Theologiae.*

Early 1300s—1517 AD
The Renaissance.

1305 AD
Papacy moved to Avignon following a dispute with Philip IV of France.

1378—1410 AD
The Great Schism—a time of division in the Roman Catholic Church due to disagreements concerning papal succession.

1382 AD (1380?)
John Wycliffe translates First English Bible. John Wycliffe was an English theologian and early proponent of reform in the Roman Catholic Church during the 14th century.

1431 AD
Joan of Arc martyred—Inspired and directed by religious visions, Joan of Arc (Jeanne d'Arc), organized the French resistance that forced the English to end their siege of Orléans (1429). Her claims of visions led to a trial for heresy which resulted in execution by burning at the stake.

1436 AD
Both printing and gunpowder invented.

1450 AD
Johannes Gutenberg pioneers the modern printing press, using rearrangeable metal letters called movable type.

1450s AD
Gutenberg Bible—first printed Bible.

1453 AD
The Christian kingdom of Constantinople finally falls to the Muslims.

1470 AD
The first parachute is sketched on paper by an unknown inventor.

1478 AD
Spanish Inquisition established by Pope Sixtus IV.

1492 AD
Some say America is first discovered.

1517—1648 AD
The Reformation—European movement aimed initially at reforming the beliefs and practices of the Roman Catholic Church.

1517 AD
Martin Luther's *95 Theses*—a document written by Martin Luther that challenged the teachings of the Catholic Church on penance, the authority of the pope, and indulgences. It sparked a theological debate that fueled the Reformation.

1525 AD
Bible translated into English from Greek by William Tyndale.

1530s AD
Gerardus Mercator helps to revolutionize navigation with better mapmaking.

1530 AD
Augsburg Confession—Primary confession of faith of the Lutheran Church and one of the Reformation. It was presented by a number of German rulers and free cities at the Diet of Augsburg.

1534 AD
Act of Supremacy passed—Henry VIII becomes supreme head of the English church.

1536 AD
John Calvin writes *The Institutes of the Christian Religion.*

1536—1541 AD
Michelangelo paints "The Last Judgment."

1549 AD
Telescopes invented.

1560 AD
Geneva Bible.

1563 AD
Foxe's *Book of Martyrs* first published.

1582—1609 AD

Douay-Rheims Bible.

1590 AD
A Dutch spectacle-maker named Zacharias Janssen makes the first compound microscope.

1596 AD
Sir John Harington describes one of the first modern flush toilets.

1600 AD
Galileo Galilei designs a basic thermometer.

1600 AD
William Gilbert publishes his great book *De Magnete* describing how Earth behaves like a giant magnet. It's the beginning of the scientific study of magnetism.

1609 AD
Galileo Galilei builds a practical telescope and makes new astronomical discoveries.

1611 AD
King James Version of the Bible.

1619 AD
Blood circulation is discovered.

1633 AD
Galileo is forced by the Catholic Church to renounce his teachings that the Earth revolves around the sun.

1643 AD
Death of Galileo.

1643 AD
Galileo's pupil Evangelista Torricelli builds the first mercury barometer for measuring air pressure.

1650s AD
Antoni van Leeuwenhoek and Robert Hooke independently develop microscopes.

1650s AD
Christiaan Huygens develops the pendulum clock (using Galileo's earlier discovery that a swinging pendulum can be used to keep time).

1648—1789 AD
The Enlightenment.

Defined: When man began to look beyond superstition and began to use his reason to discover the world. The effort to discover the natural laws which governed the universe led to scientific, political, and social advances.

1678 AD
John Bunyan's *Pilgrim's Progress* published.

1687 AD
Isaac Newton formulates his three laws of motion.

1700s AD
Bartolomeo Cristofori invents the piano.

1730s—1740s AD
The Great Awakening—A religious movement among American colonial Protestants. Key figures: Jonathan Edwards and George Whitfield.

1738 AD
John and Charles Wesley converted. They lead an Evangelical revival in England and form the Methodist church.

1741 AD
George Frideric Handel writes his most famous work, *Messiah*.

1764 AD
John Newton writes hymn "Amazing Grace."

1776 AD
British colonies in America declare independence from England.

1780 AD
The Sunday School movement is pioneered in England by Robert Raikes.

1789—1799 AD
The French Revolution.

1789—1970 AD
The Modern Church.

1794 AD
London Missionary Society founded in England, a non-denominational missionary society with missions in the islands of the South Pacific and Africa.

1800—1840s AD
Second Great Awakening.

1800 AD
Italian Alessandro Volta makes the first battery (known as a Voltaic pile).

1801 AD
The automated cloth-weaving loom is invented. The punched cards it uses to store patterns help to inspire programmable computers.

1803 AD
Henry and Sealy Fourdrinier develop the papermaking machine.

1806 AD
Humphry Davy develops electrolysis into an important chemical technique and uses it to identify a number of new elements.

1806 AD
Sir William Congreve develops long-range military rockets, based on an earlier Indian technology known as the Mysore rocket.

1807 AD
British Parliament votes to abolish the slave trade.

1807 AD
Humphry Davy develops the electric arc lamp.

1814 AD
George Stephenson builds the first practical steam locomotive.

1816 AD
Robert Stirling invents the efficient Stirling engine.

1820s—1830s AD
Michael Faraday builds primitive electric generators and motors.

1827 AD
Joseph Niepce makes the first modern photograph.

1830s AD
William Sturgeon develops the first practical electric motor.

1830s AD
Louis Daguerre invents a practical method of taking pin-sharp photographs called *Daguerreotypes.*

1830s AD
William Henry Fox Talbot develops a way of making and printing photographs using reverse images called negatives.

1830s—1840s AD
Charles Wheatstone and William Cooke, in England, and Samuel Morse, in the United States, develop the electric telegraph (a forerunner of the telephone).

1836 AD
Englishman Francis Petit-Smith and Swedish-American John Ericsson independently develop propellers with blades for ships.

1837—1901 AD
Victoria Queen of England.

1839 AD
Charles Goodyear finally perfects a durable form of rubber (vulcanized rubber) after many years of unsuccessful experimenting.

1840s AD
Scottish physicist James Prescott Joule outlines the theory of the conservation of energy.

1840s AD
Scotsman Alexander Bain invents a primitive fax machine based on chemical technology.

1844 AD
Young Men's Christian Association (YMCA) founded in London, England by Sir George Williams.

1849 AD
James Francis invents a water turbine now used in many of the world's hydropower plants.

1850s AD
Henry Bessemer pioneers a new method of making steel in large quantities.

1850s AD
Louis Pasteur develops pasteurization, a way of preserving food by heating it to kill off bacteria.

1850s AD
Italian Giovanni Caselli develops a mechanical fax machine called the *pantelegraph*.

1854 AD
Charles H. Spurgeon, only nineteen years old, called to the pastorate of London's famed New Park Street Chapel.

1854 AD
Dogma of the Immaculate conception of Mary proclaimed by the Roman Catholic church.

1855 AD
D.L. Moody converts to Christianity.

1859 AD
Charles Darwin publishes *Origin of Species.*

1860s AD
Frenchman Étienne Lenoir and German Nikolaus Otto pioneer the internal combustion engine.

1860s AD
James Clerk Maxwell figures out that radio waves must exist and sets out basic laws of electromagnetism.

1861—1865 AD
American Civil War.

1861 AD
Fire extinguishers are invented.

1863—1885 AD
Fanny Crosby writes more than 8,000 hymns including "Blessed Assurance."

1865 AD
Rev. William Booth founded Open Christian Mission in East London (in 1878 its name is changed to Salvation Army).

1867 AD
Joseph Monier invents reinforced concrete.

1868 AD
Christopher Latham Sholes invents the modern typewriter and QWERTY keyboard.

1870—1881 AD
Revised Standard version of the Bible.

1870s AD
Thomas Edison develops the phonograph, the first practical method of recording and playing back sound on metal foil.

1870s AD
Lester Pelton invents a useful new kind of water turbine known as a Pelton wheel.

1870—1871 AD
First Vatican council. Dogma of Papal infallibility proclaimed.

1871 AD
Frank Wenham, a British aeronautical engineer, invents the wind tunnel.

1876 AD
Alexander Graham Bell patents the telephone, though the true ownership of the invention remains controversial even today.

1877 AD
Thomas Edison invents his sound-recording machine or phonograph—a forerunner of the record player and CD player.

1877 AD
Edward Very invents the flare gun (Very pistol) for sending distress flares at sea.

1880 AD
Thomas Edison patents the modern incandescent electric lamp.

1880 AD
Pierre and Paul-Jacques Curie discover the piezoelectric effect.

1880 AD
First Northfield Conference led by D.L. Moody.

1880s AD
Thomas Edison opens the world's first power plants.

1880s AD
Charles Chamberland invents the autoclave (steam sterilizing machine).

1880s AD
Charles and Julia Hall and Paul Heroult independently develop an affordable way of making aluminum.

1880s AD
Carrie Everson invents new ways of mining silver, gold, and copper.

1881 AD
Jacques d'Arsonval suggests heat energy could be extracted from the oceans.

1883 AD
George Eastman invents plastic photographic film.

1884 AD
Charles Parsons develops the steam turbine.

1885 AD
English Revised Version of the Bible.

1885 AD
Karl Benz builds a gasoline-engine car.

1886 AD
Josephine Cochran invents the dishwasher.

1888 AD
Friedrich Reinitzer discovers liquid crystals.

1888 AD
John Boyd Dunlop patents air-filled (pneumatic) tires.

1888 AD
Nikola Tesla patents the alternating current (AC) electric induction motor and, in opposition to Thomas Edison, becomes a staunch advocate of AC power.

1890s AD
French brothers Joseph and Louis Lumiere invent movie projectors and open the first movie theater.

1890s AD
German engineer Rudolf Diesel develops his diesel engine—a more efficient internal combustion engine without a sparking plug.

1890s AD
Russian Konstantin Tsiolkovsky figures out the theory of space rockets.

1894 AD
Physicist Sir Oliver Lodge sends the first ever message by radio wave in Oxford, England.

1895 AD
German physicist Wilhelm Röntgen discovers x-rays.

1895 AD
American Ogden Bolton, Jr. invents the electric bicycle.

1896 AD
First Modern Olympics.

1898 AD
Nikola Tesla invents remote, radio control.

1899 AD
Everett F. Morse invents the optical pyrometer for measuring temperatures at a safe distance.

1899 AD
Gideons International founded—ministry dedicated to distributing copies of the Bible in over eighty languages and more than 180 countries of the world to those who might not otherwise encounter it.

1901 AD
Guglielmo Marconi sends radio-wave signals across the Atlantic Ocean from England to Canada.

1901 AD
American Standard Version of the Bible.

1901 AD
The first electric vacuum cleaner is developed.

1903 AD
Brothers Wilbur and Orville Wright build the first engine-powered airplane.

1905 AD
Albert Einstein explains the photoelectric effect.

1905 AD
Samuel J. Bens invents the chainsaw.

1906—1908 AD
Frederick Gardner Cottrell develops the electrostatic smoke precipitator (smokestack pollution scrubber).

1906 AD
Azusa Street Revival—Revival that took place in Los Angeles, California

and was led by preacher William J. Seymour. It launched Pentecostalism as a worldwide movement.

1906 AD
Willis Carrier pioneers the air conditioner.

1906 AD
Mikhail Tswett discovers chromatography.

1907 AD
Leo Baekeland develops Bakelite, the first popular synthetic plastic.

1907 AD
Alva Fisher invents the electric clothes washer.

1908 AD
American industrialist and engineer Henry Ford launches the Ford Model T, the world's first truly affordable car.

1909 AD
Scofield Bible.

1909 AD
German chemists Fritz Haber and Zygmunt Klemensiewicz develop the glass electrode, enabling very precise measurements of acidity.

1910 AD
Romanian-born Henri-Marie Coandă builds a simple jet plane, but it never actually flies.

1912 AD
American chemist Gilbert Lewis describes the basic chemistry that leads to practical, lithium-ion rechargeable batteries (though they don't appear in a practical, commercial form until the 1990s).

1912 AD
Titanic sinks.

1912 AD
Hans Geiger develops the Geiger counter, a detector for radioactivity.

1914—1918 AD
World War I.

1916 AD
Robert Hutchings Goddard, an American physicist, publishes influential ideas on building space rockets.

1918 AD
Billy Graham is born. Later becomes one of the most prominent evangelists in Christian history.

1919 AD
Francis Aston pioneers the mass spectrometer and uses it to discover many isotopes.

1920s—1930s AD
Frank Whittle of England and Hans Pabst von Ohain of Germany develop rival jet engines.

1920s AD
John Logie Baird develops mechanical television.

1920s AD
Philo T. Farnsworth invents modern electronic television.

1920s AD
Robert H. Goddard develops the principle of the modern, liquid-fueled space rocket.

1920s AD
German engineer Gustav Tauschek and American Paul Handel independently develop primitive optical character recognition (OCR) scanning systems.

1920s AD
Albert W. Hull invents the magnetron, a device that can generate microwaves from electricity.

1921 AD
Karel Capek and his brother coin the word "robot" in a play about artificial humans.

1921 AD
John Larson develops the polygraph ("lie detector") machine.

1928 AD
Thomas Midgley, Jr. invents coolant chemicals for air conditioners and refrigerators.

1928 AD
The electric refrigerator is invented.

1929 AD
Great Depression begins in the US.

1930s AD
Peter Goldmark pioneers color television.

1930s AD
Laszlo and Georg Biro pioneer the modern ballpoint pen.

1930s AD
Maria Telkes creates the first solar-powered house.

1930s AD
Wallace Carothers develops neoprene (synthetic rubber used in wetsuits) and nylon, the first popular synthetic clothing material.

1930s AD
Robert Watson Watt oversees the development of radar.

1930s AD
Arnold Beckman develops the electronic pH meter.

1931 AD
Harold E. Edgerton invents the xenon flash lamp for high-speed photography.

1932 AD
Arne Olander discovers the shape memory effect in a gold-cadmium alloy.

1933—1944 AD
Rise of Nazism in Germany.

1935 AD
Oswald Chamber's book *My Utmost for His Highest* published post-mortem.

1936 AD
W.B. Elwood invents the magnetic reed switch.

1938 AD
Chester Carlson invents the principle of photocopying (xerography).

1938 AD
Roy Plunkett accidentally invents a nonstick plastic coating called Teflon.

1939—1945 AD
World War II.

1939 AD
Igor Sikorsky builds the first truly practical helicopter.

1940s AD
English physicists John Randall and Harry Boot develop a compact magnetron for use in airplane radar navigation systems.

1942 AD
Wycliffe Bible Translators founded by William Cameron Townsend with the goal of making a translation of the Bible in every living language in the world.

1942 AD
Enrico Fermi builds the first nuclear chain reactor at the University of Chicago.

1945 AD
Dietrich Bonhoeffer pastor and leader of the underground church of Germany martyred.

1945 AD
US government scientist Vannevar Bush proposes a kind of desk-sized memory store called Memex, which has some of the features later incorporated into electronic books and the World Wide Web (WWW).

1945 AD
Arthur C. Clarke conceives the idea of the communications satellite, a space-based signal "mirror" that can bounce radio waves from one side of Earth to the other.

1945 AD
On August 6, 1945, during World War II (1939-45), an American B-29 bomber dropped the world's first deployed atomic bomb over the Japanese city of Hiroshima. The explosion immediately killed an estimated 80,000 people; tens of thousands more would later die of radiation exposure.

1947 AD
Dead Sea Scrolls found.

1947 AD
John Bardeen, Walter Brattain, and William Shockley invent the transistor, which allows electronic equipment to be made much smaller and leads to the modern computer revolution.

1948 AD
Formation of the World Council of Churches.

1949—1954 AD
C.S. Lewis writes *The Chronicles of Narnia.*

1949 AD
First Billy Graham Crusade, held in Los Angeles, CA.

1949 AD
Bernard Silver and N. Joseph Woodland patent barcodes—striped patterns that are initially developed for marking products in grocery stores.

1950s AD
Charles Townes and Arthur Schawlow invent the maser (microwave laser). Gordon Gould coins the word "laser" and builds the first optical laser in 1958.

1950s AD
Stanford Ovshinksy develops various technologies that make renewable energy more practical, including practical solar cells and improved rechargeable batteries.

1950s AD
European bus companies experiment with using flywheels as regenerative brakes.

1950s AD
Percy Spencer accidentally discovers how to cook with microwaves, inadvertently inventing the microwave oven.

1950 AD
All missionaries are forced to leave China.

1950 AD
Dogma of the Assumption of Mary proclaimed by the Roman Catholic Church.

1952 AD
US Congress designates a National Day of Prayer.

1952 AD
Revised Standard Version of the Bible.

1954 AD
Indian physicist Narinder Kapany pioneers fiber optics.

1955 AD
US electrical engineer Eugene Polley invents the TV remote control.

1956 AD
First commercial nuclear power is produced at Calder Hall, Cumbria, England.

1957 AD
Soviet Union (Russia and her allies) launch the Sputnik space satellite.

1957 AD
Lawrence Curtiss, Basil Hirschowitz, and Wilbur Peters build the first fiber-optic gastroscope.

1958 AD
Jack Kilby and Robert Noyce, working independently, develop the integrated circuit.

1959 AD
IBM and General Motors develop Design Augmented by Computers-1 (DAC-1), the first computer-aided design (CAD) system.

1960s AD
Douglas Engelbart develops the computer mouse.

1960s AD
James Russell invents compact discs.

1960 AD
Theodore Maiman invents the ruby laser.

1962 AD
William Armistead and S. Donald Stookey of Corning Glass Works invent light-sensitive (photochromic) glass.

1962 AD
Nick Holonyak invents the LED (light-emitting diode) while working at General Electric.

1963 AD
Ivan Sutherland develops Sketchpad, one of the first computer-aided design programs.

1963 AD
Martin Luther King Jr. leads a civil rights march in Washington, DC

1964 AD
IBM helps to pioneer e-commerce with an airline ticket reservation system called SABRE.

1965 AD
Frank Pantridge develops the portable defibrillator for treating cardiac arrest patients.

1965 AD
Amplified Bible.

1966 AD
Stephanie Kwolek patents a super-strong plastic called Kevlar.

1966 AD
Jerusalem Bible.

1966 AD
Robert H. Dennard of IBM invents dynamic random-access memory (DRAM).

1967 AD
Israel recaptures Jerusalem.

1967 AD
Japanese company Noritake invents the vacuum fluorescent display (VFD).

1968 AD
Alfred Y. Cho and John R. Arthur, Jr invent a precise way of making single crystals called molecular beam epitaxy (MBE).

1969 AD
World's first solar power station opened in France.

1969 AD
Long before computers become portable, Alan Kay imagines building an electronic book, which he nicknames the Dynabook.

1969 AD
Willard S. Boyle and George E. Smith invent the CCD (charge-coupled device): the light-sensitive chip used in digital cameras, webcams, and other modern optical equipment.

1969 AD
Astronauts walk on the Moon.

1970 AD
New American Bible.

1970 AD
Hal Lindsey writes the book *Late Great Planet Earth*.

1970s AD
Jesus Movement.

1970s – 1980s AD
James Dyson invents the bagless, cyclonic vacuum cleaner.

1970s—1980s AD
Scientists including Charles Bennett, Paul Benioff, Richard Feynman, and David Deutsch sketch out how quantum computers might work.

The Postmodern Church

1971 AD
New American Standard Bible.

1971 AD
The Living Bible.

1971 AD
Electronic ink is pioneered by Nick Sheridon at Xerox PARC.

1971 AD
Ted Hoff builds the first single-chip computer or microprocessor.

1973 AD
Martin Cooper develops the first handheld cellphone (mobile phone).

1973 AD
Roe v. Wade Supreme Court case rules in favor of legalizing abortion.

1973 AD
Robert Metcalfe figures out a simple way of linking computers together that he names Ethernet. Most computers hooked up to the Internet now use it.

1974 AD
Lausanne Congress for World Evangelization.

1974 AD
First grocery-store purchase of an item coded with a barcode.

1975 AD
Whitfield Diffie and Martin Hellman invent public-key cryptography.

1975 AD
Pico Electronics develops X-10 home automation system.

1976 AD
The Good News Bible.

1976 AD
Steve Wozniak and Steve Jobs launch the Apple I, one of the world's first personal home computers.

1977 AD
Focus on the Family founded by James Dobson.

1978 AD
New International Version of the Bible.

1979 AD
Mother Teresa wins the Nobel Peace Prize for work undertaken in the struggle to overcome poverty and distress in the world.

1980s AD
Japanese electrical pioneer Akio Morita develops the Sony Walkman, the first truly portable player for recorded music.

1981 AD
Stung by Apple's success, IBM releases its own affordable personal computer (PC).

1981 AD
The Space Shuttle makes its maiden voyage.

1981 AD
Patricia Bath develops laser eye surgery for removing cataracts.

1981 AD
Fujio Masuoka files a patent for flash memory—a type of reusable computer memory that can store information even when the power is off.

1981—1982 AD
Alexei Ekimov and Louis E. Brus (independently) discover quantum dots.

1982 AD
New King James Version of the Bible.

1983 AD
Compact discs (CDs) are launched as a new way to store music by the Sony and Philips corporations.

1987 AD
Larry Hornbeck, working at Texas Instruments, develops DLP® projection—now used in many projection TV systems.

1989 AD
New Revised Standard Version of the Bible.

1989 AD
Tim Berners-Lee invents the World Wide Web.

1990—1991 AD
Gulf War.

1990 AD
German watchmaking company Junghans introduces the MEGA 1, believed to be the world's first radio-controlled wristwatch.

1991 AD
Linus Torvalds creates the first version of Linux, a collaboratively written computer operating system.

1993 AD
The Message translation of the New Testament first published.

1994 AD
Israeli computer scientists Alon Cohen and Lior Haramaty invent VoIP for sending telephone calls over the Internet.

1995 AD
Updated New American Standard Bible.

1995 AD
Broadcast.com becomes one of the world's first online radio stations.

1995 AD
Pierre Omidyar launches the eBay auction website.

1996 AD
WRAL-HD broadcasts the first high-definition television (HDTV) signal in the United States.

1997 AD
Death of Mother Teresa of Calcutta, founder of the Missionaries of Charity. Her death on September 5 largely went unnoticed in the media because Princes Dianna was tragically killed less than a week earlier on August 31.

1997 AD
Electronics companies agree to make Wi-Fi a worldwide standard for wireless Internet.

1999 AD
Signing of the Joint Declaration on Justification by the Lutheran and Roman Catholic Churches.

1999 AD
Columbine Shooting.

2001 AD
English Standard Version of the Bible.

2001 AD
Apple revolutionizes music listening by unveiling its iPod MP3 music player.

2001 AD
Richard Palmer develops energy-absorbing D3O plastic.

2001 AD
The Wikipedia online encyclopedia is founded by Larry Sanger and Jimmy Wales.

2001 AD
Bram Cohen develops BitTorrent file-sharing.

2001 AD
Scott White, Nancy Sottos, and colleagues develop self-healing materials.

2001 AD
On September 11, terrorist suicide bombers attack the United States in New York, Northern Virginia, and Pennsylvania.

2002 AD
The Message version of the Bible.

2002 AD
iRobot Corporation releases the first version of its Roomba® vacuum cleaning robot.

2003 AD
Dan Brown writes *The Da Vinci Code*, made into a film in 2006; both the book and the film are criticized because of historical inaccuracy of the life of Jesus and history of the early Church.

2004 AD
Holman Christian Standard Version of the Bible.

2004 AD
Electronic voting plays a major part in a controversial US Presidential election.

2004 AD
Updated New Living Translation of the Bible.

2004 AD
Andre Geim and Konstantin Novoselov discover graphene.

2004 AD
The film *Passion of the Christ* is released.

2005 AD
Death of Pope John Paul II, election of Pope Benedict XVI.

2005 AD
A pioneering low-cost laptop for developing countries called OLPC is announced by MIT computing pioneer Nicholas Negroponte.

2006 AD
Together for the Gospel inaugural conference.

2007 AD
Amazon.com launches its Kindle electronic book (e-book) reader.

2007 AD
Apple introduces a touchscreen cellphone called the iPhone.

2008 AD
ESV Study Bible published.

2009 AD
World Methodist Council adopts the Lutheran/Catholic Joint Declaration on Justification.

2010 AD
The Lausanne Movement holds the Third Congress on World Evangelization in Cape Town, South Africa, 16-25 October 2010.

2010 AD
Apple releases its touchscreen tablet computer, the iPad.

2010 AD
3D TV starts to become more widely available.

2011 AD
Beatification of Pope John Paul II.

2013 AD
Resignation of Pope Benedict XVI, who is succeeded by Pope Francis.

2013 AD
Elon Musk announces "hyperloop," a giant, pneumatic-tube transport system.

2014 AD
The Tree of Life (TLV) translation of the Bible first published

2015 AD
Supercomputers (the world's fastest computers) are now a mere thirty times less powerful than human brains.

2016 AD
Three nanotechnologists win the Nobel Prize in Chemistry for building miniature machines out of molecules.

2019 AD
Google claims to have achieved "quantum supremacy"—with a quantum computer that calculates faster than a conventional one.

2020 AD
World-wide Covid-19 Pandemic. [259] An outbreak of COVID-19 was first detected in mainland China in December of 2019. The cause of this outbreak was a new virus, known as the severe acute respiratory syndrome coronavirus 2 (SARS-CoV-2). On February 12, 2020, the World Health Organization officially named the disease caused by the novel coronavirus as Coronavirus Disease 2019 (COVID-19). By early April 2020 every

continent in the world had been affected causing nearly one million diagnosed events in over 200 countries. By November 30, 2020 the virus had infected more than 62.7 million people and killed over 1.4 million worldwide, according to real-time data compiled by the Center for Systems Science and Engineering at Johns Hopkins University. [260]

Three vaccines were conditionally approved in 2021 and given in stages based on categories determined to be most at risk. Initially, receiving the vaccines were voluntary but, after almost half of the US population, for various reasons, chose not to be vaccinated after final FDA approval for all three vaccines, and variants of the disease started to appear, certain mandates were put in place. Some employers began to risk the consequences and made vaccinations a requirement of continued employment. Insurance companies began to indicate health insurance rates could cost more for unvaccinated individuals. [261]

An article published by the American Institute for Economic Research on May 1, 2020, titled "Woodstock Occurred in the Middle of a Pandemic" was written by Jeffrey A. Tucker, editorial director. In it he states that a similar flu epidemic, also with its origin in China, hit the United States in December of 1968. The flu ultimately killed approximately 100,000 people in the US, mostly over age sixty-five, and over one million worldwide. In 1969 the US population was 200 million and the average lifespan was seventy. Today there are 330 million in the US with an average life span of seventy-eight. The author's point? The 1968 outbreak was as lethal as the 2020 Covid-19 pandemic.

Some comparisons to the 1968 outbreak: Nothing was closed by force, Congress did not pass special legislation, schools mostly stayed open except for cases of absenteeism making continued classroom study impractical, and businesses remained open as well. Concerts, Woodstock in particular, went on without a hitch in August of 1969. Unemployment was not seriously impacted and the stock market did not crash. So, what was different?

The twenty-four/seven news cycle did not exist in 1968. News organizations at the time were more interested in a man landing on the moon, not with filling every waking moment with news of a virus that might kill you. The limited number of outlets at the time were not nearly as concerned about ratings in 1968. Sensationalism and drama were not written into news scripts.

To me the most disgusting overreach affecting both the science and faith community was the politization caused by Covid-19. Both science and faith were projected through a political lens. Depending on how you view this, the US Presidential election coincidentally ran parallel in 2020 with the pandemic. At a time in history when these two communities could have effectively used their internal resources to help, neither could, without strong resistance to choose a side. The public rarely received pure science concerning the effects of the disease or the ongoing efforts to cure it without that information being disseminated through a political filter. The faith community was faced with restrictions in counseling, assisting, prayer gatherings, and a sharp reduction in funds due to canceled worship where offerings were normally collected to support ministry.

While we may be able to look back on this period years from now and better understand what actually happened, we must remember that history is often written by the winner; whatever that means, as long as politics is involved.

The End of a Long, Hot Summer

When does summer end? Within world civilizations, is the end of summer simply a date on the calendar, or the aligning of certain planets, or when your favorite swimming pool closes? While it is useful to some, like Daylight Saving Time, to identify an end or a beginning of something, these designations do little to convince the weather or the sun to follow our choices for these changes. Instead, the leaves may change in late October or mid-November, or who knows? They change when they change. Then there is the small detail of hemispheres. When we are enjoying a summer breeze, certain places in Australia are freezing. Why? It is way on the other side of the Equator. Somewhere when reading this book, I would not be surprised if you questioned my use of seasons as they relate to creation.

These designations are obviously not divine inspiration. I use them because I believe in both science and faith. I grew up in both worlds. My parents were deeply religious, although my father came to the faith later than my mother. Their families came from rural Southwest Virginia coal mining communities dominated by Holiness churches. Nine years after I was born into that environment, my family moved to the mid-Atlantic coast of Florida, twenty miles south of what was then known, and now again is known, as Cape Canaveral, one of the birthplaces of rockets and outer space. I say then and now, because the space center became known as Cape Kennedy for a period of years. Everyone knew people who worked in that industry. I remember standing on the steps of my elementary school watching Alan Shephard, the first American launched into outer space on May 5, 1961. I was in the fifth grade. Science was a very big deal. So was my faith.

Neither the seasons of science nor the seasons of faith have any relationship to the made-up dates of man. I am not referring to scientifically dated rock strata or bones. I am referring to dates like December 25th or whatever date Easter is each year, or scientific general dates Carl Sagan used to love to quote, also the title of his book published in 1988, *Billions and Billions*. Within my book, I have separated the seasons based on the major changes within the faith community. Initially, I had no choice. It really doesn't matter when the Earth and everything in it began, the science community was not around, especially with the evidence they have produced since the 17th century. I also used the faith community for these four divisions because I believe science has a supportable position regarding "religion." I totally understand their position.

Moving from here I will be wading into areas that neither faith nor science can claim… because they have yet to happen… if they ever do; as prophesied or calculated, depending on which side of this discussion you support. Again, my goal is to get both sides to the table with open minds and a willingness to discuss how each can contribute to civilization's best efforts to create change where division lurks.

Welcome to the Fall of Creation—literally.

FALL

Christ's coming back at the end of the original Creation and beyond for eternity… or infinity

CHAPTER 21

WHAT HAPPENS WHEN THERE ARE NO MORE QUESTIONS TO ASK... .

... the Ultimate Understanding
... will the science community ever
reach total consensus?
... will that occur due to their
research and investigation?
... will the faith community ever become
truly trusting... with no doubt?
... what happens when many in both communities
realize they were wrong about a lot of things?
... will anyone or anything have regret at the end
of existence as we have always known it to be?

The last days will be days of apostasy: As time moves along, Scripture warns there will be a growing, universal revolt against God and authority that will manifest itself in a humanism in which man, believing the age-old lie of Satan that he can solve his own problems, becomes the center, source, and answer to all his problems. [262]

... 3 Let no one deceive you by any means; for that Day

will not come unless the falling away comes first, and the man of sin is revealed, the son of perdition...

... 11 And for this reason God will send them strong delusion, that they should believe the lie...
2nd Thessalonians 2:3 and 11 NKJV

... 3 But know this, that in the last days perilous times will come...

2nd Timothy 3:1 NKJV

Our atmosphere, proximity to the Sun, and countless other beautiful coincidences not only permit living things to survive and evolve but also thrive. And yet, here we are, sitting at desks and in coffee shops and walking down the street like it isn't some kind of extraordinary miracle. [263]

The last days will be days of globalism: Because of wars, rumors of wars, and the threat of international crises that will build in preparation for the scenario found in the Great Tribulation... [264] (For Scripture reference, please read Matthew 24 and Revelation 6-19)

And because of man's humanism, there will be a spirit of globalism. More and more the human race will look to a one-world international system to bring peace and safety... [265]

[3] *For when they say, "Peace and safety!" then sudden destruction comes upon them, as labor pains upon a pregnant woman. And they shall not escape.* 1 Thess. 5:3 NKJV

A giant asteroid crashing into our planet would instantly kill off millions of animals. But the aftermath of such an impact would be even more disastrous: Tsunamis, earthquakes, and vast clouds of dust blocking out the sun would lead to crop failure and mass extinction. Sixty-five million years ago, just such an event killed off 75 percent of species on Earth. But to really wipe life off the planet, it would take an astrophysical event so powerful that Earth's oceans would literally boil away, according to a new study.

The heat and cosmic radiation would make Earth inhospitable even to tardigrades, among the hardiest organisms ever discovered. [266]

The last days will be days of false messiahs: This hope and search for some form of a one-world system will lead to the longing and search for someone who can come on the scene of human history and mold the world into unity and peace. The last days will be days of false messiahs, a time when the world will be looking for solutions in a great world leader until such a person arises in the form of the Antichrist, the Beast of the tribulation period… [267]

5 *For many will come in My name, saying, "I am the Christ," and will deceive many.* Matthew 24:5 NKJV

(For Scripture reference please read Matthew 23-24)

3 *Let no one deceive you by any means; for that Day will not come unless the falling away comes first, and the man of sin is revealed, the son of perdition, 4 who opposes and exalts himself above all that is called God or that is worshiped, so that he sits as God in the temple of God, showing himself that he is God.* 2 Thess. 2:3-4 NKJV

13 *Then I stood on the sand of the sea. And I saw a beast rising up out of the sea, having seven heads and ten horns, and on his horns ten crowns, and on his heads a blasphemous name.* Revelation 13:1 NKJV

First the bad news: The world is going to end one day and there is nothing we can do about it. The good news: We could have a billion years to enjoy ourselves here before that happens. Humans have long been obsessed with "The End." Since the dawn of civilization, people claiming to know when the big day is coming have stirred up trouble, sparked panics and started cults. The latest apocalyptic anxiety is being drummed up by David Meade, a self-described "specialist in research and investigation" who predicts the Rapture as foretold in the Book of Revelations will begin Saturday (what a bummer Doomsday couldn't at least fall on a Monday).

> There is no scientific evidence to support Meade's theory that a "Planet X" he calls "Nibiru" is headed toward Earth—but that hasn't stopped Meade from spreading the word. There are, however, several scenarios in which humans could be wiped out. And even if we manage to avoid all of those, the sun will die one day. "A lot of people have written books about all the ways that the universe can get us," said Richard Binzel, a professor of planetary sciences at the Massachusetts Institute of Technology. Fortunately, those are all very low probability. But the one certainty is that the sun will not last forever." [268]

The Rapture or Rapio

WE SHOULD PICK up where we left off at the end of the Season of Summer. An abbreviated format of that ending might read, " Jesus Christ came back to life, met with His disciples and others on a mountain, told them to develop a church based on His death and resurrection, ascended into the air, somehow went back to Heaven, wherever that may be, and a couple of thousand years or so later, the sky in the East opened up and Jesus Christ comes riding back into our world to gather His believers and take them all back to Heaven to a place He has prepared for them where they will live for eternity with Him and His Father, the Almighty God."

Welcome to the Fall of Creation. It is prophesied that it will begin with a thing called the Rapture.

Like Creation in Genesis, the faith community believes the Rapture, or the coming back of Christ for the third time to the Earth, will be the time promised by Jesus Christ just before leaving this universe early in the 1st century. As with several other points in Christian theology, the Rapture contains a widely disputed event called the Tribulation Period. Here is how it is described in a post on the beliefnet.com website:

> *The pre-tribulation view is that the Rapture will happen before the tribulation period, the mid-tribulation view is that the Rapture will occur half-way through the tribulation period and the post-tribulation view is that the Rapture will occur at the end of the tribulation period.* [269]

This site, using Scripture, mentions the following things that will be happening just prior to the rapture while warning that Scripture cautions against trying to figure out when the Rapture will actually take place:

- Many will claim to be the Messiah who will have answers for a troubled world.
- False prophets will rise and show great signs and wonders to deceive.
- None of the above are ordinary individuals.
- People will be crying out for leaders to deliver them.
- People will seek mystics and religious leaders who claim to have a deeper knowledge.
- Disputes among nations.
- War escalating into another World War.
- Hunger and famine.
- Earthquakes.
- New diseases.

Obviously, all of these things have been occurring continuously since the beginning of time. It seems these verses are emphasizing that His return will be on just any other normal day, and not signs for the believer to use as an announcement He is about to pop out of the Easter sky.

Also, at the Rapture, Scripture speaks of a Judgment Seat of Christ. This topic, along with all of the other end-of-time illustrations in Scripture-the appearance of the antichrist, a mark on someone's forehead or right hand, a 1,000-year reign, and the number 666-are mysterious and complicated. This again takes us back to my previous thoughts regarding how God thinks: it is not the way humans think. I don't believe man is capable of understanding these Scriptures, other than that he should be ready. Below are several of these references, followed by others you may wish to look up.

10 *Because you have kept My command to persevere, I also will keep*

you from the hour of trial which shall come upon the whole world, to test those who dwell on the earth. Revelation 3:10 NKJV

[30] *Then the sign of the Son of Man will appear in heaven, and then all the tribes of the earth will mourn, and they will see the Son of Man coming on the clouds of heaven with power and great glory.* [31] *And He will send His angels with a great sound of a trumpet, and they will gather together His elect from the four winds, from one end of heaven to the other.* Matthew 24:30-31 NKJV

[20] *For our citizenship is in heaven, from which we also eagerly wait for the Savior, the Lord Jesus Christ,* [21] *who will transform our lowly body that it may be conformed to His glorious body, according to the working by which He is able even to subdue all things to Himself.* Philippians 3:20-21 NKJV

[18] *For I consider that the sufferings of this present time are not worthy to be compared with the glory which shall be revealed in us.* [19] *For the earnest expectation of the creation eagerly waits for the revealing of the sons of God.* Romans 8:18-19 NKJV

[7] *And He said to them, "It is not for you to know times or seasons which the Father has put in His own authority.* [8] *But you shall receive power when the Holy Spirit has come upon you; and you shall be witnesses to Me in Jerusalem, and in all Judea and Samaria, and to the end of the earth."* Acts 1:7-8 NKJV

[34] *"I tell you, in that night there will be two men in one bed: the one will be taken and the other will be left.* [35] *Two women will be grinding together: the one will be taken and the other left.* [36] *Two men will be in the field: the one will be taken and the other left."* [37] *And they answered and said to Him, "Where, Lord?" So He said to them, "Wherever the body is, there the eagles will be gathered together."* Luke 17:34-37 NKJV

[24] *But in those days, after that tribulation, the sun will be darkened, and the moon will not give its light;* [25] *the stars of heaven will fall, and the powers in the heavens will be shaken.* [26] *Then they will see the Son of Man coming in the clouds with great power and glory.* [27]

And then He will send His angels, and gather together His elect from the four winds, from the farthest part of earth to the farthest part of heaven. Mark 13:24-27 NKJV

17 *Then we who are alive and remain shall be caught up together with them in the clouds to meet the Lord in the air. And thus we shall always be with the Lord.* 1 Thessalonians 4:17 NKJV

5 *But concerning the times and the seasons, brethren, you have no need that I should write to you.* 2 *For you yourselves know perfectly that the day of the Lord so comes as a thief in the night.* 3 *For when they say, "Peace and safety!" then sudden destruction comes upon them, as labor pains upon a pregnant woman. And they shall not escape.* 4 *But you, brethren, are not in darkness, so that this Day should overtake you as a thief.* 5 *You are all sons of light and sons of the day. We are not of the night nor of darkness.* 6 *Therefore let us not sleep, as others do, but let us watch and be sober.* 7 *For those who sleep, sleep at night, and those who get drunk are drunk at night.* 8 *But let us who are of the day be sober, putting on the breastplate of faith and love, and as a helmet the hope of salvation.* 1 Thessalonians 5:1-8 NKJV

10 *For we must all appear before the judgment seat of Christ, that each one may receive the things done in the body, according to what he has done, whether good or bad.* 2 Corinthians 5:10 NKJV

3 *And if I go and prepare a place for you, I will come again and receive you to Myself; that where I am, there you may be also.* John 14:3 NKJV

20 *Nevertheless do not rejoice in this, that the spirits are subject to you, but rather rejoice because your names are written in heaven.* Luke 10:20 NKJV

50 *Now this I say, brethren, that flesh and blood cannot inherit the kingdom of God; nor does corruption inherit incorruption.* 51 *Behold, I tell you a mystery: We shall not all sleep, but we shall all be changed—* 52 *in a moment, in the twinkling of an eye, at the last*

> *trumpet. For the trumpet will sound, and the dead will be raised incorruptible, and we shall be changed.* 1 Corinthians 15:50-52 NKJV

Other Scripture references: *1 John 2:18 & 22, 1 John 4:1-3, 2 John 1:7, 1 Thessalonians 4:13-18, 2 Thessalonians 2:3-4, Matthew 24:24, Revelation 13:7, Revelation 13:8, Revelation 13:16-18, Revelation 14:9-12, Revelation 19:20, Revelation 20:4-6*

Before leaving this topic of the rapture we should spend a moment understanding the origin of the term. First, the term *rapture* is never used in the text of any translation of the Christian Bible. According to some sources the term was first derived from the Latin Vulgate translation of 1 Thessalonians 4:16-17 containing the phrase, "*...shall be caught up together with them in the clouds to meet the Lord in the air...*" It was also noted that Protestants normally reject the Latin Vulgate, while accepting random bits such as the Latin verb *Rapio*.

One source cites Protestant minister John Nelson Darby as one of the first to begin teaching this idea in the 1830s following the Second Great Awakening in North America. Darby studied law at Westminster School and Trinity College. He later chose to enter the diaconate of the Church of Ireland instead of practicing law. After several years he broke away from the Church of Ireland and formed The Plymouth Brethren, a conservative nonconformist evangelical Christian movement. It was at this time he began to preach this new doctrine.

A contemporary of Darby, C. I. Scofield, picked up on this new theology and expanded it to include a *millennialist* position regarding the rapture and the 1,000-year reign of Christ on His return at the end of time. This idea, due to theological disagreement, developed into three separate beliefs: Pre-Millennialism, Post-Millennialism, and Amillennialism.[270]

As a Christian believer, I have accepted how I will face the end of everything. Although quite complicated, it is somewhat outlined in the above Scriptures. I also believe that if I am wrong and there is no rapture or anything like it, at some point the universe is going to run its course and end. Either way, this is all going to go away in its present form. I don't think the faith community can tell you what will become of the universe and Earth God created. They firmly believe it was all created from nothing; or almost nothing, by an act of an almighty God. Science believes it was created from something already existing and that whatever happens, it will just transform into another form of matter; but it will not cease to exist.

An examination of the above paragraph should demonstrate that these two communities really have nothing about which to disagree. It seems all they can argue over is a distant past and a future that will not arrive until at least... well, more future was just created... there it goes again... and again... I can't keep up, neither can I predict what it will look like, so why argue over it?

In all cases, what both can do is discuss the present and prepare for the near future in our lifetimes and maybe, just maybe, the distant future. These two communities can work together, appreciating what each brings to the table. If in doing so, a believer or a non-believer decide the other position makes more sense, they can explore the other community if members of that other community are honest with them. As a believer, I am taught to simply introduce others to Christ, not "win" them over. That is the work of the Holy Spirit, who moved upon the face of the deep and darkness. These faith beliefs may or may not be something the science community can understand, but they can experience them and see if in fact it doesn't change their life. On the contrary, I believe the majority of the science community isn't trying to win anyone over to their position, they just feel everyone should understand, as they do, how things work. There is no long-term consequence to the science community if they or anyone else accepts their position or not.

Relevance

I awoke this morning six a.m. Central Time. It occurred to me, as I lay there looking at my clock turned at a ninety-degree angle to my resting head, that six a.m. was only good for me and the thousands of others who live in my time zone. Thankfully, the coffee pot kicked in at 6:20 a.m. Central Time and not Pacific Time. I didn't have an exceptionally good night's sleep since last Thursday, two days previous, I had the first of two shingles shots. I knew what the side effects were to be, so when asked in which arm I preferred to be stuck, I chose my left. I was still having trouble with my right arm from shoulder surgery a year ago. I thought I could spread the pain. Bad idea! Now I couldn't sleep on either side. What does all of this have to do with this book and the Fall Season of Creation? Let me tell you a little more about my morning.

I made breakfast and sat down in front of the TV selecting my favorite cable program, The Science Channel (I hear you, my scientist friends, it's just entertainment). They were airing a documentary on exploration in Turkey at the site of Mt. Ararat where some believe lies the remains of Noah's Ark. It was a rerun, but my thought since getting out of bed had centered on my efforts in this book as a Christian to better understand the scientific method and thus lead to a better understanding of each other's positions. Let me back up and tell you what was unsaid in these first two paragraphs.

The measurement of time, quantities, things, scientific elements, gravity, explosions, distance… just about anything humans deal with on a daily basis seem to be at the core of disagreements between science and faith. Why was it six a.m. Central Time when I woke up? It was due to the rotation of the Earth and, that some people eons ago decided to divide one rotation in to twenty-four parts. Realizing the rays of the sun didn't strike the Earth all at once, and that most humans structure their lives around daylight hours, those who designed these things decided to create zones so the people who live closer together could use relatively the same portion of the day to interact.

Why did my coffee pot begin grinding and brewing at 6:20 a.m. Central time? Because I told it to turn on at that time. It didn't intuitively know that was when I would prefer to have my coffee. At least not yet, not this model. Why did I choose to have the shot in my left arm, or for that matter, why did I decide to even have the shot? How did I know the shot was available? Why did my doctor recommend I have the shots? Why did I have the shoulder surgery a year prior to making this decision? The answer to all of these questions is measurement. Based on a period of time, and the activities during that period, my bones had developed issues requiring repair. The period of time could have been different based on different measurements of utilization and endurance. The choice of which arm was a measurement of my ability to endure pain on a scale of 1 to 10 as proposed by my doctors. The shingle's vaccination was measured in a hypodermic syringe. Too much or too little would have had bad consequences. The medicine was developed over years of testing different quantities and ingredients against results. After the first shot I had to wait two months and then have a second shot within six months. How did science know for me to wait just that long? Measurement. Time. Why do I, a Christian, part of the faith community, accept life-affecting scientific discoveries without hesitation but, while doubting and questioning their methods and accuracy of dating a rock, or the age of a dinosaur bone? So, what about the Ark? How's that for a deflection?

As I have uncovered over months and months of research for this book, and just as in the documentary mentioned above, the major disagreements centered on measurements; how far in the past, dating of regional rocks and soils, Lidar imaging of the ground under the site, how many animals were on the boat, how old was Noah, how tall was Noah, the size of the Ark, and the age of the Earth determined by science and... in the end all was compared to the position of the faith community. Archeologists, based on scientific studies, have determined the Earth is billions of years old, while some in the faith community say, based on Biblical evidence and some secular historical records, the Earth is between 6,000 and 10,000 years of age. Admittedly, a fairly large gap.

Based on relevance, I didn't question the measurements and scientific studies performed to produce a vaccine preventing me from coming down with a very painful, scary rash called shingles. I simply scheduled a time, went in, rolled up my sleeve, and told the doctor to stick me in the left arm. It didn't stop there... I went back a few months later and did it again. No questions.

There you have it—a standoff. How in the world do you resolve such a historic disagreement? The answer is in the previous sentence. Do you see it? It is the fourth word in and it is so obvious mankind has overlooked it since the first human breathed life. What could possibly be the motive of science or faith, *world*wide, to refuse to talk to each other and accept each other's point of view? Is it stubbornness? Is it science refusing to accept a God who requires obedience and praise for His creative acts? Is it faith's traditions and feeling like adjusting their position weakens the reality of the Trinity? Probably all the preceding and a lot more. Admittedly, it is complex.

Does Anything Ever Really End?

Science has made every effort to look as far away from the Earth as possible. They have seen or have developed theories about the edge of what we perceive millions of light-years straight up from the surface of the Earth. The faith community has looked to a place called Heaven. Most don't think Heaven is anywhere within the scope of what science has seen or theorized. Wherever Heaven might be, it is a long way from the Earth. So, again, how in the world do we resolve these issues between the science and faith communities? Perhaps we stop relegating the discussion to the world in which we live and the universe in which it sits. And then, we all stop holding fast to the idea that God, the Trinity, exists only in the vast universe we earthlings experience. If our existence is as large as we all know it to be, why can't it be even bigger… much bigger?

We are about to finish this journey by moving into a future filled with uncertainty, guided by at best, a coded map and endless theories. As a Christian, I believe I will go to the place from where our Creator started—Heaven. I have no idea where that is located, if location is even relevant. I have been told in Scripture a little about what it will look like by someone who never experienced in his lifetime anything to which he could compare it to form his description.

Unlike my shingles shot, science is of no help in settling my curiosity. Many in science take the position that things, and people, just end. You take your last breath and that's it, you return to dust or ashes meaning, various elements of the periodic table. Others believe a variety of things may happen; reincarnation, spirit worlds, or discovery that we never really existed in the first place. And then, here is something everyone could speculate about; other worlds, multi-universes, other… creations. Perhaps it is possible that a God who is powerful enough to create our existence, is also powerful enough to create other creations.

The only thing that stirs the imagination more than speculation about what may have existed before the Big Bang or the Creation in the Holy Book, the Bible of the faith community, is speculation concerning what happens after it all ends. The faith community looks to Scripture for an awareness of what will happen. Along with that awareness are warnings against trying to figure out when the end may come; only God knows the date and time.

One of the views of the science community regarding the end is much like a pinball machine with a large ball rolling around in space, randomly colliding and destroying other masses in the universe. One of these obstacles will eventually be the earth. Another scientific view, among thousands, looks at the universe like a machine that will one day wear out or at a minimum, just run out of fuel. But these scientific views are not the only thing science is doing about the present… or the future. That task, since November 20, 1984, was placed in the hands of two individuals, at least initially.

SETI

Looking back before November 20, 1984, there was a small project funded by NASA charged with the search for extraterrestrial intelligence. John Billingham, Chief of Life Sciences at NASA Ames Research Center, and Bernard (Barney) Oliver, formerly of Hewlett-Packard, invited Tom Pierson and others into discussions about maximizing the effectiveness of the funds and ways of putting more money into research and less into institutional overhead. These discussions led to the concept of a dedicated non-profit organization focused on research and education around the factors of the Drake Equation. The Drake Equation is used to estimate the number of communicating civilizations in our galaxy, or more simply put, the odds of finding intelligent life in the Milky Way.

This vision was realized in the 1984 incorporation of SETI (Search for Extraterrestrial Intelligence), headed originally by CEO Thomas Pierson and SETI scientist Jill Tarter. CEO Pierson passed away due to illness in 2014, and SETI was eventually folded into the larger science umbrella of the Carl Sagan Center. Today, the Institute has approximately 100 scientists as well as specialists in administration, education, and outreach. Over the years, other research disciplines have been added to the Institute's portfolio, all unified by their relevance to the search for, and understanding of, life beyond Earth. [271] Interestingly, people of faith would hold that they have more proof of extraterrestrial life than does science. The Bible is filled with a God, His Son, a Holy Spirit, angels, and other beings, none of which seem to originate on this planet. A scientific definition would consider these things "aliens." SETI has yet to find any proof of life from somewhere else or the aliens of the faith community previously mentioned.

Although it wasn't proof, an event occurring on August 15, 1977, prior to the incorporation of SETI, at a radio telescope located at Ohio State University detected a powerful, unexplainable signal while scanning the stars as part of a pre-SETI project. The radio telescope was searching for possible signals from alien civilizations. The signal appeared to be coming somewhere from the direction of Sagittarius. The signal lasted a little over one minute and was never heard again. The incident became known as the "Wow! Signal" because the computer recording the event printed a very large specific code on the continuously running paper graph. The first astronomer to see the code, Jerry Ehman, was so taken he circled the letters with a red pen and wrote beside the letters on the printout, "Wow!"[272]

This mystery went unsolved until as recent as 2016 when Astronomer Antonio Paris, who had been studying the event for a long time, suggested it may have come from a comet orbiting in the area where the "Wow! Signal" was detected. In fact, it was possibly two comets, both of which are surrounded by large hydrogen clouds.

The presence of hydrogen in this radio signal is significant because the signal would appear to be coming from a location that could have large quantities of water, since hydrogen makes up a large part of a water molecule.

> *Hydrogen atoms emit at 1420 MHz (a wavelength of 21 cm). Hydroxyl molecules, composed of one atom of hydrogen and one atom of oxygen (OH), emit at four specific radio frequencies ranging from 1612 MHz to 1720 MHz. When a hydrogen atom combines with a hydroxyl molecule, it forms a molecule of water, the most essential molecule for life as we know it. Thus, the range of frequencies from 1420 to 1720 MHz is called the Water Hole. It has been a popular frequency range for many SETI programs.*[273]

Unsurprisingly, this is not the end of the story. In May of 2015 a team of researchers using a Russian radio telescope spotted another strong signal coming from the sun-like star HD 164595 which lies ninety-four light-years away from the Earth. Like the 1977 "Wow! Signal," there was no second signal. Both were single, quick bursts. Since they did not repeat themselves, scientists are quite certain the signals did not come from a satellite, which would be on continuously. It adds credibility, but not proof, that the signals came from an intelligent source outside our solar system.

Our civilization has been sending signals into space for decades. Radio and television signals emitted from some of our very first radio and television productions are still making their way millions of light-years away from the Earth. Also, launched in 1977, both Voyager space crafts included a golden record containing messages from our planet in case another civilization discovers them. Both of the crafts have now reached interstellar space. One of the messages included is:

> *This is a present from a small, distant world, a token*
> *of our sounds, our science, our images, our music,*
> *our thoughts and our feelings. We are attempting*
> *to survive our time so we may live into yours.*
>
> —*President Jimmy Carter*[274]

Without fully admitting it, both the faith and science communities believe there was something before what is known today. Although the faith community insists everything was created from nothing, or almost nothing, in Genesis 1:1, by inference, God, the Trinity, must have existed before the Earth was without form and void and darkness was on the face of the deep. Science holds to the moment of singularity… when all things began. They too, since they do not believe something can come from nothing, by inference, are admitting there was something before the Big Bang, or whatever made us appear.

If both believe there *was* something *before* we came into existence, in a Heaven-like place or somewhere else, then it isn't a stretch for both to think there is something after our current existence ends. That is what the Fall Season of Creation is all about.

CHAPTER 22
DIRECTIONS TO THE END OF THE BEGINNING

... which term is correct, Eternity or Infinity?
... is science at odds with Faith... or Religion?
... I thought they were the same thing!

What's It All About?

THIS QUESTION, *WHAT'S It All About*, is one of the delineating points separately defining each community. The faith community will tell you God created mankind and everything in it for one purpose, to glorify Him. It seems the end for the faith community, when it arrives, will also center eternity on that same motive. The science community, instead, is always discovering, always searching for meaning in the universe. Science, by its nature, is always one discovery away from changing what it thinks about anything. Nothing is certain, nor can it be, since they are never sure of what they don't know. This textual understanding of both communities is also why each defines the future with different terms.

The faith community uses the term "eternity." Science prefers the term "infinity." Neither term has a future end, or a time-stamped beginning. They both suggest that the future is always one more day than a human can imagine, and the first day of the past was always the day before yesterday. With those parameters, let's take a look into an endless future in 360 degrees.

The Faith Community

When will the end come? Scripture says absolutely no one knows the answer to this question. Not even the one they hold up as the savior of the world.

> 36 *"But of that day and hour no one knows, not even the angels of heaven, but My Father only."* Matthew 24:36 NKJV

Will there be a warning before the end? Scripture lists several signs, but these have been occurring since the beginning of the world. The implication is that the end will come on just another average day... so be ready!

> 1 *But know this, that in the last days perilous times will come:* 2 *For men will be lovers of themselves, lovers of money, boasters, proud, blasphemers, disobedient to parents, unthankful, unholy,* 3 *unloving, unforgiving, slanderers, without self-control, brutal, despisers of good,* 4 *traitors, headstrong, haughty, lovers of pleasure rather than lovers of God,* 5 *having a form of godliness but denying its power. And from such people turn away!* 2 Timothy 3:1-5 NKJV

What will happen to all of the humans who have lived on the Earth prior to the end? Those who accepted Jesus Christ as the Messiah, after He was crucified and resurrected, and those who followed God's law before Jesus Christ was crucified and resurrected will enter into Heaven to be present with the Trinity for eternity. This is not as cut and dried as some make it appear. There are serious questions within this community. While they all agree the body has a soul, and that soul remains alive after physical death, when it leaves the deceased body and makes this eternal journey is still a frequent subject of debate. Two other issues are also still debated. One is the actual process for those who were under the Law before Jesus Christ became part of the historical lineage. The second issue deals with those who, for whatever reason, never had an opportunity to be a part of the Law or who never had an opportunity to hear about the birth, death, and resurrection of Jesus Christ before they died.

> 12 *"And behold, I am coming quickly, and My reward is with Me, to give to every one according to his work.* 13 *I am the Alpha and the Omega, the Beginning and the End, the First and the Last."* 14 *Blessed are those who do His commandments, that they may have the right to the tree of life, and may enter through the gates into the city.*
> Revelation 22:12-14 NKJV

The Science Community

When will the end come? Science testifies openly that the world, and in fact the entire universe, is going to end. Ending the universe or our solar system will likely take a very long time. For example, if our sun runs out of fuel it will implode and all of the planets around it will be consumed by the event. According to science, that will take about 5 billion years. [275]

A more relative concern involves the probability an object 100 to 450 feet across could hit the Earth. Inbound space objects are traveling so fast scientists, based on current data, would need from five to ten years notice of an impact to have enough time to possibly plan a deflection or explosion of the object. If a 450-foot-wide object struck New York City, it would erase the metropolis and all its inhabitants. After-effects would create a worldwide catastrophe. [276]

After much research, my untrained interpretation of their academic prophecy is that they don't really have a clue. The solution to the possibility of either private or government funding for the research and development of useful technology compounds any possible, practical avoidance of Earth annihilation, should a large space object intersect earth in its trajectory. As of today, even if we are given ten years notice, earthlings of any nationality have no proven, replicable system to prevent the collision.

Will there be a warning before the end? Yes, there will be a warning and if you have been paying attention to those who talk like scientists, but really are not, you are aware that Mother Nature is screaming at civilization through climate change. Documentaries on this subject often spend all their budget detailing the contemporary things mankind has done to destroy the Earth without mention of millions, and probably billions, of years of changes in the Earth's rotation, reversals of polarity, gravitational changes, and distance from the sun, and a thousand other events that have constantly changed the face of the Earth, the planets in our solar system, and the universe in general, all resulting in changes to the Earth's climate. Again, political filters. Yes, I am aware of the exploding spike in CO_2 beginning in the mid-20th century and its devastating impact on world temperature. I will address these and related issues in more detail in a bit, so please don't judge me yet...

For the faith community, there was a warning from Noah about the flood. The flood was not due to man's abuse of the air or the land, but the abuse of each other and lack of respect for God.

For the science community, people who live along the Gulf Coast are blessed with some of the most accurate weather technology in history predicting the path of hurricanes and storm surges. In spite of these warnings, people die, billions of dollars of property are damaged or destroyed... and yet people still insist on living near the coastline... below sea level. Although science can warn people about the dangers of living near the ocean, a volcano, on a tectonic plate, or in tornado alley, they still don't know the hour or the minute of disaster.

What will happen to all of the humans who have lived on the Earth prior to the end? The science community, at least those who are pure scientists, will tell you it is not the calling of a scientist to comment on this subject. In scientific study, things are... and then they are not.

If we are to know what it is all about, we will need to take a deeper dive into the stories of the science and faith community's view of eternity and infinity. Is that possible, since there is such divergence between their view of these two similar but yet very different trajectories? I believe it is probable, or at least possible, that these two communities have more in common than either wants to admit. Let's give it a go.

Eternity

What is an eternity? To a young child it is the endless few weeks before Christmas. To someone with a serious medical condition, it may be the endless time waiting on lab results. To someone who has no food, it may be the endless day waiting for the one meal they know will be served at the mission. Humans often associate eternity with short-term situations, often over things which they have some form of control. Remove the tiniest morsel of control, and the element of not knowing when it begins borders on impossible to comprehend. There you have humankind's dilemma when trying to fathom eternity, meaning… forever! In the faith community, eternity is not so much a measurement, but a quantity of life. Those of the faith community often refer to *eternal life*, or a life that never ends.

Not to be confused with the mythical discovery of Ponce de Leon's fountain of youth in St. Augustine, Florida, eternal life is believed by the faith community to be the destiny of every soul born on this Earth from the beginning of whenever until whenever finally ends on Earth and babies are no longer born. Some may struggle with this concept, but faith teaches that everyone will live forever, *where* you will spend that eternity, however, depends on a choice.

Accepting Jesus Christ as your personal Savior by believing Jesus Christ was born of a virgin, lived as a 100 percent divine—100 percent human, was crucified, died, was buried, rose from the dead and ascended into Heaven to return one day for all who believe, and then confessing they are a sinner asking for divine forgiveness, will assure that eternal place will be in Heaven with Christ, His Father God, the Holy Spirit, and multitudes of other angels and believers. Not making the decision to believe in Christ and follow His example comes with the assurance you will spend your eternity in a place called Hell. Hell is often interpreted as anywhere without the presence of God. Or, said another way, a place where nothing good ever happens.

The faith community largely holds that eternity begins when a person is first conceived as a human being as evidenced in Scripture,

> 7 *But the very hairs of your head are all numbered. Do not fear*

therefore; you are of more value than many sparrows. Luke 12:7 NKJV

And,

5 *"Before I formed you in the womb I knew you; Before you were born I sanctified you; I ordained you a prophet to the nations." Jeremiah 1:5 NKJV*

Infinity

"A long time ago in a galaxy far, far away…" [277] Nothing speaks of science fiction better than the Star Wars Movie franchise. Infinity has its own symbol, ∞. The symbol, sometimes called the *lemniscate*, was introduced by a clergyman and mathematician, John Wallis in 1655 AD. In Latin, the word lemniscate means ribbon. *Infinitas*, in Latin, means "boundless."

Long before the symbol was introduced, the concept of infinity was understood. Somewhere around the 4th to 3rd century BC in the Jain mathematical text, Surya Prajnapti assigned numbers as either enumerable, innumerable, or infinite. Anaximander, a Greek philosopher, used the work *apeiron* to refer to the infinite. Zeno of Elea, born circa 490 BC, was known for paradoxes using infinity. One of the most familiar involves a race between a turtle and Achilles. Other examples of infinity in science and math are Pi, a monkey using a typewriter, and dividing by zero. Two that are interesting and debatable are the size of the universe; creating the possibility of multiverses and second, the possibility there are different sizes of infinities. [278]

The theory of eternal inflation, which was developed in the early 1980s by Alan Guth, Andrei Linde, and others, and has led to a radically new global view of the universe. The theory is explained in a *Scientific American* online article:

"Inflation is a period of super-fast, accelerated expansion in early cosmic history. It is so fast that in a fraction of a second a tiny subatomic speck of space is blown to dimensions much greater than the

entire currently observable region. At the end of inflation, the energy that drove the expansion ignites a hot fireball of particles and radiation. This is what we call the big bang." [279]

Unlike faith, science doesn't see infinity as infinite. As you would expect, the term has been tossed about through the ages, changing and evolving based on ever-increasing complexity and related discoveries. An intellectual, peer-reviewed academic resource from iep.utm.edu titled, *The Infinite* by Bradley Dowden, California State University Sacramento, comments on the history of this term:

Looking back over the last 2,500 years of use of the term "infinite," three distinct senses stand out: actually infinite, potentially infinite, and transcendentally infinite. These will be discussed in more detail below, but briefly, the concept of potential infinity treats infinity as an unbounded or non-terminating process developing over time. By contrast, the concept of actual infinity treats the infinite as timeless and complete. Transcendental infinity is the least precise of the three concepts and is more commonly used in discussions of metaphysics and theology to suggest transcendence of human understanding or human capability. [280]

Below is the summary of his introduction:

For purposes of doing mathematics and science, the actual infinite has turned out to be the most useful of the three concepts. Using the idea proposed by Bolzano that was mentioned above, the concept of the actual infinite was precisely defined in 1888 AD when Richard Dedekind redefined the term "infinity" for use in set theory and Georg Cantor made the infinite, in the sense of infinite set, an object of mathematical study. Before this turning point, the philosophical community generally believed Aristotle's concept of potential infinity should be the concept used in mathematics and science. [281]

It should be safe to say that this area has the greatest opportunity for disagreement between faith and science. Both have a concept of a beginning that accelerates forward. However, unlike faith, science would probably disagree that this concept applies to human life and location, postulating infinity is simply, although complexly applied, to measurement. If that were true, I think it would stem not from a scientific unbelief, but again, the application of natural elements cannot not be studied in relation to feeling, emotion, belief, and certainly not faith. These are not material scientific fields of study. Perhaps that is why some scientists are also called Christian and not, as one sect claims, Christian Scientists. Such a title perhaps was never intended.

I have spent a great deal of time defining the faith community in which I live and my layperson's view of the science community and how I see the interaction of these two communities. Not covered in detail is a third community, the religious community, which could be a major barrier to any resolution between science and faith. Covering the past and present of the religious community would take a book larger than the one you are reading. For that reason alone, I will only touch on my thoughts regarding confusion between faith and religion below. I will leave the resolution up to the reader.

A Tale of Three, Not Two, Narratives

The Faith Narrative

In the Land of Eternity lived God. Not one of other gods, but *the* God, the Beginning and the End, for who else could be in charge of a place called Eternity? Some tried to imagine where God came from, but there was no place from where God could have originated since there was no other place than the Land of Eternity. God was not alone, there was a Son, and… a Spirit. These three were all one, together, yet separate. There were many angels in Eternity, some with specific duties, and one in particular who felt he could become like God. Some of the other angels agreed and they all together attempted to just become God. It did not go well for those angels and their leader. The God of the Land of Eternity cursed them all and sent them down from Eternity to live in a desolate place that was completely separate from Heaven.

At some point God, along with the Son and the Spirit, and a few others, discussed the possibility of creating beings who would take on their image but not their positions. The purpose of these new beings would be to glorify God and praise Him for Who He was and what He would do for those who obey. The only problem was that there was no place for them to live, at least not yet, in the Land of Eternity. A solution was offered and accepted.

Using the Word of the Son and the power of the Spirit, God spoke into existence a universe, a separate place where everything affected everything else and in so doing caused it all to work. God continued, his Son spoke, and the power of the Spirit flourished as waters appeared and were used to separate land. Two lights of varying shades were hung in the newly created sky. This was followed by living, breathing creatures, plants, trees, and all other things not made in his image, coming to life. God looked at all of this and called it good.

Then, the moment all had been waiting for arrived. It was time to create the very first new being made in the image of God and those who were with Him. The being was not formed out of nothing, but unlike all other creations, this being would be formed from what God had already created. After God finished, He breathed divine life into the being and the being became aware.

God placed the being in a garden and gave the being the task of naming everything God had created. God noticed that the being was lonely because there was nothing else like the human creation. Out of God's compassion, a second being was created from the first to be a companion. God gave them complete control over a special, protected garden and everything in it, except for two trees; the Tree of the Knowledge of Good and Evil and the Tree of Life. An earlier incident was about to spoil everything.

This place where God, the Son, and the Spirit created the universe and everything in it was not in the Land of Eternity, but somewhere else. This new place was either where God cast down the fallen angels and their leader, or they knew how to get to this location because the fallen leader shows up in the newly created garden. The fallen angel easily convinced the two new beings there was no harm in eating from the Tree of the Knowledge of Good and Evil and they both took a bite of the fruit. Instantly they were aware of their nakedness and hid themselves from their creator.

When God found out about their disobedience the two were sent out of the garden and never allowed to reenter because of God's concern they would also eat from the Tree of Life and live for eternity. Unlike the Land of Eternity, this new place outside the garden was not designed for beings to live there forever. The two beings were cast out into the remainder of the world created as part of the universe to make it on their own.

Part of becoming a resident of the rest of the world involved pain, work, and childbirth. They produced children, the children produced murder and from there many generations of beings were born and died. Not long after, God became totally displeased with his creation and destroyed it with a flood, except for one family.

The newly created world after the flood was still known as the Earth. God established a race of beings he called God's children. God gave them strict laws from which to govern and live. After years and years of births and deaths God chose certain beings to prophesy about the plan for the Son to be born into the midst of these beings and become one of them. The Son would teach and then die for those teachings. The Son would be raised from the dead and then return to the Land of Eternity. From there preparations would be made for those who believed that the Son was the Savior of the world to become residents.

Before the Son left the Earth after raising from the dead, he told His followers of a return some day at which time all of those who believed would be taken with the Son to the Land of Eternity to live forever in the presence of God.

That day came, and everything the Son promised came true.

No one knows what happened to the universe and everything which God originally created. Because of the joy of living in the Land of Eternity with God, no one really cared or even had memories of their past.

The Science Narrative

Billions and billions of years ago a singularity occurred… or maybe it was a Big Bang. No one understands why, or for that matter, where it came from. What is known is that it came from somewhere because something cannot come from nothing. And, since we cannot explain what happens after death, or on a larger plane, what happens to everything once all universes run out of gases needed to birth new planets and solar systems, finding out what happened before the singularity or bang is just useless work. More important is the discovery of what things are, what they are made of, and what we can combine to make new stuff that is helpful (hopefully) for the human race.

After the singularity (or bang), a lot of time, by Earth standards, passed. Find here an interpretation of what Aristotle postulated, as written in an article in ScienceDirect.com;

> *… picturing a small universe centered on the Earth and bound by the sphere of fixed stars, the Philosopher took pain to demonstrate philosophically and physically that it was necessarily eternal. For instance, a beginning of time would imply an absence of time before; but one could say before unless one had already supposed the existence of time. Likewise, a movement could not arise spontaneously: either it had existed for all eternity, or it was resulting from the action of another movement, which, itself, had existed for all eternity or was the product of a preceding movement, and so forth. And the existence of an obviously unchanging celestial world also testified to the eternity of time, because incorruptibility was by definition absolute. In On the Heavens, Aristotle thus concluded that "the heaven as a whole neither came into being nor admits of destruction, as some assert, but is one and eternal, with no end or beginning of its total duration, containing and embracing in itself the infinity of time."* [282]

As the Earth evolved, pre-life was looking for a medium in which to appear… and then water appeared on the Earth, maybe from internal sources below ground or maybe by millions of asteroids crashing into the Earth depositing water from outer space. Either way, water was already in the universe. As a result of these crashing objects from outer space billions of years ago, 4.1 billion years for reference, certain carbon elements, primarily graphite, captured in crystals of zircon created a ratio of isotopes in the graphite suggesting a biological origin. Or, since there was water, likely oceans, any life would have been a one cell organism without membrane-bound nuclei or cell organelles. If there had been a continental shelf crust made of granite, certain mineral sources may have also been available as nutrients, like phosphorus. Certain living organisms found today deep in the ocean near extremely hot hydrothermal vents are proof that these forces could have survived and created the beginning of life on Earth. [283]

From these basic formations, using billions of years for development and relying on the survival of the fittest, life flourished and developed by adapting to the immediate needs of their environment. Eventually intelligent life developed, until the civilized world of today dominated.

To understand the future, we must refer back to the beginning. If we understand that everything that exists today was formed from something that has always existed, then although it may change form, matter will always continue to exist for infinity. The only way for all matter to cease to exist is that at some point it never existed. Since science has never postulated that something can be created from nothing, an argument for all matter ceasing to exist is absurd. Therefore, there will never be an end to the world or the universe in which it sits. It may, however, transform into something else.

The Religion Narrative

Until the birth of the Son of God, in the human form of Jesus Christ, only stories and traditions existed describing the origins of everything and the purpose for mankind's presence in the universe. There was no need for a system to map out and protect the various beliefs regarding the entering of the physical world by its creator. There was only one, Judaism.

The term religion finally found its roots and reason for being in the 15th century, primarily within a group known as Catholic. A couple of centuries later, in the 1700s, there would be a powerful reformation that would become known as Protestant. The systematic study of science and religion started in the 1960s, with authors such as Ian Barbour (1966) and Thomas F. Torrance (1969) who challenged the prevailing view that science and religion were either at war or indifferent to each other. [284]

My interpretation: The war was between science and religion, not science and faith. Words have meaning. Science has no quarrel with faith because faith is not a field studied by science. Again, religion is a system, a thing. Faith cannot be looked at under a microscope or through a telescope. Although, faith is transmitted through the brain, faith is not considered a scientific field of study. Religion, on the other hand, is a system much like science. This similarity lends itself to conflict. Science only became a pure field of study after ridding itself of part-time, early religious leaders in favor of full-time individuals devoted only to the study of science.

The Missing Explanation

Faith, Science, and Religion are three very distinct and separate entities. Many through the ages have tried without success to do as a scientist would do, pour them all into a beaker and see what you get. It is my conclusion that mixing these three together, like mixing several different colors of paint, will only result in several shades of gray (or if you are from England, grey). Either way, it is not useful and if you need to match the color and repaint, it is very difficult.

Science has no language or code for the concept of good, nor a reason to do so. Science is about figuring things out, identifying and answering questions of why, not writing code for emotion. Science makes no apology nor does it ask forgiveness; when it fails, it just reboots, learning from its failure.

Religion is neither science nor faith, but a *system* of belief. Those of faith will quickly tell you religion has no saving power. A professional lobbying group called the X Club, formed by Thomas Huxley in 1864, had the sole purpose of promoting a science free from religious dogmas by removing competition from amateur scientist/clergymen. This provided an opening to the field of science for full-time professionals. It could have also created the flood of applications for grants to fund those full-time scientific professionals. Faith, on the other hand, is not a religious system nor a scientific method. [285]

Faith has no desire to *prove* anything. Nor does faith require a name, other than "faith." So, there is no reason for a rigid system of religion to protect it. Science is in no way a subset of faith; both are separate. However, science and religion are a lot alike. Both are based on a never-ending search for greater knowledge, wisdom, understanding, and control. It is often thought that religion bases its structure on the unseen while science follows only what they can touch or observe. Nothing could be further from the truth. If it were true, it would contradict the two most basic principles of science: hypothesis and theory. Science is filled with examples moving theoretical science to a level that is only later proven to exist using physical means. Examples include Pasteur's unseen pathogens, Newton's invisible gravitational force, and Einstein's imperceptible photons. [286] Religion is a system of protecting the unseen elements of faith; but religion is not faith, nor does true faith need protection. Broken down and observed, the conflicts through the years have been between science and religion, not science and faith.

That is why by grace through faith you are saved. Science nor religion can save you; they can only make your daily life better or worse. Science doesn't strive to be neutral on this matter because it does not recognize mediocrity. Science is either right or wrong. If right, it moves to the next equation. If it is wrong, it starts over. Religion reacts in much the same way as science. That can explain why there are so many different religions. Faith does none of these; actually, it does just the opposite.

Religion is a system used by some in the faith community to build a structural framework. Like scientific research, sometimes that framework must be altered, or… torn down and redesigned. The latter usually comes with much resistance. Such has been the case since the beginning of time… and the establishment of the New Testament Church in the early 1st century; e.g., circumcision vs. baptism is recorded proof.

Faith, the kind of faith expressed and demonstrated in Scripture, and in the lives of millions of living Christians around the world today, cannot be explained in hypothesis or theory. The faith community points to Hebrews 11:1 in the New Testament: "*Now faith is the substance of things hoped for, the evidence of things not seen.*" *NKJV* It is quite interesting that the author of this verse chose to call the thing that makes up faith... a *substance.* You will also note in Ephesians 2:8-9, "*8 For by grace you have been saved through faith, and that not of yourselves; it is the gift of God, 9 not of works, lest anyone should boast.*" *NKJV*

Like a chemical bond in a biological organism connecting cells, faith is a bridge allowing the grace of God to connect to humankind. Again, neither science nor religion can provide this. All of this aside, there is a problem to be addressed by science, religion, and faith: obstruction.

Each of these groups must realize what they are, who they are, and what they represent. My goal is to get each to stop creating obstacles that prevent each of the others from being successful at what they do best. Each has an inalienable right to accept the teachings of one of the other by respecting the need of each to discover and share what they have found without claiming the others to be irrelevant, mythical, or absurd. Such obstruction is not helpful or beneficial to the humanity each has proclaimed to serve.

CHAPTER 23

A SEASON UNLIKE ANY OTHER

... this is too complicated so I
will just avoid the issue.
... there's no way the world would
follow one individual!

Is it possible for one human to gain access to, and have dominion over, the entire world?

IF MORTALS CAN dream about knowledge that unlocks the universe, how much deeper could Satan dig into ways to destroy creation? Is it possible for Satan to occupy the mind of a person?

Earlier I pointed out the need to look very closely at the probability of a few things promised in the waning chapters of the New Testament regarding how civilization's end will proceed. Let's cover the first one, how is it possible for one person to rule the entire Earth.

What would you assume the following list of individuals may have in common?

- Steve Jobs
- Ben Rich
- Alan Turing

- Wernher von Braun
- John von Neumann
- Srinivasa Ramanujan Aiyangar
- Konstantin Tsiolkovsky

Science has spent much of its time, since first being known as science, trying to prove aliens exist. All the while, the faith community had already identified them. (My childhood deacons are getting restless in their graves). Yes, there will be consequences for this statement. Before erecting a stake for me, covering it with straw, and lighting a fire, let's address three areas open to discussion in the faith community and the role science has, and will, play going forward. As you consider this, please understand up front I am no more saying these individuals are Satan than did Jesus or Paul of those in their time, or others who dealt with demons who had taken over the lives of innocent people. I am, however, pointing to similarities in certain individuals who have garnered massive attention worldwide. Let's start in the present and work backward.

In this last section, called the *Fall of Creation,* it is important to look very closely at the probability of a few things promised in the waning chapters of the New Testament and predictions in scientific records regarding how civilization's end will proceed. A few faith community events would include:

- Is it possible for one human to gain access to, and have dominion over, the entire world?
- What will cause all of the predicted biblical disasters?
- What will be Satan's role in all of this?

You have most likely heard of Steve Jobs. You may also be familiar with his accomplishments over the past several years and how his inventive mind placed technology squarely in the center of our everyday lives. So much so, that a minute twitch or glitch in underlying process associated with his technology platform can cause irreversible damage, or success to you personally or economically. Steve Jobs and the Apple brand, along with a few other 19th, 20th, and 21st -century titans of technology have, in a very short period of time—compared to the entire historical age—monopolized critical centers of commerce, military, and communications.

So, to the first question above, *how does a single individual become that dominant over our lives*, you will find that none of the following individuals achieved their life's world-altering ideas or results by chance or accident. After you finish reading about these individuals, I encourage you to ask yourself these three questions:

- Is this thing we call civilization going to totally end?
- How does my answer to the first question impact me and those I love?
- Of the ideas and data presented, which one could have the greatest impact?
- Hold those answers close as you read the remainder of this book.

Seven Individuals Who Captured Long-Lasting World Attention

Steve Jobs

Born: February 24, 1955—Conceived in Homs, Syria and born in San Francisco, California without the knowledge of the biological father. The child was placed for adoption upon birth.

Raised in: San Francisco, California

Parents:

> **Biological father**, Abdulfattah al-Jandali, was Arabic growing up in Homs, Syria. He met his wife while serving as a teacher's assistant at the University of Wisconsin. While an undergraduate at the American University of Beirut, Lebanon, Steve's father was a student activist and served time in prison for his political activities.

Biological mother, Joanne Carole Schieble, a Catholic and of Swiss and German descent.

Adopted father and mother were eventually Paul and Clara Jobs. They were not the first choice, but rather a couple who held high social and financial status. The first couple soon changed their minds. The biological mother initially rejected the offer of adoption by the Jobs couple, but eventually agreed when the couple promised to send the child to college and finance his education. That turned out to be an inexpensive pledge, since the young man dropped out of college shortly after enrolling. The child grew to form a very loving, lifetime relationship with his adopted parents.

Early years: Growing up, the boy had trouble making friends with children his own age, but really enjoyed working alongside his dad building and fixing things. He turned out to be a handful when he started school. He was bored, probably due to lack of challenging educational core studies. His education was turned around by a fourth-grade teacher putting him through an advanced math curriculum. Jobs skipped the fifth grade and threatened to drop out of school in the seventh-grade if his parents didn't find a better school. They sold everything and used all their savings to buy another home in an area of Los Altos, California. It worked—a better, more challenging school and a house that would become the first office and manufacturing place for Apple computer. The house was declared a national historic site in 2013.

High School: Homestead High School had strong ties to Silicon Valley. This is where Jobs first met Steve Wozniak. Homestead High became a center of change for Jobs. He became interested in literature, electronics, girls, and like many high school students, discovered recreational drugs. By his senior year of high school Jobs had begun using LSD.

College: His friend, Wozniak, began attending the University of California, Berkeley in 1971. Jobs started attending Reed College in 1972 and dropped out that same year. His reason was he didn't want to spend his parent's money on an education that seemed meaningless to him. He instead wanted to traveled to India seeking enlightenment through the study of Zen Buddhism.

Early Career: Jobs was described as both a brain and a hippie but it was said he never really fit into either class; smart enough to be a nerd, but wasn't, and too intellectual for hippies, who just got wasted all the time. It was about this time that Wozniak designed a low-cost "blue box" to generate necessary tones to allow someone to make illegal, free long-distance telephone calls. Jobs sold the boxes and split the profits with Wozniak. Later, Jobs would credit this first "blue box" venture as the starting point of Apple's later success. He felt without this Blue Box venture nothing would have developed afterward. By 1974 he was still kind-of involved with a girl he had been seeing off and on for a few years, living in a cabin, working at Atari, and saving money for his planned trip to India.

Later Career: Became an American business magnet, industrial designer, investor, and media proprietor. Co-founded Apple, Inc. becoming its chairman and CEO. Was the chairman and majority shareholder of Pixar and a member of the Walt Disney Company's Board of Directors. Toward the end he was the founder, chairman, and CEO of NeXT. He is remembered as the pioneer of the personal computer revolution of the '70s and '80s, along with co-founder Steve Wozniak. Seven months after visiting India in 1974 Jobs, wearing traditional Indian clothing and sporting a shaved head, returned to the states.

Upon his return Jobs experimented with psychedelics, calling LSD experiences "one of two or three of the most important things [he had] done in [his] life." [287]

His Spiritual Belief: Zen Buddhism

His Inspiration: Paramahansa Yogananda

Illness: Diagnosed with a pancreatic neuroendocrine tumor in 2003

Died: October 5, 2011 of respiratory arrest related to his tumor at the age of fifty-six

Noteworthy

I will sum up this section on Steve Jobs shortly but, footnoted below are some great references if you wish to read about all of the details of what Jobs accomplished beginning in 1974. There is much, much more to the story of Steve Jobs. Most of what I have written here was taken from a simple Wikipedia search. [288] If you want to get an up-close look, I suggest you read, *Steve Jobs: The Exclusive Biography*, by Walter Isaacson. [289]

On October 16, 2011, an invitation-only memorial was held at Stanford University. Those invited were a who's-who of names at the time from all walks of life: entertainment, business, and politics. Each attendee was given a small brown box as a "farewell gift" from Jobs. The box contained a copy of the *Autobiography of a Yogi* by Paramahansa Yogananda.

In an inc.com article by Hitendra Wadhwa, founder of Mentora Institute, reporting on what one of the attendees of the memorial service, Mark Benioff, CEO of Salesforce.com, said at a TechCrunch Disrupt conference two years later regarding the gift he received at the memorial from Jobs:

> *"I look at Steve as a very spiritual person," he added. "[Steve] had this incredible realization—that his intuition was his greatest gift and he needed to look at the world from the inside out."*

The article turns to *Yogananda's* impact on society and science:

> *Yogananda's particular genius was showing the modern applicability of these ancient principles, attuning himself to an audience who aspired as much to outer success as inner growth by delivering talks on topics like "The Science of Healing" and "The Art of Getting What You Want." In that regard, he was a forerunner to 21st-century psychologists, physicians, psychotherapists, and neuroscientists who are generating powerful scientific findings on human nature and well-being—all aligned with Yogananda's teachings on consciousness, thoughts, emotions, habits, and brain-wiring.*

And then these closing thoughts:

Yogananda's teaching of universal consciousness strongly appealed to Steve Jobs, who had a self-professed hunger to "make a dent in the universe."

Since Yogananda's passing in 1952, many teachers have followed his trailblazing path to bring yoga to our world, helping make it a fixture in popular culture as it continues to take hold with young and old, the elite and the ordinary, the spiritualists and the atheists. [290]

Jobs is buried in an unmarked grave at Alta Mesa Memorial Park, the only nonsectarian cemetery in Palo Alto.

Ben Robert Rich (alias—Ben Dover)

Born: June 18, 1925 in Manila, Philippines

Raised in: The Philippians. The family fled to the US just weeks before the Japanese bombed Pearl Harbor in 1941. The family wound up in Los Angeles in 1942 where Ben became a naturalized citizen.

Parents: Isadore Reich and Annie Kupferman Reich.

Early years: The Philippians.

High School: The Philippians.

College: UCLA in aeronautical engineering, UC Berkeley, Harvard Business.

Early Career: Rich was a US engineer who conducted top secret research on advanced military aircraft while working at Lockheed Aircraft Corporation (now Lockheed Martin Corporation) under an alias, which he was required to adopt for security reasons.

Later Career: Rich was the second director of Lockheed Martin's top-secret Skunk Works program. His crowning achievement was the technology and development of the F-117 Nighthawk stealth bomber. Others included the SR-71 Blackbird, and the U-2 spy plane. [291]

His Spiritual Belief: Jewish

His Inspiration: He believed in "the existence of other intelligent life in the universe" even though he thought it likely to be distant and definitely mysterious.

Illness: Esophageal cancer

Died: January 5, 1995 in Los Angeles, California at the age of seventy.

Noteworthy

The following are excerpts from gaia.com. As a disclaimer, the site defines itself with the following statement on their site under the heading, Who Are We:

> *Answer life's deeper questions and go beyond the mainstream narrative with Gaia— a member-supported media network of truth seekers and believers empowering an evolution of consciousness. Explore over 8,000 ad-free, streaming titles that challenge modern paradigms and allow you to manifest the reality that defines your being.*

As an aeronautical engineer and considered the Father of Stealth, Rich was more of an enigma now than he was during his lifetime. Rich was the second director of Lockheed Martin's top-secret Skunk Works program. After passing away in 1995, controversy was still brewing among friends and critics and those who heard him speak publicly about whether or not he believed in UFOs and extraterrestrials. [292]

Rich is quoted as saying, "We now have the technology to take ET home," in concluding a UCLA alumni speech in 1993. Below is the text preceding this statement:

> *"We already have the means to travel among the stars, but these technologies are locked up in black projects and it would take an Act of God to ever get them out to benefit humanity... Anything you can imagine, we already know how to do."* [293]

Also questioned is the credibility of claims he stated that technology learned at Roswell (Area 51) was used to influence the development of top-secret US aircraft. These communications were said to be to his friend, John Andrews, a Lockheed engineer. And then there was the alleged deathbed confession.

James Goodall, an aerospace journalist and in-demand public speaker, became a friend of Rich. Goodall states that he had a phone conversation with Rich ten days before his death while Rich was hospitalized in the USC Medical Center in Los Angeles. Rich reportedly told Goodall:

> *"Jim, we have things out in the desert that are fifty years beyond what you can comprehend. If you have seen it on Star Wars or Star Trek, we've been there, done that, or decided it wasn't worth the effort. They have about 4,500 people at Lockheed Skunk Works. What have they been doing for the last 18 or 20 years? They're building something."* [294]

In his 1994 autobiography, *Skunk Works: A Personal Memoir of My Years at Lockheed*, Rich recounted that the technology for his crowning achievement, the F-117, virtually "fell into my lap" after a Soviet scientist openly published (1975) [295] an idea that led to the technology for the stealth fighter-bomber. [296]

Alan Mathison Turing

Born: June 23, 1912 in London, England

Raised in: England

Parents: Julius Mathison Turing and Ethel Sara Stoney

Early years: Raised by foster parents while his biological parents returned to India to continue the father's work.

High School: At age thirteen, the boy was enrolled in a socially elite private school in southern England with emphasis in Latin and Greek classic studies.

College: The young man secured a place at King's College in Cambridge to study mathematics, graduating in 1934 with a first-class honors degree.

With remarkable academic precocity he was made a fellow of the college in 1935. Turing completed a doctoral thesis at Princeton in 1938 before returning to Cambridge.

Early Career: With Britain's declaration of war on Germany in September 1939, he joined the Bletchley Park code-breakers at the Government Code and Cypher School. His wartime work earned him a high civilian award and he was named an Officer of the Most Excellent Order of the British Empire.

Later Career: Turing's fascination with computers led him to design the Automatic Computing Engine.

His Spiritual Belief: Raised Anglican—as an adult was atheist LGBT

His Inspiration: Insatiable appetite for figuring things out, especially number problems.

Illness: Suicide

Died: June 7, 1954 in Wilmslow, Cheshire at the age of forty-one

Noteworthy

He was a British mathematician and logician who made major contributions to mathematics, cryptanalysis, logic, philosophy, and mathematical biology and also to the new areas later named computer science, cognitive science, artificial intelligence, and artificial life. [297]

Parents Ethel Sara Stoney and Julius Mathison Turing had met in imperial India, where Julius' father was a senior colonial administrator.

Alan Turing, at thirteen, was enrolled in a socially elite private school in southern England with emphasis in Latin and Greek classic studies. "If he is to be solely a scientific specialist, he is wasting his time at a public school," Nowell Smith, Sherborne's headmaster, wrote to his parents. Nonetheless, he secured a place at King's College in Cambridge to study mathematics, graduating in 1934 with a first-class honors degree.

In 1936 he published the groundbreaking paper "On Computable Numbers, With an Application to the Entscheidungs Problem" (or "decidability problem"), a reference in German to a celebrated riddle that the American logician Alonzo Church had also explained.

The Government Code and Cypher School's greatest initial challenge was figuring out the method of encryption of the German Enigma device, which was invented twenty years earlier by Arthur Scherbius, a German electrical engineer. The machine was said to be capable of generating almost 159 quintillion permutations. This machine replaced a method previously developed by a Polish mathematician that had become unusable.

Using a huge contraption called the Bombe, they mimicked the operations of the Enigma machine to break its codes. "The critical factor was Turing's brilliant mechanization of subtle logical deductions," wrote biographer Hodges. [298]

In the postwar years, Turing's fascination with computers led him to design the Automatic Computing Engine. Although it was never built, Turing believed that "the computer would offer unlimited scope for practical progress toward embodying intelligence in an artificial form," Hodges wrote. [299]

In an unrelated investigation, police discovered Turing was having a physical relationship with another man. Such relationships at that time in England were considered "gross indecency." The other man was given a conditional discharge; however Turing was ordered to undergo chemical castration by taking doses of the female hormone estrogen to reduce sex drive, which entered him in the public record as a criminal. This incident would put an end to any hope of continuing his remarkable scientific journey. Two years later, at age forty-one, the motive for his apparent suicide remained unclear and left many questions.

In 2009 British Prime Minister Gordon Brown, speaking on behalf of the British government, publicly apologized for Turing's "utterly unfair" treatment. Four years later in 2013, Queen Elizabeth II granted Turing a royal pardon. [300]

(Freiherr, or Baron)
Wernher Magnus Maximilian Freiherr von Braun [301]

Born: Born March 23, 1912 in Wirsitz, Posen Province, in the German Empire

Raised in: Wirsitz, Posen Province, Germany

Parents: Freiherr Magnus Alexander Maximilian von Braun and Emmy von Quistorp

Early years: Boarding school at thirteen. At sixteen, his parents transferred von Braun to another private school where he pursued his interest in rocket engineering.

College: Graduated with a degree in mechanical engineering in 1932 at the age of twenty. He earned his doctorate degree from the Friedrich-Wilhelm University of Berlin for doctoral studies and graduated with a doctorate in physics in 1934.

Early Career: Von Braun had an ambivalent and complex relationship with the Nazi Third Reich.[302] In 1933, von Braun was working on his creative doctorate when the Nazi Party came to power in a coalition government in Germany. By the end of 1934, his group had successfully launched two liquid fuel rockets that rose to heights of two miles. He applied for membership in the Nazi Party on November 12, 1937, and was issued a membership number.

Von Braun used Goddard's plans from various journals and incorporated them into the building of the Aggregate series of rockets. The first successful launch of an A-4 took place on October 3, 1942. The A-4 rocket would become well-known as the V-2. Later, von Braun would comment: "I have very deep and sincere regret for the victims of the V-2 rockets, but there were victims on both sides… A war is a war, and when my country is at war, my duty is to help win that war." [303]

Later Career: In 1950, at the start of the Korean War, von Braun and his team were transferred to Huntsville, Alabama, his home for the next twenty years. Between 1952 and 1956, [304] von Braun led the Army's rocket development team at Redstone Arsenal, resulting in the Redstone rocket, which was used for the first live nuclear ballistic missile tests conducted by the United States. As director of the Development Operations Division of the Army Ballistic Missile Agency, von Braun, with his team, then developed the Jupiter-C, a modified Redstone rocket. [305] The Jupiter-C successfully launched the West's first satellite, Explorer 1, on January 31, 1958. This event signaled the birth of America's space program. Von Braun became NASA's first director on July 1, 1960 and held the position until January 27, 1970. [306]

His Spiritual Belief: Lutheran from birth to adulthood. In 1945 he explained his decision to surrender to the Western Allies, rather than Russians, as being influenced by a desire to share rocket technology with people who followed the Bible. In 1946, [307] he attended church in El Paso, Texas, and underwent a religious conversion to evangelical Christianity. [308] Later in life, he joined an Episcopal congregation, [309] and became increasingly religious. [310]

His Inspiration: Getting into space.

Illness: Pancreatic cancer

Died: June 16, 1977 in Alexandria, Virginia at the age of sixty-five

Noteworthy

He was the leading figure in the development of rocket technology in Nazi Germany and a pioneer of rocket and space technology in the United States. [311] How is this possible, you may ask? It seems the entire family were well-connected. One adult child served as a West German diplomat and was named as the Secretary of State in the foreign office in the 1970s. A younger brother was a rocket scientist and later senior executive with Chrysler. [312] Again, you may ask… how is this possible?

According to historian Norman Davies, von Braun was able to pursue a career as a rocket scientist in Germany due to a "curious oversight" in the Treaty of Versailles which did not include rocketry in its list of weapons forbidden to Germany. [313]

He said that he had been so influenced by the early Nazi promise of release from the post–World War I economic effects, that his patriotic feelings had increased. Yet, he also wrote that "to us, Hitler was still only a pompous fool with a Charlie Chaplin moustache" [314] and that he perceived him as "another Napoleon" who was "wholly without scruples, a godless man who thought himself the only god." [315]

At the time, Germany was highly interested in American physicist Robert H. Goddard's research. Before 1939, German scientists occasionally contacted Goddard directly with technical questions. Von Braun used Goddard's plans from various journals and incorporated them into the building of the Aggregate (A) series of rockets.

The Soviet Army was about 100 miles from Peenemünde in early 1945 when von Braun assembled his planning staff and asked them to decide how and to whom they should surrender. Unwilling to go to the Soviets, von Braun and his staff decided to try to surrender to the Americans.

In early April, as the Allied forces advanced deeper into Germany, Kammler ordered the engineering team, around 450 specialists, to be moved by train into the town of Oberammergau in the Bavarian Alps, where they were closely guarded by the SS with orders to execute the team if they were about to fall into enemy hands. On April 29, 1945, Oberammergau was captured by the Allied forces who seized the majority of the engineering team. Von Braun and several members of the engineering team, including Dornberger, made it to Austria. [316]

After the surrender, Wernher von Braun spoke to the press:

> *We knew that we had created a new means of warfare, and the question as to what nation, to what victorious nation we were willing to entrust this brainchild of ours was a moral decision more than anything else. We wanted to see the world spared another conflict such as Germany had just been through, and we felt that only by surrendering such a weapon to people who are guided not by the laws of materialism but by Christianity and humanity could such an assurance to the world be best secured.* [317]

Initially, he was recruited to the US under a program called Operation Overcast, subsequently known as Operation Paperclip. In 1952, von Braun first published his concept of a crewed space station in a *Collier's Weekly* magazine series of articles titled "Man Will Conquer Space Soon!" The space station (to be constructed using rockets with recoverable and reusable ascent stages) would be a toroid structure, with a diameter of 250 feet; this built on the concept of a rotating wheel-shaped station introduced in 1929 by Herman Potočnik in his book *The Problem of Space Travel—The Rocket Motor.*

The US Navy had been tasked with building a rocket to lift satellites into orbit, but the resulting Vanguard rocket-launch system was unreliable. In 1957, with the launch of Sputnik 1, a growing belief within the United States existed that it was lagging behind the Soviet Union in the emerging Space Race. American authorities then chose to use von Braun and his German team's experience with missiles to create an orbital launch vehicle. Von Braun had originally proposed such an idea in 1954, but it was denied at the time. [318]

NASA was established by law on July 29, 1958. One day later, the fiftieth Redstone rocket was successfully launched from Johnston Atoll in the south Pacific as part of Operation Hardtack I. Two years later, NASA opened the Marshall Space Flight Center at Redstone Arsenal in Huntsville, and the Army Ballistic Missile Agency (ABMA) development team led by von Braun was transferred to NASA. In a face-to-face meeting with Herb York at the Pentagon, von Braun made it clear he would go to NASA only if development of the Saturn were allowed to continue. [319]

On May 5, 1961 von Braun's team successfully launched Alan Shepard into space. He named his Mercury-Redstone 3 *Freedom 7.* [320] (Author's note: I have memories of standing on the outside steps of my fifth-grade elementary school in Eau Gallie, Florida looking at the white contrail of Shepard's capsule as it flew over my head.)

The Marshall Center's first major program was the development of Saturn rockets to carry heavy payloads into and beyond Earth orbit. From this, the Apollo program for crewed Moon flights was developed. Von Braun initially pushed for a flight engineering concept that called for an

Earth orbit rendezvous technique, but in 1962, he converted to the lunar orbit rendezvous concept that was subsequently realized. [321]

During Apollo, he worked closely with former Peenemünde teammate, Kurt H. Debus, the first director of the Kennedy Space Center. His dream to help mankind set foot on the Moon became a reality on July 16, 1969. Over the course of the program, Saturn V rockets enabled six teams of astronauts to reach the surface of the Moon.

He traveled to India and hoped that the program would be helpful for bringing a massive educational television project to help the poorest people in that country. [322] [323]

On March 1, 1970, von Braun was assigned the post of NASA's Deputy Associate Administrator for Planning at NASA Headquarters. After a series of conflicts associated with the truncation of the Apollo program, and facing severe budget constraints, von Braun retired from NASA on May 26, 1972.

After leaving NASA, von Braun became Vice President for Engineering and Development at the aerospace company Fairchild Industries in Germantown, Maryland, on July 1, 1972. [324]

Von Braun helped establish and promote the National Space Institute in 1975, a precursor of the present-day National Space Society and became its first president and chairman. In 1976, he became scientific consultant to Lutz Kayser, the CEO of OTRAG, and a member of the Daimler-Benz board of directors. When the 1975 National Medal of Science was awarded to him in early 1977, he was hospitalized and unable to attend the White House ceremony.

In the first half of his life, von Braun was a nonpracticing, "perfunctory" Lutheran, whose affiliation was nominal and not taken seriously. [325]

In 1945 he explained his decision to surrender to the Western Allies, rather than Russians, as being influenced by a desire to share rocket technology with people who followed the Bible. In 1946, [326] he attended church in El Paso, Texas, and underwent a religious conversion to evangelical Christianity. [327] In an unnamed religious magazine, he stated:

> *One day in Fort Bliss, a neighbor called and asked if I would like to go to church with him. I accepted, because I wanted to see if the*

> *American church was just a country club as I'd been led to expect. Instead, I found a small, white frame building... in the hot Texas sun on a browned-grass lot... Together, these people make a live, vibrant community. This was the first time I really understood that religion was not a cathedral inherited from the past, or a quick prayer at the last minute. To be effective, a religion has to be backed up by discipline and effort.* —von Braun [328]

On the motives behind this conversion, Michael J. Neufeld is of the opinion that he turned to religion "to pacify his own conscience," [329] whereas University of Southampton scholar Kendrick Oliver said that von Braun was presumably moved "by a desire to find a new direction for his life after the moral chaos of his service for the Third Reich." [330] Having "concluded one bad bargain with the Devil, perhaps now he felt a need to have God securely at his side." [331]

Later in life, he joined an Episcopal congregation, [332] and became increasingly religious. [333] He publicly spoke and wrote about the complementarity of science and religion, the afterlife of the soul, and his belief in God. [334] [335] He stated, "Through science man strives to learn more of the mysteries of creation. Through religion he seeks to know the Creator." [336] He was interviewed by the Assemblies of God pastor C. M. Ward, as stating, "The farther we probe into space, the greater my faith." [337] In addition, he met privately with evangelist Billy Graham and with the pacifist leader Martin Luther King Jr. [338]

In January 1977, now very ill, he resigned from Fairchild Industries. Later in 1977, He was, however, although too ill to attend the White House ceremony, President Gerald Ford awarded him the country's highest science honor, the National Medal of Science in Engineering. He is buried at the Ivy Hill Cemetery. His gravestone cites Psalm 19:1: *"The heavens declare the glory of God; and the firmament sheweth his handywork" (KJV).*

John von Neumann

Born: Born December 28, 1903 in Budapest, Kingdom of Hungary

Raised in: Budapest, Hungary

Parents: Neumann Miksa and Kann Margit

Early years: When he was a child, he could divide two eight-digit numbers in his head and converse in Ancient Greek. [339] He and his siblings were tutored in English, French, German, and Italian. His parents insisted he attend school with children his age; however, he was supplied special tutors in subjects he favored. He attended the elite Lutheran Fasori Evagelikus Gimnazium in 1914 until graduation.

High School: By fifteen he began the study of advanced calculus. By age nineteen, he had published two major mathematical papers.

College: University of Berlin for two years, followed by admittance to the prestigious ETH Zurich in 1923. At the same time, he entered Pazmany Peter University as a PhD candidate in Mathematics. He graduated simultaneously from the University in Budapest with a PhD in Mathematics and another PhD from ETH Zurich in Chemistry in 1926. He then went to the University of Göttingen on a grant from the Rockefeller Foundation to study mathematics under David Hilbert.

Early Career: By 1927 he had published twelve major mathematical papers. He taught at the University of Berlin in 1928. In 1929 he taught briefly at the University of Hamburg, but in October of that year he was offered a better position at Princeton University. [340]

Later Career: In the 1930s von Neumann became very interested in developing expertise in explosions, a phenomenon that is difficult to model in mathematics. This led to consultant positions with the military, eventually leading him to the Manhattan Project. His greatest contribution to the atomic bomb was the design of the explosive lenses needed to compress the plutonium core of the Fat Man weapon.

In 1950, von Neumann became a consultant to the Weapons Systems Evaluation Group, whose function was to advise the Joint Chiefs of Staff and the United States Secretary of Defense on the development and use of new technologies. In 1955, von Neumann became a commissioner of the AEC. He accepted this position and used it to further the production of compact hydrogen bombs suitable for intercontinental ballistic missile (ICBM) delivery.

Shortly before his death from cancer, von Neumann headed the United States government's top secret ICBM committee, which would sometimes meet in his home. Its purpose was to decide on the feasibility of building an ICBM large enough to carry a thermonuclear weapon. Von Neumann had long argued that while the technical obstacles were sizable, they could be overcome in time. The SM-65 Atlas passed its first fully functional test in 1959, two years after his death. The feasibility of an ICBM owed as much to improved, smaller warheads as it did to developments in rocketry, and his understanding of the former made his advice invaluable. [341]

His Spiritual Belief: Agnostic all his life, and then suddenly turned Catholic in 1930. None of the family had converted to Christianity while Max was alive, but all did afterward. [342]

His Inspiration: Literature

Illness: Either bone, pancreatic, or prostate cancer.

Died: February 8, 1957 at the age of fifty-three

Noteworthy

John von Neumann, regarded as the foremost mathematician of his time and said to be "the last representative of the great mathematicians," was a Hungarian-American mathematician, physicist, computer scientist, engineer, and polymath. His powers of recall allowed him to quickly memorize the pages of telephone directories, and recite the names, addresses, and numbers therein. [343]

He was a pioneer of the application of operator theory to quantum mechanics in the development of functional analysis, and a key figure in the development of game theory and the concepts of cellular automata, the universal constructor, and the digital computer.

He also worked on the philosophy of artificial intelligence with Alan Turing when the latter visited Princeton in the 1930s. [344]

Von Neumann believed that much of his mathematical thought occurred intuitively; he would often go to sleep with a problem unsolved and know the answer upon waking up. Ulam noted that von Neumann's way of thinking might not be visual, but more aural. [345]

During World War II, von Neumann worked on the Manhattan Project with theoretical physicist Edward Teller, mathematician Stanislaw Ulam and others, problem-solving key steps in the nuclear physics involved in thermonuclear reactions and the hydrogen bomb. He developed the mathematical models behind the explosive lenses used in the implosion-type nuclear weapon and coined the term "kiloton" (of TNT) as a measure of the explosive force generated. [346]

After the war, Robert Oppenheimer remarked that the physicists involved in the Manhattan project had "known sin." Von Neumann's response was that "sometimes someone confesses a sin in order to take credit for it." [347]

As a Hungarian émigré, concerned that the Soviets would achieve nuclear superiority, he designed and promoted the policy of mutually assured destruction to limit the arms race. [348]

Von Neumann reportedly said, "So long as there is the possibility of eternal damnation for nonbelievers it is more logical to be a believer at the end," referring to Pascal's wager. He had earlier confided to his mother, "There probably has to be a God. Many things are easier to explain if there is than if there isn't." [349]

"He was of course completely agnostic all his life, and then he suddenly turned Catholic—it doesn't agree with anything whatsoever in his attitude, outlook, and thinking when he was healthy." Father Strittmatter recalled that even after his conversion, von Neumann did not receive much peace or comfort from it, as he still remained terrified of death. [350]

Srinivasa Ramanujan Aiyangar

Born: December 22, 1887 in Madras Province an administrative subdivision (presidency) of British India

Raised in: Kumbakonam

Parents: Father: Kuppuswamy Srinivasa Iyengar; Mother: Komalatammal

Early years: In December 1889 Ramanujan contracted smallpox, but recovered. A child prodigy by age eleven, he had exhausted the mathematical knowledge of two college students who were lodgers at his home. He mastered trigonometry by the age of thirteen while discovering sophisticated theorems on his own. By fourteen he received merit certificates and academic awards that continued throughout his school career, and he assisted the school in the logistics of assigning its 1,200 students (each with differing needs) to its approximately thirty-five teachers.

High School: Though he had almost no formal training in pure mathematics, he made substantial contributions to mathematical analysis, number theory, infinite series, and continued fractions, including solutions to mathematical problems then considered unsolvable. When he graduated from Town Higher Secondary School in 1904, Ramanujan was awarded the K. Ranganatha Rao prize for mathematics by the school's headmaster.

College: He received a scholarship to study at Government Arts College, Kumbakonam, but was so intent on mathematics that he could not focus on any other subjects and failed most of them, losing his scholarship in the process. He later enrolled at Pachaiyappa's College in Madras. There he passed in Mathematics, choosing only to attempt questions that appealed to him and leaving the rest unanswered, but performed poorly in other subjects such as English, Physiology, and Sanskrit. Ramanujan failed his Fellow of Arts exam in December 1906 and again a year later. Without an FA degree, he left college and continued to pursue independent research in mathematics, living in extreme poverty and often on the brink of starvation. Seeking mathematicians who could better understand his work, in 1913 he began a postal partnership with the English mathematician G. H. Hardy at the University of Cambridge, England. Recognizing Ramanujan's work as extraordinary, Hardy arranged for him to travel to Cambridge. In his notes, Hardy commented that Ramanujan had produced groundbreaking new theorems, including some that "defeated me completely; I had never seen anything in the least like them before." [351]

Early Career: He tried to interest the leading professional mathematicians in his work, but failed for the most part.

Later Career: While on his death-bed in 1920, Ramanujan wrote a letter to his mentor, English mathematician G. H. Hardy. The letter described several new functions that behaved differently from known theta functions, or modular forms, and yet closely mimicked them. "No one was talking about black holes back in the 1920s when Ramanujan first came up with mock modular forms, and yet, his work may unlock secrets about them," Ono says. Expansion of modular forms is one of the fundamental tools for computing the entropy of a modular black hole. Some black holes, however, are not modular, but the new formula based on Ramanujan's vision may allow physicists to compute their entropy as though they were. [352]

His Spiritual Belief: Ramanujan's first Indian biographers describe him as a rigorously orthodox Hindu. He credited his acumen to his family goddess, Namagiri Thayar (Goddess Mahalakshmi) of Namakkal. He looked to her for inspiration in his work and said he dreamed of blood drops that symbolized her consort, Narasimha. Later he had visions of scrolls of complex mathematical content unfolding before his eyes. He often said, "An equation for me has no meaning unless it expresses a thought of God." [353] Hardy, cited above, remembers Ramanujan remarking that all religions seemed equally true to him.

His Inspiration: Unbreakable equations

Illness: First thought to be tuberculosis, but later diagnosed in 1994, after another review of his medical records, as hepatic amoebiasis, a curable disease.

Died: April 26, 1920 in Kumbakonam, Madras Province at the age of thirty-two

Konstantin Tsiolkovsky

Born: September 17, 1857 at Izhevskoye, Ryazan Governorate, Russian Empire

Raised in: A log house on the outskirts of Kaluga, about 120 miles southwest of Moscow

Parents: Father: Makary Edward Erazm Ciołkowski, Mother: Mariya Ivanovna Yumasheva

Early years: Due to hearing problems connected to a bout with scarlet fever, he was not admitted to any school. He spent three years attending a Moscow library where he studied by himself. [354]

High School: As a teenager he began to speculate about space travel. At sixteen, he traveled to Moscow for three years studying chemistry, mathematics, astronomy, and mechanics. He learned to listen to lectures with the use of an early "ear trumpet." Upon learning that his son was going hungry and being overworked, the elder Tsiolkovsky called his son home to Vyatka in 1876.

College: Cut off from most central learning and teaching centers, the young man, now married and holding a teaching position at a local school did his own experimentation and research. He worked out a very complicated equation related to the kinetic theory of gasses but after sending it to the Russian Physico-Chemical Society in St. Petersburg he was informed it had already been solved twenty-five years earlier. That same Society, impressed by the young man's intellectual independence, invited him to become a member. [355]

Early Career: Despite his lack of early educational opportunities, he became a prominent Russian research scientist in aeronautics and astronautics who pioneered rocket and space research and the development and use of wind tunnels for aerodynamic studies. He was also among the first to work out the theoretical problems of rocket travel in space. [356]

Later Career: Without original funding, Tsiolkovsky built the first wind tunnel ever in Russia and studied the varying effects of wind on dirigibles and airplanes leading him to design advanced aircraft designs. The period of 1900 to 1915 proved to be the worst of his life including the suicide of his son in 1902 and a devastating flood in 1908 destroying much of his accumulated work over the previous years. The Academy of Sciences did not recognize the value of his aerodynamic experiments followed in 1914 by the Aeronautical Congress in St. Petersburg dismissing, through indifference, his models of an all-metal dirigible. In the final eighteen years of his life, Tsiolkovsky continued his research, with the support of the Soviet state, on a wide variety of scientific problems. His contributions on stratospheric exploration and interplanetary flight were particularly noteworthy and played a significant role in contemporary astronautics. On November 9, 1921, the council of the People's Commissars granted him a pension for life in recognition of his services in education and aviation. [357]

Most space launch vehicles trace their heritage to ballistic missiles developed for military use during the 1950s and early '60s. Those missiles in turn were based on the ideas first developed by Konstantin Tsiolkovsky in Russia, Robert Goddard in the United States, and Hermann Oberth in Germany. [358]

It is difficult to find alternatives to chemically fueled rocket propulsion for lifting mass out of Earth's gravity. One concept, originally advanced by Konstantin Tsiolkovsky in 1895, is a "space elevator"—an extremely strong cable extending from Earth's surface to the height of geostationary orbit or beyond. This concept, far-fetched as it may seem, has been the subject of serious preliminary research. [359]

His Spiritual Belief: Came from a Roman Catholic, Russian Orthodox family

His Inspiration: Space travel

Illness: Cancer

Died: September 19, 1935 in Kaluga, Russia at the age of seventy-eight

Let's summarize the above stories regarding these seven individuals who drew their first and last breath on this Earth and what they had in common.

I selected them based purely on the unique impact they had on civilization. Or more specifically, on how they came up with ideas that completely changed the course of civilization and the attention given them by the world, not just those with whom they interacted.

Not being initially aware of intimate details about these seven individuals, I was surprised at some of the commonalities:

- Almost all were born somewhere other than the US or had parents living in another country.
- Several died before their time due to various illnesses.
- Most did not profess a sincere belief in God other than a casual involvement in religion; if that. One, late in life, became evangelical.
- Several had dysfunctional families.
- Few came from wealthy means.
- All apparently possessed an almost savant understanding of education, especially math. I find it pivotal, based on these commonalities, that much of the Revelation of John at the end of the New Testament is saturated with numbers. Equally, both Hebrew and Greek language link their alphabets to numbers and the symbolism they represent. [360]
- All have been accepted by the world as highly contributing figures in history. All gained enormous notoriety around the world with little, if any, personal effort to do so.
- Several had ties or encounters with the country of India, a country with a large Buddhist following.

What do you take away from the seven stories above? That Steve Jobs was possessed by Satan after being caught up in Zen? Not hardly. How does one man, after slaughtering English citizens under the command of the man recognized by much of the world as pure evil, Adolph Hitler, become the head of the leading space agency, NASA, run by the United States? How do men, in some cases with very little formal education, become savants in mathematics, physics, and chemistry? What this may demonstrate is how one individual, a future antichrist figure, could dominate humankind in the last days, and I stress "could" because none of the above examples was the antichrist.

These unlikely individuals, or savants, could be held up as men who turned all society on its ear by doing and creating things that changed the world view for all creation. In some cases, creating the possibility of world annihilation by shifting leverage to a very few who could demand submission of all civilization. And, for the faith community, most all of these seven individuals were heavily connected to mathematics. Numbers play a large role in the Book of Revelations and Hebrew and Greek languages. Only one of the seven held fast to an abiding faith in Christianity. The others are held in high esteem even though they had very little attachment to the Christian faith, or in some cases, any faith at all.

What Does Gravity Have to Do with It?

Above I have summarized the life and contribution of seven individuals, all of which left civilization with something that did not exist before they were born. All of whom, one way or the other, were in the scientific field. Before leaving these seven, I want to blend in the story of one other individual who is considered a scientist, theologian, and mystic.

If you could ask most any true scientist what they felt to be the most important element of our universe, the likely answer would be gravity. If that were the result, it would follow that the most influential scientist related to gravity would have been, Sir Isaac Newton, considered the father of the laws of gravity. At the beginning of an online article in 2016, i24news.tv used the following quote near the beginning of their article: "Sir Isaac Newton is mostly known as the pre-eminent scientist who fathered the laws of gravity. But a collection of his manuscripts held by the Israeli National Library reveals a new facet to the 17th-century British scholar, one which is not only generally not known, but also somewhat surprising. Introducing Isaac Newton, the mystic."

This article reveals a side of Newton rarely published or discussed. Dr. Milka Levy Rubin, the curator of the Humanities collection at the National Library of Israel, told i24news, "He kept them secret because he was a devout Christian and he was concerned that if people would see the scripts, he would be considered a heretic." According to the article, Newton used parts of the Book of Daniel from the Christian Bible, and other Old Testament texts, to calculate that the end of the world will occur *no sooner than* 2060. Unlike other Doomsday prophets, Newton saw this date as a beginning of a new era rather than the apocalypse. This *new era* will be discussed later under the heading, New 21st Century Scientific Era—specifically, Newton's new era containing very new developments regarding… gravity.

The article summarizes Newton's Christian position: "Newton was ahead of his time even in his religious outlook by being close to the Evangelical church, which became popular only after his death and had a different vision to the predominant Anglican Church, whose traditional view at the time was that the Jews were doomed to wander the earth, symbolizing the victory of Christianity over Judaism." This prediction is even more fascinating because he also predicted "…the return of the Jews from exile to the land of Israel"—an event that took place more than 200 years after his death. And according to Levy-Rubin, the great scientist was 'actually one of the first Evangelists.'" [361]

New 21st Century Scientific Era

Back to the new era alluded to earlier by Sir Isaac Newton. We all know how fast technology is developing. Well, things even as important as gravity aren't escaping this rush to double knowledge every eighteen months or so. If you have yet to hear the word, LIGO, or Laser Interferometer Gravitational-Wave Observatory, let's get you up to speed... literally.

In simple terms, if that is possible, LIGO represents man's ability to, since the late 1990s monitor and measure incredibly massive objects colliding violently in space. These collisions send out ripples in space-time that reverberate through the cosmos for billions of years. Einstein first theorized these gravitational waves 100 years earlier. While several scientists continued to theorize about their existence, in 1990 the National Science Foundation approved the assembly of two LIGO detectors: one in Hanford, Washington and another in Livingston, Louisiana.

The detectors were so sensitive they had to be redesigned between 2010 and 2014 to eliminate such false readings as ravens pecking on the pipes. The redesign was successful and in September 2015, the observatory began picking up the signature of its first gravitational waves. On Feb. 11, 2016, the finding was made public, with physicists announcing that they had detected the collision of two black holes twenty-nine and thirty-six times more massive than the Sun, respectively, that occurred nearly 1.3 billion years ago.

In January 2020, LIGO detected a second neutron star smashup that involved colossal objects with a combined mass 3.4 times that of the Sun. Such weighty neutron stars have never before been seen in telescopes and push the size limit of what should *theoretically* be possible for such objects, leaving scientists to scratch their heads over how those stars could have been created.[362]

Adding to these less than decade-old findings are additional LIGO facilities being completed around the world in recent years, leaving little doubt that centuries-old attitudes, understandings, and beliefs will likely evolve. In addition to the two original sites, locations now include Italy, Germany, Japan, and India. [363]

Okay, hold up! What does this have to do with the end of time? Great discussion topic. Let's take a look at how this is affecting the science community.

Raymond Bergner, a doctor of Psychology at Illinois State University, wrote the following on June 3, 2017 under the title, "Intelligent Design: Maybe True, Maybe False, But Not Absurd":

While there is disagreement about its implications, there is little disagreement among physicists today that our universe is "fine-tuned" both for existing in its present form and for bringing about life forms. Various physical parameters, among them the value of the strong nuclear force, the charge of the electron, and the rate of expansion of the universe in the first second after the big bang, all have a wider range of theoretically possible values. However, only an extremely tiny fraction of these values, and these allowing for essentially zero deviation, allow for such things as the existence of atoms, the formation of stars, the clumping together of matter to form planets and galaxies, and ultimately the origination of life forms. This being the case, the scientific consensus is that our universe is an extraordinarily unlikely one. The realization of each of these values, taken alone, is extraordinarily improbable. The fact that so many of them conveniently have precisely the necessary value represents such an incomprehensible unlikelihood that, Stephen Hawking, himself, an avowed atheist and opponent of intelligent design, refers to our universe as "an apparent miracle."[364]

From the Wall Street Journal:

Here's the story: The same year Time *featured the now-famous headline, the astronomer Carl Sagan announced that there were two important criteria for a planet to support life: The right kind of star, and a planet the right distance from that star. Given the roughly octillion—1 followed by 27 zeros—planets in the universe, there should have been about septillion—1 followed by 24 zeros—planets capable of supporting life.*

There's more. The fine-tuning necessary for life to exist on a planet is nothing compared with the fine-tuning required for the universe to exist at all. For example, astrophysicists now know that the values of the four fundamental forces—gravity, the electromagnetic force, and the "strong" and "weak" nuclear forces—were determined less than one millionth of a second after the big bang. Alter any one value and the universe could not exist. For instance, if the ratio between the nuclear strong force and the electromagnetic force had been off by the tiniest fraction of the tiniest fraction—by even one part in 100,000,000,000,000,000—then no stars could have ever formed at all. Feel free to gulp.

Multiply that single parameter by all the other necessary conditions, and the odds against the universe existing are so heart-stoppingly astronomical that the notion that it all "just happened" defies common sense. It would be like tossing a coin and having it come up heads 10 quintillion times in a row. Really?

Fred Hoyle, the astronomer who coined the term "big bang" said that his atheism was "greatly shaken" at these developments. He later wrote that "a common-sense interpretation of the facts suggests that a super-intellect has monkeyed with the physics, as well as with chemistry and biology. The numbers one calculates from the facts seem to me so overwhelming as to put this conclusion almost beyond question."

Theoretical physicist Paul Davies has said that "the appearance of design is overwhelming" and Oxford professor Dr. John Lennox has said "the more we get to know about our universe, the more the hypothesis that there is a Creator… gains in credibility as the best explanation of why we are here." [365]

Any questions? If you have read this far, I hope you can see why I feel it is past time for science and faith to find a quiet room, sit at a table and with open minds discover what they have in common.

Global Communications and Growth of Technology

Further strengthening the possibility of one-man domination are 21st -century, worldwide, instant communications. When considering how one individual could rule the world, we can probably agree that it would not have been a possibility until the age of worldwide communications. You cannot control something with which you cannot communicate.

On a site called, The Emerging Future [366] you will find a set of data and charts that will, at a minimum, make you take a deep breath and possibly momentarily numb your brain. The site postulates that computers will double their capabilities every twelve to eighteen months, including the information technologies that use them. What does that mean? According to their chart that capability will increase thirty-two times over a short time span from today's systems. How does that affect you and I who have difficulty just figuring out changes to our laptops and phones with a simple program update?

More to the point, using The Emerging Future's (TEF's) charts, those increases become uncontrollable over ten, twenty, thirty, forty, and fifty years due to the human brains inability to keep up that pace. Here is a sample of this site's interpretation of these advancements.

Ten Years—So just how good is your ability to predict the changes that will occur when our capabilities become increased hundreds of times more each year? [367]

Twenty Years—Eighteen to twenty years out, technological advancements will be hundreds of thousands to a million times more advanced. That makes our first fourteen years of exponential growth seem flat-lined (no progress), when in fact, it will be 4,000 times more advanced than today. We currently have regenerative medicine in clinical trials and consumer wireless computer brain interfaces for $300. If we can do this today, what will it be like in twenty years, when technology is a million times better, and more importantly… faster? [368]

Thirty Years—This is the time period that "The Singularity" is supposed to occur. This means that bio, nano, robotic and computer technology will become so rapid, so advanced, and so profound that today's limited understanding does not allow us to describe, within reason, what life will be like. In the early 2040s, the rapid pace of improved changes will be hundreds of millions of times faster than today as each year passes. The only way humans will be able to keep up or even know about these changes will be with virtual assistants and computers inside us, and around us. Evolution will bring about artificial general intelligence (AGI). Without the fear of death, poverty, boredom, disease, pain, or even maintenance, preoccupation with self-interests may diminish or even dissolve. "Living" may become simply a spontaneous series of events. [369]

Forty Years—Forty years out, with technology a trillion times more advanced than today, we will be so far away from our current knowledge base that it is pure guesswork as to what will be going on. [370]

Fifty Years—Fifty years out, technology that is a quadrillion-times more advanced than today will give us the ability to *perceive* a quadrillion. There will be an explosion of highly intelligent biological, non-biological, micro, nano, virtual, mixed, and morphing life forms colonizing the solar system and beyond. Life spans will develop into life continuums. [371]

With this kind of data projections, it may be that the eventual ruler of the entire Earth isn't a human at all, but rather some form of technology, or whatever it is called in 2070. Considering that possibility adds more emphasis to the number 666 mentioned in the Book of Revelation.

With this exponential leap in technology, the earlier look at eight specific individuals should come into focus. With this monolithic leap in technology only an infinitesimal number of individuals would have the ability to understand and control its power.

Control of Currency and Living Conditions

The other item needed for one person to rule is control; control of monetary means, in whatever form that takes and, control over living conditions. The current move to the "cloud" of everything we do, and the coming exponential increase of technology, could clearly place both of these conditions under the control of some central entity… or person! As of this date in June 2021, four single corporations—Face Book, Twitter, Google, and Amazon—each headed by a single individual (Google is co-owned), were making corporate decisions affecting the lives of hundreds of millions of people and trillions of dollars. This control and these dollars were and are wildly affecting the world economy. Many small countries' entire economies aren't as large as these five CEOs individually:

Face Book—Mark Zuckerberg—— $115,700,000,000 USD [372]

Twitter—Jack Dorsey—$13,000,000,000 USD [373]

Google—Larry Page and Sergey Brin, co-owners—$100,000,000,000 USD Plus [374]

Amazon—Jeff Bezos $191,000,000,000 USD [375]

Other individuals include Elon Musk—$166,000,000,000 USD and Bill Gates—$129,700,000,000 USD. [376]

Let's look at some other statistics supporting the notion of a single individual being recognized as a world leader. The following individuals, worldwide, are being followed on leading social media platforms by an astounding number of human beings: [377] [378] [379] [380] [381]

Cristiano Ronaldo—Portuguese—480,920,000—Soccer (football) (All Social Media Platforms Combined)

Ariana Grande—USA—301,800,000—Singer (Instagram and Twitter only)

Lionel Messi—Barcelona—276,940,000—Soccer (football) (All Social Media Platforms Combined)

Neymar—Brazilian—260,660,000—Soccer (football) (All Social Media Platforms Combined)

Dwayne Johnson (The Rock) —USA—216,000,000—Movie Star (Instagram Only)

Labron James—USA—154,800,000—NBA Basketball (All Social Media Platforms Combined)

Barack Obama—Past US President—129,200,000—Politics (Twitter only)

David Beckham—ex-English—120,270,000—Soccer (football) (All Social Media Platforms Combined)

Shakira—US—109,000,000—Singer (Facebook Only)

Will Smith—US—104,000,000—Movie Star (Facebook Only)

Vin Diesel—US—103,000,000—Movie Star (Facebook Only)

Donald Trump—Past US President—88,783,411—Politics (Twitter Only, 1-8-2021)

Odell Beckham, Jr. —US—14,300,000—NFL Football (All Social Media Platforms Combined)

Tom Brady—US—9,500,000—NFL Football (All Social Media Platforms Combined)

For comparison:

Bob G. Shupe—US—562—Business Owner, author, good guy (Facebook Only)

Having researched and mentioned all of the above very public, and in most cases, well-known individuals, I must confess I don't personally think the individual will come from any of those sources. I also don't believe they will be a well-known politician or religious figure, as we define those vocations. What I believe more likely is that a figure will surface proposing a blend of faith and science. Yes, you read that correctly... faith and science blended together. But Bob, isn't that what you are trying to accomplish with this book? Isn't your goal to get these two groups together? Yes and... no.

The goal of this book is for both faith and science to understand and recognize each other for what they do and what each represents. I am not pushing for a Christian scientist, but rather scientists who are or become Christians. Anything other than those results will only produce an Antichrist candidate. Do such Antichrist candidates exist today? Yes. Consider if you will, Li Hongzhi. [382] Just to be clear, I did not say Li Hongzhi is the Antichrist. Just as the other eight individuals mentioned earlier, Li Hongzhi has the ability to attract worldwide attention.

There have been other similar human figures through history, but they are either dead, in jail, or being detained in an institution. While aggregating these would-be leaders, it appears they have learned from each other's failures. It also seems that these characters become more and more international.

Falun Gong

Li Hongzhi, mentioned above, is very much alive and last reports have him living in Dragon Springs, a high-security compound in upstate New York near the small town of Deer Park. Before giving a quick rundown of this compound, let's take a look at the practice, or belief in, Falun Gong.

Falun Gong, in Chinese called "Discipline of the Dharma Wheel," is also known as Falun Gong and Falun Dafa. One term is the belief and the other is the teaching of that belief. The two terms however are often intermingled. Falun Gong is an offshoot of *qigong* or, "Discipline of the Vital Breath." And you thought I had already used every possible version of the book title.

The beginnings of this tradition developed in the early 1950s by members of the Chinese medical establishment as a means of promoting traditional Chinese medicine in modern socialist China. With great apprehension the Chinese Communist Party (CCP) eventually allowed the practice, placing it alongside the science of biomedicine. After being banned between 1966-1976, it was accepted again in the late 1970s with a purported lab discovery of *gigong* energy associating the practice with scientific and economic development. Overnight the practice became "scientific."

Over the next twenty-year period, the 1980s and 1990s, hundreds of millions of Chinese were drawn to a wide variety of gigong groups. Certain individuals prospered monetarily while other government and military figures saw an opportunity for this to become a *Chinese Science* that would expand human potential and propel China to the forefront of the developed world. Most advocates of qigong however, regarded it as a self-cultivation practice with a scientific basis and not as a religion, which was heavily regulated in China. It was in 1992, after the practice had again fallen into disfavor with China's political authorities, Li Hongzhi founded Falun Gong in an effort to restore the flagging fortunes of the dying movement. Estimates of his followers varied greatly, ranging from 2 to 60 million.

Li's efforts fell on hard times in 1995, causing him to move his operations to the United States. [383] Back to Deer Park, upstate New York.

It will not surprise you that Li's compound, Dragon Springs, is listed as a tax-exempt religious organization. The compound is located on a heavily secured 400-acre plot. According to sources who spoke to NBC, internet access is restricted inside the compound, relationships are arranged, and conventional medicines are frowned upon. Falun Gong's founder, Li Hongzhi, lives there, although he has not been seen for years. Surrounded by forests and guarded by armed security, Dragon Springs has been declared a refuge for persecuted Falun Gong practitioners. Few outsiders are permitted inside. [384]

Several of Li's followers have fled China to other countries seeking refuge from spiritual persecution. One of the reports outlining in detail, with references, was published by the United Kingdom. It is easy to read and very informative. Please check the reference section of this book for access. [385]

Dragon Springs, fighting the reluctance of neighboring communities, has grand plans to expand the facility and the programs and offerings they develop, including living space to accommodate 500 instead of the current 100 people. Falun Gong is a system that combines Buddhism, mysticism, and exercise, but it also touches on aliens and ethnic separation.

The compound houses an academy of arts which acts as a "feeder school" to Shen Yun, which is a type of theatrical dance performance. Or, according to the New Yorker's Jia Tolentino, it's "essentially religious-political propaganda." Shen Yun is a non-profit organization, and in 2016 had about $75 million worth of assets and brought in $22 million in revenue. Chinese political experts view Shen Yun as part of Falun Gong's elaborate and well-put-together public relations plan. The act has anti-Communist messages and China's ruling party sees it as a propaganda tool meant to subvert their authority.

Li teaches that practicing Falun Dafa will transform bodies into high-energy matter, and even assists some people to fly to heaven. Li also believes in aliens. In a *Time* interview in 1999, he said, "If aliens are not to replace human beings, society will destroy itself on its own." That same year he told the *Washington Post* he was not some sinister mastermind, but more of "an accidental prophet." One of his teachings was that people should never get involved in politics. "We have never interfered in government and never done anything wrong. We are all law-abiding citizens," Li said. Despite its stated political stance, a consortium of Falun Gong members owns the *Epoch Times*, a multi-lingual newspaper Li has referred to as "our media."

"The drinking water is polluted. No matter how we try to purify it, it cannot return to its original purity," Li said in a *Time* interview. Despite his focus on the environment, locals are concerned Dragon Springs ruined the neighboring land when in 2019, Dragon Springs requested further expansion to build a 100,000-gallon-per-day wastewater plant, which neighbors fear will ruin surrounding waterways. Once the water is treated it will be discharged into local rivers, like Basher Kill and Neversink River. [386]

What I just described is a human who has attracted tens of millions of followers from around the world. He did it using a lifestyle that is faith/science/religion or basically whatever you want to believe in, as long as you devote your entire self and all you have to one individual. This *thing*, whatever it is, belongs in the Steve Jobs category of *self-realization*. This is contrary to what both true faith and science stand for; a God Who loves those who have faith in Him, and a science that is constantly in search for new discovery, always looking away from themselves.

With these thoughts and facts, I believe that if not now, at least in a few Earth years, it will be possible for one human, or one device to rule the world.

CHAPTER 24

A SEASON UNLIKE ANY OTHER
(CONTINUED)

... now, the world coming to an end; I can see that.

Climate Change, What Does Man Have to Do With It?

THE SECOND THING that has to occur at the End of Days is worldwide disaster. I hesitate to comment on this subject because it is, and has been, more political and militant on both sides of the debate than science and faith. However, based on the world ending as some in both the science and faith communities contend, this issue must at least be broached. I will attempt to comment on both sides. My position is not whether climate change is real or who or what is causing it, my reasoning is that something is going to end this planet. These are just other possibilities.

As I stated earlier, the disasters listed in Scripture are not unique events and there are secular records of them occurring throughout humanities known history. Recorded events are not just in the Bible but also documented in secular history. Biblical events occur within the 6,000 to 10,000 years some in the faith community believe to be the age of the Earth. Scientific records are limited only by mankind's discovery of writing and archeological evidence. The difference between these historical and biblical events may be that these future events will be happening all over the Earth, and all at the same time.

If that is the intent, then yes, that could be a unique event for modern humankind. However, since most, if not all of these disasters may be acts of nature, can we assume they will only be the result of man's inhospitality toward our alien occupation of the Earth? And then there is the nuclear factor, allowing several countries to initiate global destruction. This cannot be ruled out based on destructive elements described in Scripture. Those who wrote those Biblical books had no point of reference to describe such an event, or to assume it was humanly initiated, even if God revealed the scenes to them.

As an aside, an interesting challenge for the science community would be to explain how a human, over 2,000 years ago, could come up with these concepts, including an extensive knowledge of disease and destruction. Science *and* the faith community might also consider and explain how the writer of Genesis, although not knowing what to call it, understood the concept of Pangaea and plate tectonics; since this ancient writer speaks of all land being in one place and it is then later separated into land masses separated by water. Sorry for the detour. Consider it homework.

Let's begin our wade through the weeds with a few definitions of *climate change*:

Merriam-Webster— *significant and long-lasting change in the Earth's climate and weather patterns.* [387]

Cambridge—*changes in the world's weather, in particular the fact that it is believed to be getting warmer as a result of human activity increasing the level of carbon dioxide in the atmosphere.* (And on down the page)—*changes in the world's weather, particularly an increase in temperature, thought to be caused by things such as carbon dioxide in the atmosphere.* [388]

Lexico—*a change in global or regional climate patterns, in particular a change apparent from the mid- to late- 20th century onwards and attributed largely to the increased levels of atmospheric carbon dioxide produced by the use of fossil fuels.* [389]

Collins—*changes in the Earth's climate, especially the gradual rise in temperature caused by high levels of carbon dioxide and other gases.* [390]

We could go on, but those quoted above are several great resources. Regardless of the given definition, early on each of these sites indicates that humans are the central cause of these current changes. The common denominator in all definitions includes the term *carbon dioxide,* except Merriam. Yes, Merriam gets around to it, but I find their initial pure definition a much better place to begin. Why? They do not include a central blame as if nothing else could be a contributor. In addition, climate change is controlled, not manipulated, by much more than temperature.

It could be that temperature fluctuation is the result of a chain of events, not a reaction to a sole instigator. If someone has an agenda, it is convenient to ignore the entire situation and blame one thing, as if fixing a simple singular issue will correct a complex problem. This is especially true when the single culprit contains a monetary and therefor economic motive. Here are a few questions for an open-minded discussion:

- Has this happened before in the history of the universe?
- Are there records showing evidence of it happening and what caused it?
- What other levers beside carbon dioxide affect our weather patterns, and does humanity have the ability to, in any way, alter those additional components?
- Based on recorded research, are the current levels of CO_2 vastly greater and more concentrated in recent history than across the rest of our developed historic research?

Is This the First Time the Earth has Seen Drastic Changes in Temperatures, Hot or Cold?

First let's address the common denominator currently at the center of this topic. The world's countries emit vastly different amounts of heat-trapping gases into the atmosphere as demonstrated in the graphs and charts contained in the referenced site. This data was compiled by the International Energy Agency, which estimates carbon dioxide (CO_2) emissions from the combustion of coal, natural gas, oil, and other fuels, including industrial and non-renewable municipal waste.

The site ranks the top twenty highest emitters of annual carbon dioxide in 2018 (their most recent available data for this report). The chart from which these numbers is drawn was originally designed on July 16, 2008 and updated on August 12, 2020.

The Union of Concerned Scientists USA (UCSUSA) chart shows most of the countries on Earth and what percent of CO_2 each contributes to the atmosphere. By volume, in this report, the US contributes 15 percent. Countries over which the US has little, if any control—China, India and other Asian countries—collectively contribute 46 percent. In addition, the category they list as "rest of the world" is 21 percent making a total of 67 percent. The page closes with the following comment: [391]

These uneven contributions to the climate crisis are at the core of the challenges the world community faces in finding effective and equitable solutions to global warming.[392]

As an update to the 2018 report above the following figures became available during the editing of this book. A report published on May 6, 2021 by New York-based research firm Rhodium Group found:

China alone was responsible for over 27 percent of total global emissions, with the United States being a distant second at 11 percent. India was ranked third with 6.6 percent, followed by the 27-country bloc of the European Union at 6.4 percent.

China's emission reached nearly 14.1 gigatons of carbon dioxide equivalents in 2019, more than triple 1990 levels and a 25 percent increase over the past decade, according to the report. Global emissions in 2019 topped 52 gigatons. [393]

This was a fair assessment of the CO_2 issue. Now let's look at the rest of the story and how relevant it is today, if that is the case. The following data is from a site labeled and referenced as owlcation.com. Below is their "about us" statement on the site:

Who We Are: Welcome to Owlcation! We are a site created by educators and experts on topics related to education as a place to share expertise and knowledge about all things academic. We offer articles ranging from the ten most important moments in the French Revolution to the different species of owls in Florida. If you're looking for information that is education or academia related, we are here for you!

Our Purpose and Goal: Our goal at Owlcation is to be the best place online to find high-quality, informative-style content. We are also a platform for enthusiasts and aficionados to create original, useful, in-depth articles about topics they know and love. In most cases, we pair our authors with professional editors to create well-written, media-rich, and informative pieces.

Our Team: We are a diverse group of engineers, product and community advocates, moderators, and editors that are passionate about writing and online know-how. In addition to our official team, we are a tight-knit community of thousands of writers and enthusiasts.

The site lists an approximate temperature range 420,000 years ago based on replicable scientific studies using deep-sea sediment cores in Antarctica substantiating changes in temperature going back 450,000 years largely conformed to changes in the Earth's orbit.

Low: 41°F

High: 63°F

Range: 22°F

Atmospheric CO_2 less than 33 ppmv

Occurrences in the last 10,000 years:

- Hunter-gathers became farmers
- Plant and animal species were domesticated
- Civilizations were developed
- Global mean temperatures probably did not vary by more than 2°F in any 100-year period

They state that the surface temperature of the Earth tends to go up rapidly and then settle down again in cycles of roughly 100,000 years as shown in UN Environment Program (UNEP) charts. These charts trace the temperature differences over the last 420,000 years compared to the present time, defined as the year 1950.

In 1950, according to NASA's Goddard Institute for Space Studies [394], the mean surface temperature of the Earth was 57° Fahrenheit. This would mean the Earth's absolute temperature (as opposed to its temperature change) during the last 420,000 years varied from a low of about 41°F to a high of about 63°F, a range of approximately 22°F.

Although this range is no more than what most of us experience in the course of a year's time going from summer to winter, UNEP refers to these temperature changes as "very significant" and to the Earth's climate during most of these years as "unstable." Toward the bottom of the range the temperature was cold enough for glaciers to increase in size and at the top was warm enough for glaciers to be reduced by melting. The colder years are referred to as periods of glaciation and the warmer years as periods of interglaciation.

Enter the Milankovitch Theory of Earth's Temperature Changes. [395] Milutin Milankovitch, born May 28, 1879 and passing away December 12, 1958 in Austria-Hungary, was a Serbian mathematician and geophysicist best known for his work that linked long-term changes in climate to astronomical factors affecting the amount of solar energy received at Earth's surface. [396] Milankovitch, like other scientists before him, observed that the Earth's orbit around the Sun was irregular in three particular aspects: Eccentricity, Obliquity, and Precession. His theory was largely ignored until 1976, a little less than twenty years after his death.

The elements of his theory are said to have accounted for respectively, 50 percent, 25 percent and 10 percent of temperature change. Milankovitch's theory is now accepted as the best explanation of climate change "on time scales of tens of thousands of years." And the theory suggests that it is about time for Earth to begin a new long-term cooling cycle.

The summary of the site's analysis is as follows: [397]

> *Now, however, the recent surge in greenhouse gas emissions has trumped the Milankovitch Cycles. The number one cause of the current upward trend in Earth's temperature is the increase in atmospheric carbon dioxide, not an increase in sunlight. And the increase in carbon dioxide is caused primarily by human activity, notably the burning of fossil fuel (oil, natural gas and coal), not by the temperature increase— although in another positive feedback mechanism*

the temperature increase may help to increase the amount of carbon dioxide in the atmosphere.

That is as far as I want to wade into this political debate. I do encourage you to study these and other references before getting trapped on one side or the other of this discussion. What can we learn from this disagreement applicable to discussions between science and faith? I think plenty.

If you read any dialogue or information related to Satan you will clearly see it is always about Him or whatever "it" is. Satan has a lot of control over a human who does not trust and depend on faith. The only thing Satan has to fight God is its mouth or wherever the words come from. The two most-leveraged words Satan places in a human-to-human discourse are *always* and *never.* I find myself using these words in hurtful discussions with my wife, and to be fair, her words to me as well. It is only when one of us realizes one or both of these words dominate our discourse do we begin to work through the issue with mutual understanding. As an example, I may be told that, "you never make the bed" or I tell her, "you always want to watch the Hallmark Channel." Sorry Hallmark folks but some men don't like to be caught crying.

The truth is I do occasionally make the bed, and my wife has been known to watch a TV program where things blow up. Bottom line, most of us are bad about not blending in both sides of the story. That is the case with climate change and the discussion, or lack of, between science and faith.

What if someone stated that all of the glacier's melting is a result of a combination of historical events and natural causes. Admitting that if we do fix the current global pollution issue, over which the USA has only a 11 percent ability to contribute, that the Earth will still continue to warm in long-term cycles as it has for hundreds of thousands of years. As opposed to, we need to stop mining coal and building gas burning motors if mankind is going to save the Earth.

Personally, the last part of that analogy bothers me most. To me this phrase is right out of Satan's playbook. You see, humans cannot save the Earth. We cannot fix it. Nature heals itself. If we decide to not shoot someone in the head, did we save their life? Regarding climate we need to put the gun away, but realize it isn't just one gun. Removing one gun from the head of an individual surrounded by a gun-toting mob does not end the possibility of one of the other several participants pulling the trigger.

Since I am writing this book as a citizen of the United States of America it seems appropriate to comment on our country's role in the end-of-days dialogue. This will not take long. There is nothing in Scripture that speaks of a country like ours. Personally, I don't think this country will be around whenever this end-of-days thing occurs.

To the point of this book, what if science stopped demanding that nothing about the faith community's beliefs could be true, since by the science community's own admission they cannot disprove it? What if the faith community stopped denying replicable data that doesn't seem to support their understanding of the history of the universe and mankind? What if both sides stopped using the words never and always?

What if... based on faith's belief in an all-powerful, almighty, all-knowing God, and science understanding of the vastness and enormity of the universe, both consider that this universe and literally everything in it is not all there is to consider? Even if what we know is all there is, both have to admit there are still things of which we are not aware, much less understand, contained in what we think we know. Who in the faith or science community is able to proclaim there isn't anything else outside what we know and have experienced? Both the faith and science community, maybe without realizing, agree on this matter in the following way. Faith expresses there could be more since God can create anything, anytime. Science says the only reason this is all there is, is because they have yet to discover and prove there is more. As an old Staples office supply ad campaign used to proclaim with a battery powered button that, when pushed, activated a voice proclaiming, "That was easy!"

Disease, not a Meteor, Ends Civilization?

Even if disease doesn't end all life on Earth, it can certainly disrupt it and create a sense that one person has the knowledge and resources to save a subservient sector of the population. There have been multiple diseases over time across the globe that have come close, and in some cases annihilated entire civilizations.

According to a History.com editorial, the first suspected outbreaks began at the end of the hunter-gather days and the beginning of the agrarian life some 10,000 years ago. [398] Not having names for the diseases that resulted in death in ancient times makes them no less devastating. They were likely early versions of malaria, tuberculosis, influenza, smallpox, and others.

This same article gives a timeline of several major events:

430 BC—Athens—a probable typhoid fever pandemic passed through Libya, Ethiopia, and Egypt crossing the Athenian walls. As much as two-thirds of the population died.

165 AD—Antonine—possible early appearance of smallpox beginning with the Huns and then spreading to the Germans who passed it on to the Roman Empire. Ending in about 180 AD it took the life of the then Roman Emperor, Marcus Aurelius.

259 AD—Cyprian—a plague causing city-dwellers to flee to the country to escape the disease which led to its spread. Possibly starting in Ethiopia, it spread through Northern Africa, then Rome, followed by Egypt and then northward eventually reaching Britain in 444 AD.

541 AD—Justinian—first appearing in Egypt, then Palestine, and the Byzantine Empire it eventually reached much of the Mediterranean area. Oddly enough the disease is credited with creating an apocalyptic atmosphere spurring the rapid spread of Christianity. Believed to be the first significant appearance of the bubonic plague spread by rats and fleas, it killed about fifty-million people, or 26 percent of the world population.

11th Century—Leprosy—this disease grew into a pandemic in Europe in the Middle-Ages. The major resource for controlling the disease was isolation of the victim. It was believed to be a punishment of God leading to moral judgment and ostracization. It is today known as Hansen's Disease.

1350 AD—Black Death—responsible for the death of one-third of the world population. This was a second outbreak of the Bubonic Plague. Possibly starting in Asia and then spreading by caravans to Sicily, ultimately entering Europe where it spread rapidly. It actually caused France and England to call a truce to their war. It also halted the Vikings exploration of North America.

1492 AD—Columbian Exchange—with the arrival of the Spanish in the Caribbean smallpox, measles, bubonic plague and others were passed to native populations devastating the islands with as many as ninety percent dying throughout the area. Christopher Columbus arrived on the island of Hispaniola with a population of around 60,000. By 1548 there were about 500 survivors. In 1520 the Aztec Empire was destroyed by a smallpox infection.

Research in 2019 even concluded that the deaths of some 56 million Native Americans in the 16th and 17th centuries, largely through disease, may have altered Earth's climate as vegetation growth on previously tilled land drew more CO_2 from the atmosphere and caused a cooling event. [399]

1665 AD—Plague of London—the Bubonic Plague led to the death of 20 percent of the London population. The worst of the plague tapered off in 1666 about the time of the Great Fire of London.

1817 AD—1st Cholera Pandemic—This was the first of seven Cholera pandemics that would occur over the next 150 years. It originated in Russia killing over one million people. It was passed to British soldiers who brought it to India killing millions more. The disease eventually spread to Spain, Africa, Indonesia, China, Japan, Italy, Germany, and America.

1855 AD—Third Plague Pandemic—Starting in China and moving to India and Hong Kong, the bubonic plague claimed 15 million people. It was considered active until 1960 when cases dropped below a couple of hundred.

1875 AD—Fiji Measles Pandemic—Visiting Australia at the invitation of Queen Victoria a royal party was infected with the measles. Upon their return to Fiji the royal party and the police who greeted them further spread the disease to the island population. One-third of the island population perished, a total of approximately 40,000.

1889 AD—Russian Flu—Starting in Siberia and Kazakhstan, followed by Moscow and then to Finland and Poland, ending with a spread into Europe at-large and finally crossing the ocean to North America and Africa; 360,000 died of this first significant flu outbreak.

1918 AD—Spanish Flu—The avian-borne flu resulted in 50 million deaths worldwide. Starting in Europe, then the United States and parts of Asia, and then around the world.

1957 AD—Asian Flu—Starting in Hong Kong, then China, followed by the United States and England, eventually worldwide. Caused an estimated 1.1 million deaths globally.

1981 AD—HIV/AIDS—Destroying a person's immune system, the disease causes death from infection; 35 million people have died from the disease since inception.

2003 AD—SARS—Severe Acute Respiratory Syndrome possible started with bats, spread to cats and then humans in China followed by twenty-six other countries world-wide. Although not as devastating a death toll as other pandemics, SARS was a wake-up call to improve outbreak responses used to fight H1N1, Ebola, and Zika.

2019 AD—COVID-19—The disease moved through 114 countries in three months infecting 118,000 people. Without a vaccine, the virus spread to nearly every country in the world. By December 2020 it had infected more than 75 million people and at the time caused the deaths of 1.6 million people worldwide. At the time, more than 500,000 new cases were being reported each day.

Through May 2021, there were approximately 156,077,747 cases reported worldwide with 3,256,034 deaths. [400] Several vaccines were developed around the world, three in particular which were rushed through using FDA Emergency Use Authorization language. These vaccines, along with proper hygiene, social distancing, and masks helped significantly slow the spread of the disease after almost a year and a half. By September 2021 all three vaccines were completely FDA and CDC approved for those who had received shots before and after government approval. However, at that time, tens of millions of qualified individuals, for various reasons, refused to have the vaccination shots, causing a spike in infection from emerging variants.

The COVID-19 pandemic disrupted the life of almost everyone in the US and around the world. All sectors of the economy, education, health, and many others were significantly impacted and even disrupted. Having lived through an ordeal that is still not over, and very much fresh on the minds of everyone, demonstrates just how fragile life is on this Earth and just how little it takes to swing dependence on one, or a few, individuals. If those are the wrong individuals, or individual, it is clear how an end-of-days scenario could occur. Add to a pandemic a global natural disaster(s), taking over the control of the "cloud" or the internet, dismantling the infrastructure, placing faith in personalities—not leaders, global nuclear war, and billions of dollars in cash in the hands of one individual makes the End of Days, in the mind of the faith or science community… a reality.

All this end-of-days orchestra needs is a conductor. Someone… some being… something sitting in the dark shadow of all history. That would be the third factor in determining all of the ingredients for the faith community's version of the End of Days. Anyone want to meet Satan?

CHAPTER 25
A SEASON UNLIKE ANY OTHER
(CONTINUED)

... Satan? Only a fool would believe that

What is Satan's role in all of this?

AND THE THIRD aspect of the End of Days...

Regardless of how it occurs, the faith community believes something is going to happen and some part of that event will involve Satan, the Deceiver. The more important question may be, what is the easiest target for Satan to control? Let's explore that briefly, remembering that original sin dealt with Satan deceiving Eve and Adam regarding their prohibition against eating from the Tree of Knowledge; at a minimum an interesting correlation.

Humans are the obvious target for Satan as he plans his end-of-days celebration. The eight scientists we examined earlier had the ability to understand immensely complex concepts, sometimes at an extremely young age. Some of their thoughts and inventions were capable of changing humankind forever. Sometimes those changes were for good and sometimes they provided devices that could end the world. Nevertheless, that knowledge did not come from nature's norms. These sources could more easily be manipulated by an evil force looking for weaknesses. A couple of examples could be an overactive yearning for self-realization and reliance on a belief having no recognized God. An example is Buddhism, which recognizes anything one wishes their god to be. [401]

Have you ever heard the phrase, "Satan is alive and well in the world today?" How about, "That person is just pure evil?" Do you believe the first phrase? Regarding the second phrase, do you think it possible that an individual could be the pure personification of Satan Himself?

Do you remember these words and Who spoke them: "Get behind Me Satan"? How about this one? "And this she did for many days. But Paul, greatly annoyed, turned and said to the spirit, 'I command you in the name of Jesus Christ to come out of her.' And he came out that very hour." And one more, "Be sober, be vigilant; because your adversary the devil walks about like a roaring lion, seeking whom he may devour."

You can find all of the above, beginning with "Do you remember these words…" in Scripture with a simple Google search. The faith community talks a lot about the devil, Satan, the evil one, Lucifer; but sometimes I feel like these beliefs are compared to a fairytale or a Dungeons and Dragons game. Most often when I hear these references I am not struck with fear or a feeling of awe and power. You see, Satan is not a wimp. Nor is he stupid; he is just not… God. Satan rules the lives of billions of humans, and has for centuries, since the beginning of time. According to Scripture, he will drag many of these souls into an eternity of separation from God and all that is good. My point? Satan, like God, Christ, the Holy Spirit, and legions of angels who have been on and in this Earth are… aliens; meaning not of this world.

It goes without saying that the science community will have no comment concerning Satan. As a practical matter, if God cannot be explained, the idea of an evil Satan is unthinkable. What science needs to possibly think through is, what part of humankind's natural development created evil? On the other hand, the faith community might not be any more successful in explaining Satan's ultimate role in this entire ordeal. After all, it is possible to believe in God and not Satan or Hell. In fact, many Christians take this very position based on a Barna Group reference later in this section.

In a world where there is a God and not a Satan, how do we explain

Scripture's insistence within its pages of the fallen angel, especially during the last days? It would appear the entire purpose of the End of Days is the separation of believers and non-believers. Since there is very little chance of there being a single human present on Judgment Day who is perfect, there must be a system involving a decision to follow good or evil and, the end result based of that decision. There are some who have studied this proposed End-of-Times situation who have developed some very creative descriptions of Satan's activities.

One of many internet sites containing End-of-Times interpretation in preachitteachit.org hosted by its founder, Dr. Roger Barrier. He is proclaimed to be a well-known author and sought-after conference speaker. His site says he has taught thousands of pastors in thirty-two countries. [402] I am not qualified to refute or prove his teachings. That isn't my point. Rather than an opinion, I have a question.

As Jesus Christ was leaving those on the mountain during the ascension and returning to His Father in Heaven, He clearly said no one, not even He, knew when His return would take place. If the time of His returning is of such insignificance, why is it so important for us, in the Book of Revelation, to speculate about a description of various beasts, candles, and horns? It would seem the most important aspect of all Scripture points to a decision—believe or don't believe.

I often tell folks that I know when the end of time is going to take place. Try telling that to someone, or a group, and after a dramatic pause measure the reaction. Actually, I think my theory is as valid as any other. Look closely at the question, "When will the end of time occur?" What is the most important word in the question? At first glance you might say the word, "when" because that is what you are trying to determine. But what has to happen before the event has any meaning? Time must stop. What time? Earth time, there is no other… that we understand. When we say that Jupiter rotates on its axis every 9.9 hours—whose hours? It is automatically translated into our Earth hours. No one on Jupiter cares. Sorry, that was a little rude.

Forgive my stating this again—for the third time—as I have earlier in this book but, I believe it bears repeating. My simple explanation is that the end of time will be when I exhale my last breath. At that moment time stops for me. Whatever happens after that will not be measured by a watch; no matter how accurate it may be. Earth time is allusive. Once I die, once you die, we will all be at *the* Judgment. This concept is why it was just as important in the 1st century as it is in the 21st century to have made a decision, based on the teachings of Jesus Christ. Neither you nor I have until civilization's end of time to decide. So there always is and always has been a sense of urgency.

Before dismissing this possibility of a single person dominating the world, we should ask when Satan stopped possessing people? Is evil possession only true if it is recorded in the Bible? Are we in the 21st century immune to the power of Satan? Maybe so when many believers of Christ in the world no longer believe or teach about the existence of a Satan. Consider the following results of a survey taken in 2009 by the Barna Group, a nationally recognized research organization:

> *Four out of ten Christians (40 percent) strongly agreed that Satan "is not a living being but is a symbol of evil." An additional two out of ten Christians (19 percent) said they "agree somewhat" with that perspective. A minority of Christians indicated that they believe Satan is real by disagreeing with the statement: one-quarter (26 percent) disagreed strongly and about one-tenth (9 percent) disagreed somewhat. The remaining 8 percent were not sure what they believe about the existence of Satan.* [403]

Back in Chapter 23, I first posed the following three questions concerning the end of everything according to the faith community:

- Is it possible for one human to gain access to, and have dominion over the entire world?
- What will cause all of the predicted biblical disasters?
- What will be Satan's role in all of this?

Based on my research I believe all three are plausible. I have attempted to address each of these faith assertions and introduce them in a logical format. When summing up the science community view of what they may call mass extinction, the end will come sooner or later either by the hand of man… or by nature. The big differences are in relation to what happens after the end. For faith it is some form of eternal heaven or hell. For science it is some other existence but, with a whole new cast of characters and beings. The difference between these two endings is that one will be in another existence with everything currently known disappearing. The other ends with everything that is known occupying the same space and becoming something else within the same place.

So, what is missing in each? The story of faith is missing replicable details while science fails to recognize a higher power. Can faith find replicable details? Can science discover a higher power? Yes, they can. And where can they find these missing links?… Perhaps by talking to each other, not at and about each other. Sounds like an excellent subject for a book…

CHAPTER 26
IN CLOSING

Something Personal to My Believer Companions

LIKE SOME OF you, I grew up in a faith that had no room for what at the time was considered light-hearted comments about Scripture. This comes from a now grown-up kid who, when much younger I and a few friends enjoyed sitting on the back pew singing, "I was sinking deep in sin… wheeeeeey." Yes, there were consequences. I came to understand, as I grew older, that the position of some of my elders wasn't so much a position of truth as it was their interpretation of reverence and respect for the Bible. While that approach is admirable, my memory of their statements seemed to instill in me their religiosity, not a respect for what God was trying to tell me about faith through His Holy word and, more importantly through His actions… what He did.

In Malachi 3:10 KJV He says to "test" Him, "*Bring the whole tithe into the storehouse, that there may be food in my house. Test me in this," says the Lord Almighty, "and see if I will not throw open the floodgates of heaven and pour out so much blessing that there will not be room enough to store it.*"With this permission from God… *to test Him*, and the imagination He breathed into humankind at Creation, I want to leave you with these thoughts and a challenge, opening your mind to possible responses you may be asked when discussing your faith with an unbeliever. My apologies to the well-meaning deacons in my childhood church sixty-five or so years ago.

An article on abc.net.au titled "Are Religion and Science always at Odds? Here are Three Scientists that Don't Think So" by Anna Salleh, begins with this opening salvo. The father of the Big Bang Theory was actually a Catholic Priest. The pioneer of modern genetics was an Augustinian monk, and the decoder of the human genome, Dr. Francis Collins, converted from atheism to Christianity in his late twenties has been the Director of the National Institutes of Health (NIH) since 2009.

One of the scientists interviewed, Dr. Jennifer Wiseman, an astrophysicist, is quoted as saying, "Those of us who study astronomy can tell you the numbers and all of that, but comprehending it is something totally different." Dr. Wiseman is credited with the discovery of the comet Wiseman-Skiff in 1987. She also stated, "You have to look at biblical literature from the perspective of when it was written, the original audiences, the original languages, the original purposes... the message that was meant to be conveyed by it," she says. "The Bible's not a science text." [404]

Time Magazine later brought together two scientists in 2006 to debate the subject of science vs. God. Each had produced a book that same year. Evolutionary biologist Richard Dawkins published *The God Delusion*, maintaining that post-Darwinian scientific advances have rendered any belief in God irrational and unnecessary. Dr. Collins, mentioned above, also published a book in 2006 titled, *The Language of God: A Scientist's Evidence for Belief.* In the debate, Dr. Collins, after listening to endless contempt for religious people, makes two very unscientific comments.

Explaining the difference between the two scientists' approach, Dr. Collins remarked, "The difference is that my presumptions of the possibility of God and therefore the supernatural is not zero, and yours is." Later, "I think we don't do a service to dialogue between science and faith to characterize sincere people by calling them names... Atheists sometimes come across as a bit arrogant in this regard, and characterizing faith as something only an idiot would attach themselves to is not likely to help your case." [405]

These same thoughts, and others, are included in a new book by David Horowitz titled, *Dark Agenda, The War to Destroy Christian America*, which I encourage you to obtain and read. [406]

Promises Kept

I promised you back in Chapter 15 that I would provide a platform on which to construct dialogue between the science and faith communities. It wasn't until I asked a friend to read my incomplete manuscript and give me his opinion about the direction of this book that I realized my inadequacy to fulfill that promise. His words were succinct and powerful: "Bob, you aren't a scientist." He was right. It is that simple. Any view I have stated was colored by my faith. For that reason, what you have read previously in this book was adjusted to this opinion. Not only did I need to be cautious not to appear to be speaking for the science community, or for that matter the faith community which also has wide diversity but, in the end, these are my opinions only; and a few hundred footnotes.

With that background, I make the following observations which may be helpful prior to, during, or after formal and informal discussions:

- Each individual or group should state how they see themselves in light of purely accepted definitions of whom they claim to represent.
- For two individuals or two groups to have a discussion, both should come to the table with open minds. After all, that is what a discussion is about, relinquishing your position long enough for the other position to provide a reason that it might be wrong to some degree or more often, improperly applied to a situation.
- Unlike the Dawkins – Collins's debate mentioned earlier, care should be taken not to mischaracterize anyone or brand an individual or group of people by using derogatory labels.
- Those in discussion may need to approach the event, not attempting to prove the other wrong, but listening for value that may change how they perceive the issue.
- Each person or group should listen more than they talk. Intentionally try to out-listen each other.
- Only then can each one state what they have in common regarding the issue being discussed.

- No decisions are made or positions taken. Instead, each one respects the other's thoughts and agrees to think about what has been discussed. It is understood each will research anything discussed that was not clear or well-documented.
- Another time is set to discuss again, using the same outline.

People who believe they are the center of their universe and insist they are right about certain things should probably be ignored and would likely make poor participants in the platform just mentioned. I have heard those in science, faith, and religion who I felt were the center of their universe. Admittedly, that is my opinion and I could be wrong. I hope I am. What could affect my opinion is relativity. Maybe I didn't consider what they were and just concentrated on who they were. See how easy it is to admit you were wrong?

I mentioned earlier about writing styles. I intentionally left a lot of blanks in this end-of-days narrative. I encourage you to fill in a lot of details based on how you see this end happening. I don't think you will be able to say it will not end. Even if you don't believe in a higher power, in the God of Christianity, you surely know it isn't *if* you die, but, *when* you die. The only thing left after you pass away is what you leave behind. Even if there is no end after you die, that stuff you left will still disappear.

In 1983, at the Templeton Prize award ceremony in London, Alexander Solzhenitsyn, a survivor of the Soviet Gulag concentration camp, began his remarks with the following memorable words:

"More than half a century ago, while I was still a child, I recall hearing a number of older people offer the following explanation for the great disasters that had befallen Russia: Men have forgotten God; that's why all this has happened."

Continuing, "Since then I have spent well-nigh fifty years working on the history of our Revolution… But if I were asked today to formulate as concisely as possible the main cause of the ruinous Revolution that swallowed up some 60 million of our people, I could not put it more accurately than to repeat: Men have forgotten God; that's why all this has happened." [407]

In working through this ageless debate between the seen and the unseen, between evidence and faith, it does indeed come down to a decision. It doesn't matter if you are in the science camp or the faith camp, you must make a decision regarding what happens after everything is over and gone. In the end, it is easy for the faith community to explain where the concept of *decision* came from. Not so much for those in the science community. My point?

It's time to talk.

Ecclesiastes 3
NKJV

[1] *To everything there is a season, A time for every purpose under
heaven:* [2] *A time to be born, And a time to die; A time to plant, And
a time to pluck what is planted;* [3] *A time to kill, And a time to heal;
A time to break down, And a time to build up;* [4] *A time to weep,
And a time to laugh; A time to mourn, And a time to dance;* [5] *A time
to cast away stones, And a time to gather stones; A time to embrace,
And a time to refrain from embracing;* [6] *A time to gain, And a time
to lose; A time to keep, And a time to throw away;* [7] *A time to tear,
And a time to sew; A time to keep silence, And a time to speak;* [8] *A
time to love, And a time to hate; A time of war, And a time of peace.*

[9] *What profit has the worker from that in which he labors?* [10] *I have
seen the God-given task with which the sons of men are to be occu-
pied.* [11] *He has made everything beautiful in its time. Also He has
put eternity in their hearts, except that no one can find out the work
that God does from beginning to end.* [12] *I know that nothing is better
for them than to rejoice, and to do good in their lives,* [13] *and also
that every man should eat and drink and enjoy the good of all his
labor—it is the gift of God.* [14] *I know that whatever God does, It shall
be forever. Nothing can be added to it, And nothing taken from it.
God does it, that men should fear before Him.* [15] *That which is has
already been, And what is to be has already been; And God requires
an account of what is past.*

[16] *Moreover I saw under the sun: In the place of judgment, Wickedness
was there; And in the place of righteousness, Iniquity was there.* [17]
*I said in my heart, "God shall judge the righteous and the wicked,
For there is a time there for every purpose and for every work."* [18] *I
said in my heart, "Concerning the condition of the sons of men, God
tests them, that they may see that they themselves are like animals."*
[19] *For what happens to the sons of men also happens to animals; one
thing befalls them: as one dies, so dies the other. Surely, they all have
one breath; man has no advantage over animals, for all is vanity.* [20]

All go to one place: all are from the dust, and all return to dust. 21
Who knows the spirit of the sons of men, which goes upward, and
the spirit of the animal, which goes down to the earth? 22 *So I per-*
ceived that nothing is better than that a man should rejoice in his
own works, for that is his heritage. For who can bring him to see what
will happen after him?

ENDNOTES

1 *https://www.umcdiscipleship.org/resources/history-of-hymns-breathe-on-me-breath-of-god*
2 *https://www.businessinsider.com/how-many-days-can-you-survive-without-water-2014-5*
3 *http://www.transweb.org/faq/q3.shtml*
4 *https://www.simpletoremember.com/articles/*
5 *https://www.eternalanswers.org/where-does-god-come-from/*
6 2 Corinthians 5:17 NIV
7 *https://www.abarim-publications.com/YomProblem.html*
8 *https://www.discovermagazine.com/the-sciences/could-the-big-bang-be-wrong*
9 Revelation 1:8 and 21:6
10 Exodus 3:14, John 8:58
11 *https://www.discovermagazine.com/the-sciences/could-the-big-bang-be-wrong*
12 *https://www.smithsonianmag.com/science-nature/scientists-discover-oldest-known-dinosaur-152807497/*
13 *https://www.gotquestions.org/Trinity-Bible.html*
14 *https://www.everystudent.com/forum/trinity.html*
15 John 4:24
16 *http://www.nbcnews.com/id/42154769/ns/technology_and_science-science/t/did-god-have-wife-scholar-says-he-did/#.XijGBGhKhhE*
17 *https://www.theguardian.com/commentisfree/belief/2011/jul/06/gods-wife-heavenly-mother-mormon-church*
18 Ephesians 5:25-27
19 John 3:16 NKJV
20 *https://www.vocabulary.com/dictionary/begotten*

21 *https://www.compellingtruth.org/only-begotten-son.html*
22 John 3:16 NIV
23 Luke 1:34-35
24 Genesis 1:26-27
25 *https://www.christianity.com/wiki/angels-and-demons/are-angels-male-or-female.html*
26 *https://www.christianity.com/theology/theological-faq/how-did-lucifer-fall-and-become-satan-11557519.html*
27 *https://www.biblegateway.com/passage/?search=Revelation+12%3A7-12&version=NKJV*
28 Revelation 12:12b
29 *https://www.huffpost.com/entry/can-we-measure-the-earths_b_14748994*
30 *https://www.christianity.com/bible/bible.php?ver=niv&q=matthew+25:41*
31 *https://www.theguardian.com/science/2017/oct/29/scott-kelly-astronaut-interview-space-younger-twin-endurance*
32 *https://www.quora.com/How-many-different-versions-of-the-Bible-is-there-in-circulation*
33 *https://www.thomasnelson.com/p/mysteries-of-the-messiah/*
34 Genesis 1:1
35 *https://www.bibleodyssey.org/en/people/related-articles/was-the-gospel-of-thomas-gnostic*
36 The Gospel of Thomas – Saying 50
37 *https://www.livescience.com/13363-7-theories-origin-life.html*
38 *http://scienceviews.com/dinosaurs/dinofacts.html*
39 *https://creation.com/agnosticism*
40 *https://www.abc.net.au/news/2014-09-02/smart-atheism-has-a-creation-myth,-too/5712050*
41 *https://stacker.com/stories/3543/50-space-terms-understanding-universe*
42 https://ng.gensdecoeur.com/23845-25-terrifying-facts-that-prove-how-fragile-life-is-on.html
43 *https://www.nasa.gov/audience/foreducators/5-8/features/F_How_Big_is_Our_Universe.html*

44 *http://www.bbc.com/earth/story/20160610-it-took-centuries-but-we-
now-know-the-size-of-the-universe*

45 *https://www.grc.nasa.gov/WWW/K-12/Numbers/Math/
Mathematical_Thinking/how_long_is_a_light_year.htm*

46 *https://earthsky.org/human-world/
how-to-spot-the-international-space-station*

47 *https://www.thoughtco.com/speed-of-the-earth-1435093*

48 *https://spaceplace.nasa.gov/review/dr-marc-earth/earth-rotation.html*

49 *https://fullerstudio.fuller.edu/
say-dinosaurs-christian-physicist-reflects-origins-life/*

50 *https://science.howstuffworks.com/environmental/earth/geology/dino-
saur-bone-age1.htm*

51 *https://www.christianity.com/wiki/bible/what-does-the-bible-say-
about-dinosaurs.html*

52 IBID

53 *https://www.livescience.com/3945-history-dinosaurs.html*

54 *https://www.history.com/topics/pre-history/dinosaurs-an-introduction*

55 *https://www.history.com/news/prehistoric-ages-timeline*

56 *https://www.history.com/news/prehistoric-ages-timeline*

57 *https://www.timesofisrael.com/vast-and-developed-9000-year-old-set-
tlement-uncovered-near-jerusalem/#gs.fvscf0*

58 *https://archive.archaeology.org/0811/abstracts/turkey.html*

59 *https://www.usgs.gov/faqs/did-people-and-dinosaurs-live-same-
time?qt-news_science_products=0#qt-news_science_products*

60 *https://www.usgs.gov/special-topic/water-science-school/science/
water-you-water-and-human-body?qt-science_center_objects=0#qt-
science_center_objects*

61 *http://www.jesuscreated.org/Biblical/Before_Creation.php*

62 *https://blog.logos.com/2017/11/
god-wasnt-alone-created-world-says-bible/*

63 "How to pray… a simple guide for normal people" by Pete Greig,
by Tyndale House Publishers, 10-1-2019

64 *https://www.quora.com/What-was-Einsteins-viewpoint-when-he-
said-The-more-I-study-science-the-more-I-believe-in-God*

65 *https://www.vanderbilt.edu/AnS/physics/brau/H182/Term%20 papers%20'02/Benjamin.htm*
https://newrepublic.com/article/115821/ einsteins-famous-quote-science-religion-didnt-mean-taught
66 *https://www.history.com/ news/6-things-you-may-not-know-about-the-dead-sea-scrolls*
67 *http://www.ldolphin.org/sewell/sewellchron.html*
68 *https://www.gotquestions.org/gap-theory.html*
69 *https://www.hebrew4christians.com/Scripture/Parashah/Summaries/ Bereshit/bereshit.html*
70 *https://biblehub.com/lexicon/matthew/5-48.htm*
71 Revelation 3:15-16
72 *https://www.livescience.com/61734-weak-force-alien-universe.html*
73 *https://now.northropgrumman.com/ when-quantum-mechanics-and-relativity-collide/*
74 *https://www.newscientist.com/article/mg24031982-200-theres-a-glitch-at-the-edge-of-the-universe-that-could-remake-physics/*
75 *https://www.guinnessworldrecords.com/world-records/ smallest-living-organism-/*
76 *https://www.forbes.com/sites/jimhandy/2011/12/14/ how-big-is-a-nanometer/#343d14996fb0*
77 *https://futurism.com/the-byte/giant-magellan-telescope-construction*
78 *https://www.rmg.co.uk/stories/topics/ james-webb-space-telescope-vs-hubble-space-telescope*
79 *https://www.jwst.nasa.gov/content/about/launch.html*
80 *https://www.forbes.com/sites/startswithabang/2017/05/27/ask-ethan-what-does-the-edge-of-the-universe-look-like/#33b210814a16*
81 IBID
82 *https://spaceplace.nasa.gov/oort-cloud/en/*
83 *https://nineplanets.org/oort-cloud/*
84 *https://solarsystem.nasa.gov/solar-system/oort-cloud/in-depth/*
85 *https://www.forbes.com/sites/startswithabang/2016/06/17/ is-dark-matter-required-for-life-to-exist/#b7050db38795*
86 *https://time.com/77676/why-science-does-not-disprove-god/*
87 *https://www.sciencemag.org/news/1998/10/fly-gene-stretches-life-span*

88 *https://www.rep.routledge.com/articles/thematic/universal-language/v-1/sections/adamic-language*
89 *https://qz.com/1680404/the-human-genome-project-is-complete-what-about-the-rest/*
90 *https://sepetjian.wordpress.com/2012/06/12/anticipated-discoveries-of-the-bible-circumcision-the-8th-day/*
91 *https://medlineplus.gov/circumcision.html*
92 Genesis 32:28
93 Genesis 50:26
94 *https://en.wikipedia.org/wiki/Moses*
95 *https://www.history.com/news/7-influential-african-empires*
96 *https://en.wikipedia.org/wiki/History_of_the_Americas*
97 *https://knowindia.gov.in/culture-and-heritage/ancient-history.php*
98 *https://www.timemaps.com/history/china-1500bc/*
99 *https://www.chabad.org/library/article_cdo/aid/1935026/jewish/A-History-of-the-Jewish-Monarchy.htm*
100 *https://www.learnreligions.com/herod-the-great-enemy-of-jesus-christ-701064*
101 *https://www.blueletterbible.org/faq/don_stewart/don_stewart_678.cfm*
102 *https://www.learnreligions.com/introduction-to-the-pentateuch-p2-248895*
103 *https://www.churchofjesuschrist.org/study/ensign/1982/04/five-empires-of-the-ancient-near-east-a-historical-backdrop-of-1-kings-to-matthew?lang=eng*
104 *https://www.babycentre.co.uk/a25013415/developmental-milestones-smell*
105 Moore, Keith L. *Essentials of Human Embryology*. Toronto: B.C. Decker Inc, 1988, p.2 from, *https://www.princeton.edu/~prolife/articles/embryoquotes2.html*
106 *https://www.knowablemagazine.org/article/living-world/2020/how-do-body-parts-grow-their-right-sizes*
107 *https://www.sciencedaily.com/releases/2018/05/180516101418.htm*
108 *https://www.christiancourier.com/articles/723-matthew-2-19-20-the-death-of-herod-the-great*

109 *https://www.blueletterbible.org/faq/don_stewart/don_stewart_1312.cfm*

110 *https://www.quora.com/Is-there-any-evidence-that-John-the-Baptist-existed*

111 *https://catholicphilly.com/2013/03/catholic-spirituality/did-jesus-baptize-anyone-lectors-and-the-shehe-problem/*

112 N. T. wright and Michael F. Bird co-wrote a book, *The New Testament You Never Knew*, in 2018 published by Zondervan

113 *https://en.wikipedia.org/wiki/18th_century*

114 *https://en.wikipedia.org/wiki/19th_century*

115 *https://www.thoughtco.com/major-wars-and-conflicts-20th-century-1779967*

116 *https://en.wikipedia.org/wiki/List_of_wars:_1990%E2%80%932002*

117 *https://www.usatoday.com/story/money/2019/06/24/most-important-event-the-year-you-were-born/37164307/*

118 *https://en.wikipedia.org/wiki/List_of_wars:_2003%E2%80%93present*

119 *https://www.britannica.com/event/Afghanistan-War*

120 Luke 2:22-24 NKJV

121 *https://www.apocryphicity.ca/2014/04/26/new-book-on-the-infancy-gospel-of-thomas/*

122 *http://www.earlychristianwritings.com/infancythomas.html*

123 *https://www.livescience.com/38014-physical-evidence-jesus-debated.html*

124 *https://www.theatlantic.com/magazine/archive/1986/12/who-do-men-say-that-i-am/305710/*

125 *https://www.vatican.va/archive/ccc_css/archive/catechism/p122a3p1.htm*

126 *https://www.franciscanmedia.org/born-jewish-jesus-was-the-first-christian/*

127 *https://stphilipinstitute.org/2020/02/07/the-incarnation-of-jesus-christ/*

128 *https://equip.sbts.edu/journals/journal-of-theology/a-model-of-jesus-christs-two-wills-in-view-of-theology-proper-and-anthropology/*

129 https://equip.sbts.edu/journals/journal-of-theology/a-model-of-jesus-christs-two-wills-in-view-of-theology-proper-and-anthropology/

130 https://equip.sbts.edu/journals/journal-of-theology/a-model-of-jesus-christs-two-wills-in-view-of-theology-proper-and-anthropology/

131 https://stphilipinstitute.org/2020/02/07/the-incarnation-of-jesus-christ/

132 https://www.ucc.org/beliefs_jesus-christ-is-both-human

133 https://www.bookreviewsandmore.ca/2018/08/the-biggest-lie-in-history-of.html

134 https://wearefaith.org/blog/christ-jesus-or-jesus-christ/

135 https://www.etymonline.com/word/Incarnation

136 https://www.ewtn.com/catholicism/teachings/pius-ixs-ineffabilis-deus-defining-the-immaculate-conception-153

137 https://www.smithsonianmag.com/science-nature/here-are-eight-species-bending-rules-reproduction-180962676/

138 https://redeeminggod.com/god-cannot-look-upon-sin/

139 https://www.crossway.org/articles/10-things-you-should-know-about-the-incarnation/

140 http://www.thywordistruth.com/questions/qa086.html#.XvpdTChKhhF

141 https://www.versebyverseministry.org/bible-answers/who-baptized-john-the-baptist/#:~:text=In%20summary%2C%20John%20the%20Baptist%20was%20never%20under,is%20now%20practiced%20in%20the%20New%20Testament%20church.

142 https://www.learnreligions.com/baptism-of-jesus-by-john-700207

143 http://www.givemechrist.com/2016/10/why-jesus-was-40-days-in-the-wilderness/

144 https://bibleresources.americanbible.org/resource/jesus-and-wilderness

145 https://www.catholic.com/qa/why-did-jesus-need-to-be-driven-by-the-spirit-into-the-desert

146 https://www.christianbook.com/testament-never-knew-and-study-guide/9780310085294/pd/085294

147 https://www.learnreligions.com/temptation-of-jesus-bible-story-summary-700067

148 *https://www.livescience.com/56341-where-did-satan-come-from.html*

149 Jerry Walls, "Heaven, Hell and Purgatory: Rethinking the Things That Matter Most" (Brazos Press, 2015)

150 Elaine Pagels, "The Origin of Satan" (Random House, 1995).

151 *https://greekherald.com.au/culture/oldest-olive-tree-in-the-world-is-located-in-crete/*

152 *https://www.ncbi.nlm.nih.gov/pmc/articles/PMC2810702/#:~:text=Hematohidrosis%20also%20known%20as%20Hematidrosis,%5B1%5D%20Manonukul%20et%20al.*

153 https://www.trusting-in-jesus.com/Arrest-of-Jesus.html

154 *http://www.jewishencyclopedia.com/articles/1554-annas*

155 *https://christianity.stackexchange.com/questions/75552/what-is-the-basis-for-believing-that-40-lashes-would-be-deadly-especially-in-ro*

156 *https://www.christianity.com/wiki/jesus-christ/medical-account-of-the-crucifixion.html*

157 *https://en.wikipedia.org/wiki/Via_Dolorosa*

158 *https://www.apu.edu/articles/the-science-of-the-crucifixion/*

159 *https://www.apu.edu/articles/the-science-of-the-crucifixion/*

160 *https://www.evangelicaloutreach.org/crucifix.htm*

https://www1.cbn.com/medical-view-of-the-crucifixion-of-jesus-christ

161 *https://www.catholicculture.org/culture/library/view.cfm?recnum=4743*

162 *https://www.jw.org/en/library/books/bible-teach/origin-cross-symbol-pagan/*

163 *http://jbeard.users.rapidnet.com/bdm/Cults/science.htm*

164 *https://www.tomorrowsworld.org/commentary/the-crucifixion-darkness-what-happened-at-the-death-of-christ*

165 *https://www.bbc.com/news/blogs-magazine-monitor-27067136*

166 *https://bible.org/question/where-did-christ-go-after-he-died-and-he-rose-dead*

167 *https://bible.org/article/where-was-jesus-spirit-when-his-body-was-tomb*

168 *https://www.bibleref.com/1-Peter/3/1-Peter-3-19.html*

169 *https://www.mindonjesus.com/jesus-three-days-in-the-tomb/*

170 *https://carm.org/where-did-jesus-go-after-he-died-cross*

171 *https://www.gotquestions.org/where-was-Jesus.html*

172 *https://www.christianity.com/wiki/jesus-christ/medical-account-of-the-crucifixion.html*

173 *https://bible.knowing-jesus.com/topics/Blessed*

174 *https://www.christiantoday.com/article/life-after-death-what-did-jesus-do-between-his-resurrection-and-ascension/82998.htm*

175 *https://www.amazon.com/Case-Christ-Journalists-Personal-Investigation/dp/0310339308*

176 *https://www.alternet.org/2019/03/how-case-for-christ-author-lee-strobel-fabricated-his-best-selling-story/*

177 *https://leestrobel.com/response*

178 *https://www.beliefnet.com/ilovejesus/features/what-did-jesus-do-after-his-resurrection.aspx*

179 *https://www.desiringgod.org/articles/split-rocks-open-tombs-raised-bodies*

180 *https://www.crosswalk.com/faith/spiritual-life/what-does-amen-mean-and-why-do-we-say-it.html*

181 *https://bible.org/seriespage/41-ascension-luke-2431-acts-11-11*

182 *https://www.catholic.com/magazine/online-edition/did-the-ascension-really-happen*

183 *https://www.biblestudytools.com/dictionary/theophilus/*

184 *https://www.learnreligions.com/saint-andrew-apostle-541673*

185 *https://www.britannica.com/biography/Saint-Peter-the-Apostle*

186 *https://www.learnreligions.com/profile-of-apostle-james-701062*

187 ibid

188 ibid

189 *https://www.mountcarmelblessedsacrament.com/the-apostles-after-the-ascension-of-jesus/*

190 *https://www.mountcarmelblessedsacrament.com/the-apostles-after-the-ascension-of-jesus/*

191 *https://raychoi.org/2012/06/01/what-happened-to-the-12-disciples-after-the-resurrection-and-ascension/*

192 ibid

193 *http://www.aboutbibleprophecy.com/p156.htm*

194 *http://saintsresource.com/james-the-less-apostle/*

195 *https://raychoi.org/2012/06/01/what-happened-to-the-12-disciples-after-the-resurrection-and-ascension/*

196 ibid

197 *https://raychoi.org/2012/06/01/what-happened-to-the-12-disciples-after-the-resurrection-and-ascension/*

198 ibid

199 *https://raychoi.org/2012/06/01/what-happened-to-the-12-disciples-after-the-resurrection-and-ascension/*

200 ibid

201 ibid

202 *https://www.gcs.edu/mod/book/view.php?id=4475&chapterid=70*

203 *https://www.ucg.org/the-good-news/profiles-of-faith-james-half-brother-of-jesus*

204 *https://www.newadvent.org/cathen/02300a.htm*

205 *https://en.wikipedia.org/wiki/Barnabas*

206 *https://www.leadershipresources.org/who-is-barnabas-in-the-bible-10-things-to-know/*

207 *https://www.britannica.com/biography/Saint-Barnabas*

208 *https://crossexamined.org/wrote-gospel-mark/#*

209 *https://crossexamined.org/wrote-gospel-mark/#*

210 *https://www.learnreligions.com/john-mark-author-of-the-gospel-of-mark-701085*

211 *https://www.gotquestions.org/life-Silas.html*

212 *https://www.christianity.com/bible/people-of-the-bible/st-timothy-pauls-associate-11629587.html*

213 *https://bible.org/seriespage/who-were-timothy-and-titus*

214 *https://www.bible.ca/history/eubanks/history-eubanks-5.htm*

215 *https://ca.thegospelcoalition.org/columns/detrinitate/the-first-twenty-years-what-happened-to-the-church-immediately-after-jesus-died/*

216 *https://christianpublishinghouse.co/2018/05/30/who-was-gamaliel-that-taught-saul-of-tarsus/*

217 Just a note that Paul, in Galatians the first chapter, discloses a side trip to Arabia at some point.

218 *https://www.thegospelcoalition.org/article/no-saul-the-persecutor-did-not-become-paul-the-apostle/*

219 These two links are excellent dissertations about the background of the disagreement and its resolution. *https://bible.org/seriespage/23-jerusalem-council-gospel-defined-and-defended-acts-151-35*

https://biblewise.com/bible_study/characters/the-jerusalem.php

220 *https://thebiblemadeplain.com/why-did-paul-shave-his-head-in-cenchrea/*

221 Much of the research on the life of Paul came from Scripture and the links below. The dates listed are a collection of estimates by religious scholars. As you will see from these references, writers interpret Paul's Missionary Journeys differently and the actual number of journeys, the motive and effect of those who label four and five in particular. *https://www.biblestudy.org/apostlepaul/timeline-life-of-paul-from-first-missionary-journey-to-death.html https://www.conformingtojesus.com/charts-maps/en/paul%27s_letters-journeys_chart.htm https://www.understandchristianity.com/timelines/chronology-pauls-ministry/ https://www.friendshiptours.com/the-3-missionary-journeys-of-the-apostle-paul/*

222 *https://www.britannica.com/biography/Saint-Paul-the-Apostle#ref259967*

223 *https://www.britannica.com/biography/Saint-Luke*

224 *https://www.covenantepc.org/blog/luke-the-historian/*

225 *https://overviewbible.com/saint-luke/*

226 *https://www.learnreligions.com/luke-gospel-writer-and-physician-701087*

227 *https://www.britannica.com/biography/Saint-Luke*

228 *https://www.goodreads.com/book/show/40589736-the-new-testament-you-never-knew-study-guide*

229 *https://www.theatlantic.com/magazine/archive/1986/12/who-do-men-say-that-i-am/305710/*

230 *https://plato.stanford.edu/entries/religion-science/*

231 *https://plato.stanford.edu/entries/religion-science/*

232 *https://answersingenesis.org/world-religions/religion-what-where-how-does-bible-view-it/*
233 *https://www.oed.com/viewdictionaryentry/Entry/161944#:~:text=5.-,a.,a%20code%20of%20living%2C%20esp.*
234 *https://www.intellectualtakeout.org/blog/nobody-knows-where-word-religion-comes/*
235 *https://plato.stanford.edu/entries/religion-science/*
236 Guessom, Nidhal, 2009, Islam's Quantum Question: Reconciling Muslim Tradition and Modern Science, London and New York: Tauris.
237 *https://www.ancient.eu/religion/*
238 *https://www.ewtn.com/catholicism/library/first-religion-of-mankind-10914*
239 *https://probe.org/the-origin-of-mans-religions/*
240 *https://www.thegreatcoursesdaily.com/the-religious-impact-of-the-printing-press/*
241 Clark, Kelly, J., 2014, Religion and the Sciences of Origins. Historical and Contemporary Perspectives, New York: Palgrave MacMillan.
242 *http://symbionticism.blogspot.com/2013/04/the-history-of-science-when-was-word.html?m=1*
243 *http://symbionticism.blogspot.com/2013/04/the-history-of-science-when-was-word.html?m=1*
244 *http://symbionticism.blogspot.com/2013/04/the-history-of-science-when-was-word.html?m=1*
245 *https://plato.stanford.edu/entries/religion-science/*
246 *https://plato.stanford.edu/entries/religion-science/*
247 *https://www.huffpost.com/entry/accommodationism-and-why-_b_715915*
248 *https://theconversation.com/yes-there-is-a-war-between-science-and-religion-108002*
249 *https://americanhumanist.org/what-is-humanism/war-science-religion/*

250 *https://www.sciencedirect.com/science/article/pii/S163107131730072X#ack0005*

251 *https://www.sciencedirect.com/science/article/pii/S163107131730072X#ack0005*

252 *https://www.explainthatstuff.com/timeline.html* ; *http://historyworld.net/timesearch/default.asp?conid=1062&gtrack=mtoph* ; *http://www.sciencetimeline.net/1651.htm* ; *https://www.christianity.com/church/church-history/timeline-important-dates-in-ad-christian-history-11542876.html* ; https://en.m.wikipedia.org/wiki/Religion#:~:text=The%20Greek%20word%20threskeia%2C%20which,well%20into%20the%20medieval%20period.

253 *https://www.christianity.com/church/church-history/timeline-important-dates-in-ad-christian-history-11542876.html*

254 *https://www.christianity.com/church/church-history/timeline-important-dates-in-ad-christian-history-11542876.html*

255 *https://www.christianity.com/church/church-history/timeline-important-dates-in-ad-christian-history-11542876.html*

256 *https://www.pbs.org/wgbh/pages/frontline/shows/religion/why/legitimization.html*

257 *https://en.wikipedia.org/wiki/Constantine_the_Great_and_Christianity*

258 *http://www.historyworld.net/wrldhis/PlainTextHistories.asp?ParagraphID=bwd*

259 *https://www.aier.org/article/woodstock-occurred-in-the-middle-of-a-pandemic/*

260 *https://www.mycentraloregon.com/2020/11/30/coronavirus-live-updates-hospitalizations-of-covid-19-patients-hit-all-time-high-in-us/*

261 *https://www.webmd.com/vaccines/covid-19-vaccine/covid-vaccines-compared*

262 *https://bible.org/article/prophecies-birth-christ*

263 *https://www.sciencealert.com/8-terrifying-ways-the-world-could-actually-end*

264 *https://bible.org/article/prophecies-birth-christ*

265 *https://bible.org/article/prophecies-birth-christ*

266 *https://www.sciencemag.org/news/2017/07/what-it-would-take-kill-all-life-earth*

267 *https://bible.org/article/prophecies-birth-christ*

268 *https://www.usatoday.com/story/tech/2017/09/22/world-definitely-going-end-just-probably-not-saturday/689803001/*

269 *https://www.beliefnet.com/faiths/christianity/what-the-bible-really-says-about-the-rapture.aspx*

270 *https://d2y1pz2y630308.cloudfront.net/3064/documents/2019/1/Week%2012%20-%20The%20Rapture-Left%20Behind%20%20or%20Left%20OUT%20of%20Scripture.pdf*

271 *https://www.seti.org/about*

272 *https://www.popularmechanics.com/space/solar-system/a26767/wow-signal-mystery-solved/*

273 *https://www.seti.org/seti-institute/project/details/seti-observations#:~:text=Hydrogen%20atoms%20emit%20at%201420%20MHz%20%28a%20wavelength,frequencies%20ranging%20from%201612%20MHz%20to%201720%20MHz.*

274 *https://en.wikipedia.org/wiki/Voyager_Golden_Record*

275 *https://www.space.com/14732-sun-burns-star-death.html*

276 *https://www.businessinsider.com/nasa-asteroid-simulation-reveals-need-years-of-warning-2021-5*

277 *https://starwars.fandom.com/wiki/Opening_crawl*

278 *https://www.thoughtco.com/infinity-facts-that-will-blow-your-mind-4154547*

279 *https://www.scientificamerican.com/article/multiverse-the-case-for-parallel-universe/*

280 *https://iep.utm.edu/infinite/*

281 Ibid

282 *https://www.sciencedirect.com/science/article/pii/S163107131730072X#ack0005*

283 *https://www.livescience.com/57942-what-was-first-life-on-earth.html*

284 *https://plato.stanford.edu/entries/religion-science/*

285 *https://hyperleap.com/topic/X_Club*

286 *http://acidrayn.com/2011/05/18/comparing-similarities-between-science-and-religion/*

287 Ibid

288 *https://en.wikipedia.org/wiki/Steve_Jobs*

289 *https://www.amazon.com/dp/B005J3IEZQ/ref=dp-kindle-redirect?_encoding=UTF8&btkr=1*

290 *https://www.inc.com/hitendra-wadhwa/steve-jobs-self-realization-yogananda.html*

291 *https://www.britannica.com/biography/Ben-R-Rich*

292 *https://www.gaia.com/article/ben-rich-lockheed-martin-and-ufos*

293 *https://www.blueblurrylines.com/2014/10/ben-rich-area-51-taking-et-home.html*

294 *https://www.secretprojects.co.uk/threads/lockheed-skunk-works-ben-rich-comments-revealed.5048/*

295 *https://en.wikipedia.org/wiki/Pyotr_Ufimtsev*

296 *https://www.biblio.com/skunk-works-by-rich-ben-r-janos-leo/work/56433*

297 *https://www.britannica.com/biography/Alan-Turing*

298 *https://www.turing.org.uk/publications/dnb.html*

299 Ibid

300 ibid

301 https://en.wikipedia.org/wiki/Wernher_von_Braun

302 Neufeld, Michael J. (20 May 2019). "Wernher von Braun and the Nazis". American Experience: Chasing the Moon. PBS. Retrieved 24 July 2019.

303 Neufeld, Michael J. 2008. Wernher von Braun: Dreamer of Space, Engineer of War. Vintage

304 "Wernher von Braun | Encyclopedia of Alabama". Encyclopedia of Alabama. Retrieved 27 March 2016.

305 "Reach for the Stars". TIME Magazine. 17 February 1958. Archived from the original on 21 December 2007.

306 "Photos: Wernher von Braun, Space Pioneer Remembered". Space.com. 2012. Retrieved 15 February 2019.

307 Neufeld, Michael J. (2007) Wernher von Braun: Dreamer of Space, Engineer of War, Knoff, New York

308 Mallon, Thomas (October 22, 2007) "Rocket Man", The New Yorker, Access date: January 8, 2015.

309 Mallon, Thomas (October 22, 2007) "Rocket Man", The New Yorker, Access date: January 8, 2015.
310 Stuhlinger, Ernst & Ira Ordway, Frederick. 1994. Wernher von Braun, crusader for space: a biographical memoir. Krieger Pub,
311 Wernher von Braun: History's Most Controversial Figure?, Al Jazeera
312 Spires, Shelby G. (27 June 2003). "Von Braun's brother dies; aided surrender". The Huntsville Times.
313 Davies, Norman (2006). Europe at War 1939–1945: No Simple Victory. London: Macmillan.
314 Spangenburg & Moser. 2009. Wernher von Braun, Revised Edition. Infobase Publishing
315 See Ward (2005), Chapter 5: "Encounters with Hitler."
316 "vonBraun". Archived from the original on April 28, 2015. Capture of Werner von Braun by the 324th Regiment Anti-tank Company
317 Arts & Entertainment, Biography (1959–1961 series). Mike Wallace, television biography of Wernher von Braun, video clip of the press statement.
318 Matthew Brzezinski (2007) Red Moon Rising: Sputnik and the Hidden Rivalries That Ignited the Space Age, Henry Holt, New York
319 "Stages to Saturn – The Saturn Building Blocks – The ABMA Transfer". NASA.
320 West 2017
321 "Concluding Remarks by Dr. Wernher von Braun about Mode Selection for the Lunar Landing Program" (PDF). Lunar Orbit Rendezvous File. NASA Historical Reference Collection. 7 June 1962.
322 Spangenburg & Moser. 2009. Wernher von Braun, Revised Edition. Infobase Publishing.
323 See: Dr. Wernher von Braun talks about ATSF satellite project
324 ibid
325 Neufeld, Michael J. (2008) Wernher von Braun: Dreamer of Space, Engineer of War, Vintage.

326 Neufeld, Michael J. (2007) Wernher von Braun: Dreamer of Space, Engineer of War, Knoff, New York
327 Mallon, Thomas (October 22, 2007) "Rocket Man", The New Yorker, Access date: January 8, 2015.
328 ibid
329 Walker, Mark (2008) "A 20th-Century Faust" Archived April 2, 2015, at the Wayback Machine, American Scientist, Access: January 8, 2015
330 Oliver, Kendrick (2012) To Touch the Face of God: The Sacred, the Profane, and the American Space Program, 1957–1975, p. 23, Johns Hopkins University Press
331 Oliver, 2012,
332 Mallon, Thomas (October 22, 2007) "Rocket Man", The New Yorker, Access date: January 8, 2015.
333 Stuhlinger, Ernst & Ira Ordway, Frederick. 1994. Wernher von Braun, crusader for space: a biographical memoir. Krieger Pub,
334 von Braun, Wernher (1963) "My Faith: A Space-Age Scientist Tells Why He Must Believe in God", (February 10, 1963) The American Weekly, New York: The Hearst Corporation.
335 See von Braun's speeches in The voice of Dr. Wernher Von Brain: An Anthology. Apogee Books Publication; ed. by Irene E. Powell-Willhite: These touch "a variety of topics, including education, the cold war, religion, and the space program".
336 See the same article by von Braun, Wernher, published as "Science and religion", in Rome Daily American, September 13, 1966. Available in New Age Frontiersn (Oct. 1966) United Family, Vol-II, No. 10
337 See "The Farther We Probe into Space, the Greater my Faith": C.M. Ward's account of His Interview with Dr. Warner von Braun (1966) Springfield, MO: Assemblies of God,
338 Ward, Bob (2013) Dr. Space: The Life of Wernher von Braun, Ch. 1: "The Accursed Blessing", Naval Institute Press
339 *https://en.wikipedia.org/wiki/John_von_Neumann*
340 Ibid
341 Ibid

342 ibid

343 *Macrae, Norman* (1992). *John von Neumann: The Scientific Genius Who Pioneered the Modern Computer, Game Theory, Nuclear Deterrence, and Much More.* Pantheon Press.

344 *Macrae, Norman* (1992). *John von Neumann: The Scientific Genius Who Pioneered the Modern Computer, Game Theory, Nuclear Deterrence, and Much More.* Pantheon Press.

345 Blair, Clay, Jr. (February 25, 1957). "Passing of a Great Mind"

346 *https://en.wikipedia.org/wiki/John_von_Neumann*

347 ibid

348 ibid

349 *Blair, Clay, Jr.* (February 25, 1957). *«Passing of a Great Mind»*

350 Pais, Abraham (2006). J. Robert Oppenheimer: A Life. Oxford: Oxford University Press

351 *https://en.wikipedia.org/wiki/Srinivasa_Ramanujan*

352 *https://www.science20.com/news_articles/divine_patterns_ramanujans_magical_mind_gets_math_formula-99186*

353 Kanigel, Robert (1991). *The Man Who Knew Infinity: a Life of the Genius Ramanujan.* New York: Charles Scribner's Sons.

354 *https://simple.wikipedia.org/wiki/Konstantin_Tsiolkovsky*

355 IBID

356 *https://www.britannica.com/biography/Konstantin-Eduardovich-Tsiolkovsky*

357 IBID

358 *https://www.britannica.com/technology/launch-vehicle#ref959057*

359 *https://www.britannica.com/technology/launch-vehicle/Launch-bases#ref1263765*

360 *https://menorah-bible.jimdofree.com/english/structure-of-the-bible/alphabets-and-numerical-values/*

361 *https://www.i24news.tv/en/news/culture/103742-160222-isaac-newton-scientific-pioneer-mystical-dreamer*

362 *https://www.space.com/LIGO-Laser-Interferometer-Gravitational-Wave-Observatory.html*

363 *https://www.ligo.caltech.edu/page/ligo-sister-facilities*

364 *https://mindmatters.ai/2021/01/why-intelligent-design-of-the-universe-is-not-an-absurd-idea/*

365 *https://www.wsj.com/articles/eric-metaxas-science-increasingly-makes-the-case-for-god-1419544568*

366 *http://theemergingfuture.com/speed-technological-advancement.htm*

367 *http://theemergingfuture.com/speed-technological-advancement-ten-years.htm*

368 *http://theemergingfuture.com/speed-technological-advancement-twenty-years.htm*

369 *http://theemergingfuture.com/speed-technological-advancement-thirty-years.htm*

370 *http://theemergingfuture.com/speed-technological-advancement-forty-years.htm*

371 *http://theemergingfuture.com/speed-technological-advancement-fifty-years.htm*

372 *https://www.google.com/search?q=What+is+the+wealth+of+Face+Book%27s+owner&rlz=1C1CHBF_enUS768US768&oq=What+is+the+wealth+of+Face+Book%27s+owner&aqs=chrome..69i57j33i22i29i30l2.26034j1j15&sourceid=chrome&ie=UTF-8*

373 *https://www.google.com/search?q=What+is+the+net+worth+of+twitter%27s+owner&rlz=1C1CHBF_enUS768US768&sxsrf=ALeKk00kRMOK-ESL_Qe4Kz-jR6wezwtSNw%3A1620667808290&ei=oG2ZYLiOEYjIsQXt2qXADg&oq=What+is+the+net+worth+of+twitter%27s+owner&gs_lcp=Cgdnd3Mtd2l6EAw6BwgAEEcQsAM6BwgjELACECc6-BAghEAo6BggAEAcQHjoICAAQBxAKEB5Qm-gVWLLEF2CT4RdoAXABeACAAV6IAeoRkgECMjiYAQCgAQGqAQdnd3Mtd2l6y-AEIwAEB&sclient=gws-wiz&ved=0ahUKEwi43ouI0r_wAhUIZK-wKHW1tCegQ4dUDCA4*

374 *https://www.businessinsider.com/google-larry-page-sergey-brin-net-worth-billionaire-wealth-bloomberg-2021-4#:~:text=Google%20founders%20Larry%20Page%20and,8%20centibillionaires%-20in%20the%20world&text=The%20Google%20cofounders%20Larry%20Page,pandemic%20as%20tech%20stocks%20soar.*

375 *https://www.google.com/search?q=What+is+the+net*

+worth+of+amazon%27s+owner&rlz=1C1CHBF_
enUS768US768&biw=1280&bih=648&sxsrf=ALeKk02XASNej
2gyH46UR8O-yiEm8bzGRQ%3A1620668333012&ei=rG-ZYP
T8PJDp_QaVw6nYBQ&oq=What+is+the+net+worth+of+amaz
on%27s+owner&gs_lcp=Cgdnd3Mtd2l6EAwyBAgAEB4yBggAE-
AgQHjoHCAAQRxCwAzoGCAAQBxAeOggIABAIEAcQHjoICA
AQCBANEB46BAgAEA06BggAEA0QHjoFCAAQhgNQgDVYm
2Ng_npoAnACeACAAZEBiAHECJIBAzUuNZgBAKABAaoBB
2d3cy13aXrIAQjAAQE&sclient=gws-wiz&ved=0ahUKEwj0gqa
C1L_wAhWQdN8KHZVhClsQ4dUDCA4

376 IBID

377 https://footgoal.pro/top-10-football-players-with-most-fans-in-the-world-in-2021/#:~:text=So%2C%20who%20is%20the%20number,social%20media%20official%20accounts'%20followers.

378 https://www.sportskeeda.com/nfl/top-10-nfl-players-social-media-followers

379 https://socialmediadata.com/most-followed-people/

380 https://www.sportskeeda.com/basketball/5-active-nba-players-social-media-following

381 https://socialblade.com/twitter/user/realdonaldtrump

382 https://www.businessinsider.com/dragon-springs-falun-gong-upstate-new-york-compound-photos-2019-9

383 https://www.britannica.com/topic/Falun-Gong

384 https://www.businessinsider.com/dragon-springs-falun-gong-upstate-new-york-compound-photos-2019-9

385 https://www.justice.gov/eoir/page/file/1347746/download

386 IBID

387 https://www.merriam-webster.com/dictionary/climate%20change

388 https://dictionary.cambridge.org/dictionary/english/climate-change

389 https://www.lexico.com/definition/climate_change

390 https://www.collinsdictionary.com/dictionary/english/climate-change

391 https://owlcation.com/stem/Earths-Temperature-Brief-History-of-Recent-Change

392 https://www.ucsusa.org/resources/each-countrys-share-co2-emissions

393 https://www.theepochtimes.com/china-emits-more-green-

house-gas-than-developed-countries-combined-report_3806400.html?utm_source=ChinaDaily&utm_medium=email&utm_campaign=2021-05-09
394 *https://www.giss.nasa.gov/research/news/20140121/*
395 *https://www.nature.com/scitable/knowledge/library/milankovitch-cycles-paleoclimatic-change-and-hominin-evolution-68244581/*
396 *https://www.britannica.com/biography/Milutin-Milankovitch*
397 ibid
398 *https://www.history.com/topics/middle-ages/pandemics-timeline*
399 *https://www.history.com/topics/middle-ages/pandemics-timeline*
400 *https://news.google.com/covid19/map?hl=en-US&gl=US&ceid=US%3Aen*
401 *https://www.bbc.co.uk/religion/religions/buddhism/ataglance/glance.shtml*
402 *https://preachitteachit.org/ask-roger/detail/which-nations-are-involved-in-end-time-events/*
403 *https://www.barna.com/research/most-american-christians-do-not-believe-that-satan-or-the-holy-spirit-exist/*
404 *https://www.abc.net.au/news/science/2018-05-24/three-scientists-talk-about-how-their-faith-fits-with-their-work/9543772*
405 *http://inters.org/dawkins-collins-cray-science*
406 *https://horowitzfreedomcenterstore.org/products/dark-agenda*
407 *https://www.templetonprize.org/laureate-sub/solzhenitsyn-acceptance-speech/*

Made in United States
Cleveland, OH
04 November 2026

47643614R00297